# Mechanical Tolerance Stackup and Analysis

# MECHANICAL ENGINEERING
A Series of Textbooks and Reference Books

*Founding Editor*

## L. L. Faulkner

*Columbus Division, Battelle Memorial Institute
and Department of Mechanical Engineering
The Ohio State University
Columbus, Ohio*

**Additional Volumes in Preparation**

*Mechanical Engineering Software*

*Spring Design with an IBM PC*, Al Dietrich

*Mechanical Design Failure Analysis: With Failure Analysis System Software for the IBM PC*, David G. Ullman

# Mechanical Tolerance Stackup and Analysis

**Bryan R. Fischer**
*Advanced Dimensional Management*
*Sherwood, Oregon, U.S.A.*

MARCEL DEKKER, INC.          NEW YORK · BASEL

**Library of Congress Cataloging-in-Publication Data**
A catalog record for this book is available from the Library of Congress.

**ISBN: 0-8247-5379-8**

This book is printed on acid-free paper.

**Headquarters**
Marcel Dekker, Inc., 270 Madison Avenue, New York, NY 10016, U.S.A.
tel: 212-696-9000; fax: 212-685-4540

**Distribution and Customer Service**
Marcel Dekker, Inc., Cimarron Road, Monticello, New York 12701, U.S.A.
tel: 800-228-1160; fax: 845-796-1772

**Eastern Hemisphere Distribution**
Marcel Dekker AG, Hutgasse 4, Postfach 812, CH-4001 Basel, Switzerland
tel: 41-61-260-6300; fax: 41-61-260-6333

**World Wide Web**
http://www.dekker.com

The publisher offers discounts on this book when ordered in bulk quantities. For more information, write to Special Sales/Professional Marketing at the headquarters address above.

# Preface

Every product manufactured today is subject to variation. Typically, the manufacturing process is the source of this variation. From the peaks and valleys of integrated circuits in the microscopic regime to the buttons on the cell phone in your pocket, to the large steel structures of dams and bridges in the macroscopic regime, no product or part is immune from variation and its sources. Understanding this variation and quantifying its effect on the form, fit, and function of parts and assemblies are crucial parts of the mechanical design process.

Tolerances are engineering specifications of the acceptable levels of variation for each geometric aspect of a component or assembly. Although tolerances are typically specified on engineering drawings, it is becoming popular for tolerances to be defined in a CAD file as attributes of a three-dimensional solid model. Whether explicitly specified on a drawing or as part of a CAD model, tolerances indicate the variation allowed for part and assembly features.

Tolerances may be used to control the variation allowed for individual feature geometry, such as form and size, or to control the geometric relationship between part and assembly features, such as orientation and location. Tolerance Analysis and Tolerance Stackups are the tools and techniques used to understand the cumulative effects of tolerances (accumulated variation) and to ensure that these cumulative effects are acceptable.

There are two methods used to specify tolerances: traditional plus/minus tolerancing and geometric dimension and tolerancing (GD&T). This text covers both techniques. GD&T and its principles are discussed in depth, as the point of Toleranc... and tolerancing scheme ...'s is ultimately to prove that a dimensioning required geometric conditi... The only way to precisely specify the minus tolerancing is still co...ough the use of GD&T. Although plus/ perform Tolerance Stackups o... used, and this text discusses how to part of the goal of this book is to ...and assemblies based on plus/minus, much better system. ...e reader understand why GD&T is a

This volume presents the back... niques required to solve simple and ...d material and step-by-step techniques. Using these techniques, the design engin...ex tolerance analysis problems. related parts and assemblies will satisfy ...n ensure that the form and fit of turning, inspection, assembly and service pers...el can use these techniques to intended function. Manufac- troubleshoot problems on existing designs, to ...ify th... their in-process steps will meet the desired objective, or even to find w...ys to improve performance and reduce costs.

In-depth coverage of worst-case and statistical tolerance analysis techniques is presented in this text. Worst-case techniques are covered first, followed by statistical techniques, as the statistical techniques follow the same steps. In-depth derivation and development of the mathematical basis for the applicability of the statistical method are not included in this text. The reader is directed to appropriate references on the subject in the bibliography.

Although the text is primarily devoted to the solution of one-dimensional Tolerance Stackups, two-dimensional and three-dimensional methods are discussed as well.

Because all tolerance analyses and stackups are truly three-dimensional, the problem solver is forced to frame the problem in such a manner as to facilitate a one-dimensional solution. Simplification and idealization of the problem are required. The text discusses the rules and assumptions encountered when simplifying tolerance analysis problems. Any assumptions used as a basis for a particular solution must be presented with the results of the Tolerance Stackup.

Tolerance Analysis is part art and part science. To effectively solve a Tolerance Analysis problem, the design engineer must first understand the problem, set the problem up in a manner that will yield the desired result, solve the problem, and report the information in a way that can be easily understood by all parties involved. The last two steps are essentially one and the same: using the techniques in this book, solving the Tolerance Analysis problem, and creating a report that can be shared or communicated with others happen concurrently. This book presents the Advanced Dimensional Management approach to Tolerance Analysis, which yields consistent and easy-to-understand results.

The importance of a standardized approach to solving Tolerance Analysis problems cannot be overstated. Equally important is the need to communicate the results of a Tolerance Stackup. Rarely (if ever) is a Tolerance Stackup done without the need to share the results or to convince someone else to make a change. Again, the techniques in this text help ensure that the problem will be solved correctly and that the results will be understood by all parties involved. Chapter 13 presents the techniques for developing and formatting a standardized Tolerance Stackup Sketch. Chapter 14 presents the techniques for entering data into a standardized Tolerance Stackup Report Form. Almost every Tolerance Stackup performed must be shared with others to get their concurrence. A clearly written and properly formatted report is essential to communicate the results and get the desired response.

As stated above, Tolerance Analysis is an art, and it requires practice to become an effective problem solver. Using the techniques presented in this book, the readers will be on the path to understanding and effectively solving their tolerancing problems.

This book is intended for the following audiences: technology and engineering students, drafters, designers, CAD operators, technicians, and engineers; manufacturing, assembly, inspection, quality, and service personnel; and anyone who needs to solve Tolerance Analysis problems. This text is also useful to consultants and trainers of GD&T, tolerance analysis, and tolerance stackups.

I would like to thank my mother for her guidance and encouragement, constantly telling me that I could do anything I put my mind to. I would like to thank my wife Janine for her support and assistance with many of the figures in this text. My stepfather, who worked as a land surveyor for many years, helped to nurture my interest in geometry and to develop my problem-solving skills. I would also like to thank my coworkers, clients and fellow subcomittee members over the years. It is from you that I have learned the most, and have developed greater awareness of the problems faced in industry. Thank you all.

Bryan Fischer

# Contents

# Mechanical Tolerance Stackup and Analysis

# 1

## Background

Tolerancing, Tolerance Analysis, and Tolerance Stackups have been around in one form or another for a long time. Sometime in the past, it became necessary to determine whether a collection of parts would fit together before they were manufactured. A design team may have needed to know how thin a part feature could become during manufacture, to make sure the part would remain strong enough to work. They may have needed to know how large a hole could be and how far it could be from its nominal position to make sure there was enough surface contact to properly distribute the load from a fastener. Perhaps the manufacturing team needed to understand why an assembly of parts that met the drawing specifications did not fit together at assembly. By performing Tolerance Analysis and Tolerance Stackups, these and many other important questions about the design can be answered. Indeed that is why Tolerance Analysis and Tolerance Stackups are done—to provide answers to questions. The techniques in this text will help you, the reader, understand your tolerancing problems, answer your tolerancing questions and solve your tolerancing problems.

How can the designer ensure that parts will fit together at assembly? Better yet, how can the designer ensure that imperfect parts will fit together at assembly, as all parts are imperfect? How much imperfection or variation is allowable? Does it matter if a part is manufactured a bit larger than nominal, and the mating part is manufactured smaller than nominal? What if both parts are manufactured on the small side and mating holes in each part are slightly tilted or out of position? What affects the performance of the assembly more—variation in size or variation in position? What happens to a feature on one part if a surface on the mating part is tilted? These questions all lead to a *Tolerance Stackup*.

Tolerance Analysis and Tolerance Stackup techniques have evolved over time, increasing in complexity to meet the increasingly complex needs of the products they study. Interestingly, a change in manufacturing philosophies is likely the primary reason that Tolerance Stackups are so important today. Design tools and techniques have changed, the design community and the manufacturing community have become separate entities, and the need to clearly and unambiguously communicate design requirements between the two has driven the need for Tolerance Stackups.

The need for and the ability to design ever more complex parts and assemblies, the need to guarantee fit at assembly, and the need to guarantee interchangeability of parts have contributed greatly to the widespread need for Tolerance Analysis. Complexity is an interesting issue—it is difficult to determine if a complex design will satisfy its objectives even when all parts are at their nominal state. Throw in variation and the problem can be overwhelming. Through the application of standardized Tolerance Analysis techniques, such as the ones presented in this text, the problem can be reduced to a more manageable form and solved. The need for interchangeable parts and the need for parts that will fit without rework or adjustment at assembly can only be ensured by Tolerance Analysis. These factors are hallmarks of modern manufacturing philosophies, and the only way to ensure these goals are achieved is through the proper use of Tolerance Analysis techniques.

Tolerance Analysis can be found today in nearly all manufacturing industries, from the very small geometry found in integrated circuits to the very large geometry found on rockets, the Space Shuttle and the International Space Station. Anywhere that parts must fit together, anywhere the possibility of accumulated variation may cause a problem, or merely needs to be understood and quantified, Tolerance Analysis and Tolerance Stackup techniques are being used.

Although it is not the only environment where Tolerance Analysis is needed, Tolerance Analysis has found its greatest application in mass production, where interchangeability of blindly selected parts is essential. Just-in-time manufacturing increases the demand for parts that absolutely must fit at assembly, as it is much less likely today to have a stock of spare parts waiting in the warehouse. Tolerance Analysis is the only way to ensure that the tolerances specified on drawings will lead to parts that fit.

Tolerance Analysis is equally beneficial in research and development (R&D) and for one-of-a-kind components and assemblies, as there is no other way to ensure that the accumulated variation of individual part features is functionally acceptable. Whether it is the fit of a robotic end effector on a robotic arm, a cover that fits over an enclosure, a clearance hole that must allow a fastener to pass, or the location of a bracket in an assembly, Tolerance Analysis is the only way to guarantee that parts will fit together at assembly.

For millennia, mankind has been designing and manufacturing parts, assemblies, and structures. Early on, the person that designed an assembly was also responsible for manufacturing the parts that made up the assembly. Indeed, the design wasn't complete until final assembly, where many parts were ground, scraped, drilled, bent, and modified to match the mating parts. Such assemblies worked well enough, but they were one of a kind. It was common for all the parts in such an assembly to be custom fit. This sort of custom, craftsman-oriented manufacturing philosophy was necessary back in the days before automated, high-precision manufacturing machinery. It was the only way that the craftsman could be sure the parts could be assembled. Such assemblies, however, presented a huge problem in terms of cost, time (both production time and assembly time), and maintenance.

As all parts were essentially one of a kind, and required a great amount of labor by the craftsman, the cost was high. It also took a long time to manufacture such an assembly, as only a few could be made at any given time. Lastly, and this is perhaps the greatest problem, was the problem of replacement parts. There was no easy way to replace a part that malfunctioned or broke in service. Because the most or all of the parts had been custom fit at assembly, there was no guarantee that a replacement part pulled off the shelf would work without more drilling, grinding, and modification. Machinery designed and manufactured using these methods is subject to extended downtime when a failure occurs. The iteration and the extra labor required in getting the replacement part to match the mating parts just so and the downtime in the use of the machinery can lead to potentially large profit losses.

Over time, designs matured. This year's new design borrowed bits and pieces from previous designs, improving upon earlier approaches. As designs became more complex, designers developed into a specialized group, with skills and talents unique to their craft. Likewise, as the processes and methods used to manufacture the parts and assemblies became more complex, the craftsmen that made the parts also became more specialized, evolving into a distinct, highly skilled and talented group. Eventually, the person that designed the part and the person that made the part were different people. No longer was it satisfactory for just the designer to know what was required of the parts making up the assembly—the ideas and requirements for the parts and assemblies had to be communicated to someone else, communicated to the craftsmen who were manufacturing the parts and assemblies.

During this transition, it became evident that drawings were needed to define what was to be made, to communicate the designer's ideas to those manufacturing the component. Drawings had to be dimensioned, as all geometric information had to be specified on the drawing, in a related document or conveyed verbally. Over time it became obvious that the best way to ensure the part or assembly being manufactured satisfied the needs of the designer was to completely dimension the drawing.

Today, virtually all manufactured items are defined by engineering drawings. Among other things, engineering drawings describe or define the geometric form and size of all geometric features on a part; equally important, engineering drawings also describe or define the relationship between part and assembly features, including their relative orientation and location.

There are two components to the definition of part geometry: description of the nominal state and description of the allowable variation. The model itself, or the two-dimensional (2D) drawing geometry itself in the case of drawings, provides a description of the nominal state. Dimensioning is an extension of the description of the nominal state, as dimensioning typically represents the nominal condition of part geometry. Tolerancing is a description of the allowable variation for each part feature and between part features. Together these provide a complete description or definition of part geometry and its allowable variation.

Every feature on a part should be fully dimensioned and toleranced, which includes each feature's form, size (as applicable), orientation, and location relative to the rest of the part. The dimensioning system may use traditional plus/minus dimensions and tolerances, it may use Geometric Dimensioning and Tolerancing (GD&T), or it may use a combination of both systems. Although all of these methods are common in industry today, it is the author's opinion that the absolute best method to use is GD&T—it is by far the clearest, most accurate, and least problematic method to describe the dimensioning and tolerancing requirements for parts and assemblies. If the designer's goal is to completely and unambiguously define the allowable geometric relationships between all part features, to guarantee that the part geometry will satisfy its functional requirements at assembly, then GD&T is the only way to go. The problems and vagaries of the plus/minus system are numerous.

Accurate Tolerance Analysis can only really be done on parts and assemblies dimensioned and toleranced using GD&T—there are far too many inconsistencies and assumptions required to validate parts dimensioned and toleranced using the plus/minus system alone. That said, this text covers Tolerance Analysis using the plus/minus system and GD&T, as many companies still resist making the move to GD&T and continue to use to the familiar plus/minus system.

Today, many complex features are implicitly dimensionally defined by the mathematical data in a three-dimensional (3D) computer-aided-design (CAD) solid model file. What is a complex feature you may ask? An even better question is, "What is a feature?" According to ASME Y14.5M-1994*, a *feature* may be a surface, a hole, a slot, the surface of an impeller, the helical surface of a screw thread, any distinctly discernable portion of a part. Simply put, a feature is a surface. The surface of the impeller, the helical surface of the screw thread, the surface of an airfoil, the nose cone of a rocket or the surface of an automobile fender are examples of complex features. Such complex features are difficult (if not impossible) to fully dimension using the familiar rectangular or polar coordinate dimensioning systems used on most engineering drawings.

All features are composed of an infinite set of points. The difference between a simple feature and a complex feature can be thought of as being related to the number of dimensions required to completely define the surface: the greater the number of dimensions required to define the surface, the more complex the surface. A simple feature such as a plane is easy to dimensionally define using rectangular coordinates, as all its points lie in a single plane. Often only a single dimension completely defines the surface. Cylinders, widths (opposed parallel planes), and spheres are also simple features. They are called *features of size*, and they are controlled by rule #1 in the ASME Y14.5M-1994 Standard. These features, unlike other features, are defined by a single size dimension. All points of a perfectly cylindrical surface are equidistant from an axis; all points of a perfect width are equidistant from a center plane; all points of a perfectly spherical surface are equidistant from a center point. That makes these features easy to completely dimension on an engineering drawing.

Features such as polygons, e.g., squares and hexagons, are actually comprised of several features of size at angles to one another. To completely dimension a polygon requires more than one dimension, which differentiates a polygonal or bounded feature from a feature of size. Complex features such as the surface of an automobile fender or the hull of a ship present a great challenge in dimensioning, as all the points lie on one or more complex warped surfaces.

Historically such surfaces have been dimensioned using rectangular or polar coordinates, where a finite set of points are dimensioned in three dimensions, developing an $(x,y,z)$ Cartesian coordinate system of sorts. As mentioned earlier, a surface is constituted of an infinite set of points. To fully dimension a surface such as the fender would require an infinite number of dimensions on the drawing. Obviously this is impractical, and even ridiculous. Historically, an adequate, representative set of points was dimensioned, enough points to describe the overall shape of the surface. This is a three-

*ASME Y14.5M-1994 Dimensioning and Tolerancing Standard. New York, New York: The American Society of Mechanical Engineers, 1995.

dimensional adult version of "connect the dots" that we enjoyed so much as children. The surface in between the dimensioned points was a bit of a problem, as it was undefined. Here craftsmanship took over, and a note may have been added to the drawing to "blend" the surface, to create as smooth a transition between the dimensioned points as possible. Although imprecise, this method worked well enough. The problems it presented were outweighed by the difficulty or impossibility of completely dimensioning the surface.

Today parts and assemblies are designed using computers. Computer-aided design and drafting (CADD or CAD) programs are mathematically precise, employing algorithms based on an IEEE double floating-point precision standard. Such programs are precise to 16 places, and the three-dimensional shapes that are modeled using these programs can be considered to be completely dimensionally defined within the CAD system. Using such a CAD system, an operator can obtain as much dimensional information about a surface as required, as all points on the surface are mathematically defined.

Today it is common in many industries to eliminate most or all of the dimensions on drawings of such complex shapes. In fact, some companies are eliminating most or all of the dimensions from all their drawings, regardless if the part geometry is simple or complex. These drawings contain one or more notes instructing the person using the drawing to get the dimensional information for the part features directly from the 3D CAD model. This approach works well as long as everyone that needs to obtain dimensional information from the drawing has access to the correct CAD program. Such drawings are called *math-data-based drawings, model-based drawings, limited-dimension drawings*, or other names.

Indeed the fantastic 3D solid modeling CAD tools available to designers today allow ever more complex geometry to be designed. In many cases, the 3D solid model dimensional data representing the part is electronically transferred directly to a computer-based manufacturing center, and the drawing is not even consulted for dimensional information. The manufacturing computers are programmed to make the part described by the CAD solid model. Likewise the 3D solid model dimensional data is electronically transferred into a computer-based inspection tool, such as a coordinate measuring machine (CMM), and the inspection computers are programmed to inspect the part described by the CAD solid model.

It is very important to recognize that something is still missing, however. The 3D solid model data merely represents the part's nominal or as-designed geometry—it is analogous to the dimensions on a drawing, as it is truly dimensional data. The model only tells the dimensioning half of the story—the tolerances must still be specified.

Every feature of a part must be completely toleranced, requiring one or more tolerances to define its limits of acceptability. This leads to an important question: What is a Tolerance?

## Definitions of a Tolerance

Option 1: A Tolerance specifies how closely to the nominal (or exact) location, size, form, or orientation that a feature on a part must lie.

Option 2: A Tolerance specifies the range of acceptable deviation for a feature on a part.

Option 3: According to *Merriam Webster**, "The allowable deviation from a standard; *especially*: the range of variation permitted in maintaining a specified dimension in machining a piece."

Final Answer: A Tolerance is the specified amount a feature is allowed to vary from nominal. This may include the location, size, form, or orientation of the feature as applicable.

Manufacturing processes are used to make every feature on a part. For example, on a machined part as in Fig. 1.1, the surfaces are milled, the holes are drilled, and the groove is cut using a milling cutter. Each manufacturing process is capable of attaining a certain level of accuracy and precision. One process may be capable of greater accuracy and precision than another, such as drilling a hole with and without a drill bushing or reaming or boring that same hole. A sheet metal part stamped using an automated process is typically more accurate and precise than the same part produced using a manual process. Tolerances should be chosen that are achievable using a chosen manufacturing process.

Manufacturing processes are often measured in terms of their precision and accuracy. Precision is a measure of how repeatable a process is, how closely it can hit the same point, regardless of where that point is. For instance, Kevin may be a bad shot, but if he consistently misses the target and hits the same wrong spot, he is precise Accuracy is a measure of how close to the chosen target a process can get. For instance, Sandra may be a good shot, and if she hits the bull's-eye or near the bull's-eye, she is accurate.

As an example, consider throwing darts at a dartboard. Precision is a measure of how closely grouped all the darts are. Accuracy is a measure of how closely a dart is to the bull's-eye. Accuracy and precision are shown in Fig. 1.2.

Every feature on every part is subject to variation. No feature can be made perfectly—all manufactured parts are understood to be imperfect replicas of the part defined on the drawing. If the drawing specifies that a dimension shall be 8.000-in., it must also specify how much variation is acceptable. Consider the following examples:

---

* *Webster's Ninth New Collegiate Dictionary*. Springfield, Mass.: Merriam-Webster, Inc., 1984. p. 1241.

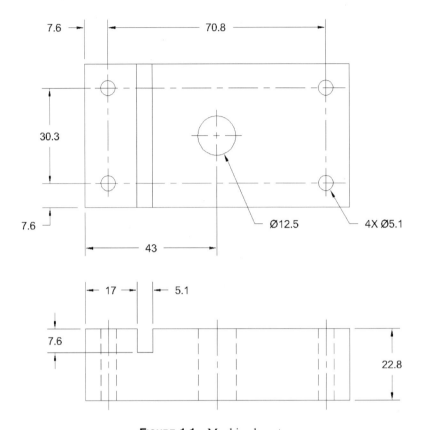

**FIGURE 1.1** Machined part.

## Example 1.1

A machinist sets up a part on her machine and removes metal, approaching the 8.000-in. dimension. She measures the part and sees it is 8.002 in. (See Fig. 1.3). She realizes that if she takes one more cut on the part it will remove 0.003 in., 0.001 in. more than the 0.002 in. of material remaining above the 8.000-in. dimension. So her choice is 7.999 or 8.002 in., unless she changes to a different, more precise process. Looking at the drawing, she sees that the tolerance for the dimension is ±0.005 in. She realizes that the part is within tolerance as it is.

What if the drawing didn't have any tolerances specified? The machinist to would have to guess how closely she had to make the part. Perhaps she arbitrarily decides that ±.010 in. is close enough, and machines the surface down to 8.008 in. and stops. The part then goes to inspection, and the inspector arbitrarily decides that the dimension should be within 0.001 in.,

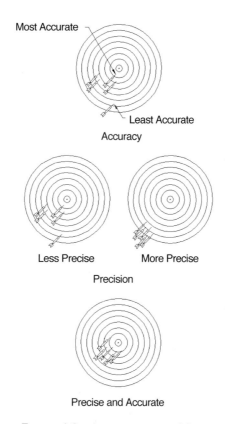

**FIGURE 1.2** Accuracy vs. precision.

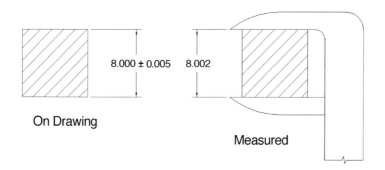

**FIGURE 1.3** As-produced part.

and rejects the part. He calls the machinist, who then calls the engineer and asks what he thinks he can live with. Not wanting to throw parts away, the engineer calls the inspector and tells him to accept the part. However, the design required the surface to be within $\pm.005$ in. By trying to reduce scrap and keep everybody happy, the engineer has accepted a bad part.

Obviously, the designer should never leave the responsibility of determining how accurately a part must made or how closely a tolerance must be held to someone else. The only person who understands the functional requirements of the part is the designer, and it is the designer's responsibility to determine, calculate, and communicate the limits of acceptability. These limits of acceptability are the tolerances specified on the drawing, on a formally referenced document or in a company standard. If the tolerances that apply are the default tolerances in the title block, the designer must ensure that those tolerances are acceptable. Whether the tolerances are explicitly specified or implicitly specified, they must be verified to work.

### Example 1.2

A first-article or prototype sheet metal part such as the one shown in Fig. 1.4 is stamped using a die. Many thousands of parts are to be made using this die. The part is inspected after stamping, and it is found that two holes are located 0.5 mm from nominal and the 90° angle between the flanges is actually 91.5°.

Consulting the drawing, the inspector sees that the holes must be located within $\pm1$ mm, and the flanges must be within $\pm1°$. The holes are within tolerance, but the flanges are out of tolerance. The part is rejected, and the die is modified to bring the parts within specification.

If the drawing did not have any tolerances specified, the die maker would have to guess how accurately to make the die, the press operator would have to guess if the die was functioning properly, and the inspector would have to guess if the parts were within specification. Of course, these determinations would be made independently, without knowing what the engineer determined was necessary for the design to function. Again, the drawing must specify the tolerances, so everyone using the drawing works to the same specifications.

The confusion and costly waste of time resulting from missing specifications could have been avoided if the engineer had done his job up front and carefully specified the tolerances on the drawing. In fact, those responsible for preparing drawings today must apply tolerances to all dimensions, whether directly or indirectly (explicitly or implicitly), in the form of plus/minus or geometric tolerances.

### Example 1.3

In this example, it is agreed that a part will be manufactured directly from 3D CAD model geometry data. The 3D CAD model geometry will be exported

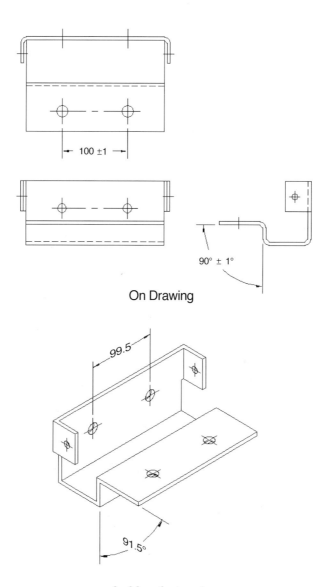

On Drawing

As Manufactured

**FIGURE 1.4** Sheet metal part.

from the CAD program directly into a computer numerically controlled (CNC) manufacturing program. The steps of the manufacturing process will be programmed and built around the CAD geometry. Additionally, a coordinate measuring machine (CMM) will be used to inspect the part. Again, the 3D CAD model geometry will be exported from the CAD program directly into the CMM computer, and the steps of the inspection routine will be programmed and built around the CAD geometry.

Because the manufacturing and inspection processes are automated and will use the CAD model geometry directly, the designer decides not to add any dimensions or Tolerances to the drawing at all. The designer understands that the dimensional data exported to the manufacturing and inspection programs completely defines the part, and no additional dimensions are required. However, the designer misses an important point.

A 3D CAD solid model accurately and completely defines the nominal part geometry—the model represents the perfect part, without variation. That is only half the problem. Without specifying tolerances it is impossible to know the limits of acceptability and whether the as-produced part is within those limits. Without any tolerances specified, no one can differentiate between a good part and a bad part, so it makes no sense to inspect the part. Obviously this is unacceptable.

The designer decides to rely on the manufacturing process capabilities to determine the allowable tolerances. He calls the shop and asks the manufacturing representative about the processes and the capabilities of their machinery. For the part in question, the manufacturing representative tells him the machine is accurate to ±0.005 and repeatable to within ±0.008. The designer is now happy that the burden of tolerancing the part has been lifted, calls the inspection shop, and tells them all the features on a part will be within ±0.005 and all the parts will be within ±0.008.

There are still some problems. The design manager learns of what is transpiring and calls the designer. She asks the designer "Why didn't you specify tolerances for the part?" The designer explains his position. The design manager explains that the tolerances must be formally stated on the drawing or in a related document to be legally binding. Still looking for a shortcut, the designer puts a note on the drawing as follows "Tolerances on all part features shall match process capability of ACME mill #123 in Bldg. A." He is happy and feels he has nailed it. He has tied the tolerances to the capability of the exact machine that will be used to manufacture the part.

Again his supervisor calls with more questions. She asks "Did you tolerance the drawing?" He explains what he did and his justification. She asks if he can tell her exactly what the limits are for a particular dimension; say the distance between two parallel faces. He reverts to what he was told about the process capability and tells her "The tolerance on that dimension is

±0.005, and the variation part to part is ±0.008. She asks him if he obtained formal SPC data from the shop for that exact part on that exact machine to verify their capabilities. He says "No" and tells her that the values were from the operator's manual that came with the machine when it was new.

The design manager explains that the capability information that came with the machine when it was new is only a starting point, and that there are many other sources of variation that add to these initial values. She also explains that merely adding a note to the drawing stating that the tolerances are tied to the manufacturing process is legally inadequate, as the process could change, and in fact will change over time. So no limits were actually defined. The designer grumbles and tells the design manager "C'mon, the parts that come off that machine always work—it's a very accurate machine. Why bother with Tolerancing the parts anyway?"

The design manager explains that the parts made on a particular machine may work and that they may satisfy their functional requirements. The problem, she says, is that the limits are not defined and with that comes several more problems. First, without defined tolerances, it is nearly impossible to do a Tolerance Stackup—the only way a Tolerance Stackup can be done in such a situation is to guess or assume values for the tolerances. Second, if it was decided to change the process and allow an outside vendor or another machine shop to make the parts, the process capabilities would be different, which would lead to different tolerances. Indeed, since the tolerances are not defined, there would be no way to tell a good part from a bad part.

Now frustrated, the designer, still looking for the shortcut, changes the note to read "Tolerances on all part features: ±0.008." He believes this captures 100% of the parts and that he has done his job. Again the design manager calls. She tells him that she has seen the updated drawing, has read the note, and has several other issues. First, it is apparent that the designer has not determined functional tolerance values, tolerances that when even at their worst-case will still allow the part to function. The designer has merely resorted to picking a global tolerance that can be manufactured. It is important that the tolerances are achievable by manufacturing, but it is more important the part will function properly. Furthermore, although he has defined linear limits of acceptability with the ±0.008 tolerance on every feature, he has not adequately defined the relationship of the features to one another. The angular relationship between the features is undefined. The designer points to the default angular tolerance in the title block and adds another note to make it apply to the CAD model geometry. This gets him close, but the specifications are still very ambiguous and subject to multiple interpretations.

The ambiguity problem can only be solved using GD&T. The part must be staged or setup for inspection. Part features must be related to one another clearly and unambiguously, and GD&T is the only way to do it. Finally he relents, takes the time, and applies GD&T to the drawing. He still doesn't explicitly state basic dimensions on the drawing, as he relies on the CAD model geometry for the basic dimensions. However, now the tolerances, specifically GD&T, are clearly stated on the drawing. It is now exactly clear what the tolerances mean and how they relate to the part. GD&T is a mathematically precise method of dimensioning and tolerancing, and it is appropriate to use such a precise method in this digital context. The tolerance zones created by the GD&T specifications are precisely located in space relative to their datum reference frames, and the rules of GD&T explain exactly where the tolerance zones are relative to one another. Finally, by using GD&T the designer has done his job, having completely and unambiguously toleranced the drawing. He was able to take a shortcut by using the CAD model geometry and avoid adding basic dimensions to the drawing. It just took him a while to understand that the dimensions are only half the story.

This example brings several issues to light. Automated and semiautomated manufacturing and inspection processes are prevalent in industry today, and many firms want to take advantage of the increased accuracy and potential savings they offer. It is important to understand what can and what cannot be eliminated from the drawing in such scenarios or whether a drawing is needed at all. It is a good idea to consult with one of the firms that specialize in streamlining documentation for automated manufacturing and inspection processes, such as Advanced Dimensional Management™. Such firms can help make sure the part and its limits of acceptability are completely defined and that everyone has access to the information they need.

It should be noted that the best way to tolerance features is by using Geometric Dimensioning and Tolerancing (GD&T). GD&T is the only way to ensure that everyone interprets the dimensioning and tolerancing specifications the same way.

GD&T is covered in Chapter 9 of this text. Advanced Dimensional Management offers several GD&T courses tailored to specific needs. Most importantly, their GD&T courses reinforce and compliment the ideas in this text, as the GD&T is presented and the tolerancing schemes developed are verified using the Tolerance Analysis techniques in this text.

It is one thing to splatter and sprinkle Tolerancing on a drawing so that it looks good; so that it appears that an adequate tolerancing job was done. It is quite another thing to do it right, to make sure that the tolerances will work, that they will not lead to parts that don't fit or function, that they can be achieved using the intended manufacturing and assembly processes, and that all downstream users of the specifications understand them.

The designer's goal should be the latter, as the decisions made during the design of a part live throughout the lifecycle of the part. Solutions to design problems such as tolerancing should be long-term solutions.

# 2

## Dimensioning and Tolerancing

This chapter provides a brief review of types and formats of dimensions and tolerances in both U.S. inch and metric formats.

### TYPES OF DIMENSIONS

Dimensions specify the nominal size, location, orientation, and form of part features. Every feature on a part, either individually or as part of a pattern, must be dimensioned. Historically dimensions have been included on drawings, as a dimensioned drawing was the only means available to describe a part. Today, most drawings are generated using 3D computer aided design (CAD) solid modeling software, and many of those drawings do not include dimensions. Part geometry in CAD files is defined mathematically and is often referred to as *mathematically defined* or *model data*. CAD math data drawings are often used by companies that can read the three-dimensional computer data directly into their manufacturing and inspection systems, thus reducing the necessity of a fully dimensioned drawing. Specific dimensions may be obtained by measuring or querying the part model using the CAD software or a similar program.

There are several types and formats of dimensions. Figure 2.1 includes examples of the various types of linear, polar, radial, and diametral dimensions. It also shows two examples of chain dimensions, one with the dimensions chained completely across the part and another with an overall dimension and one of the chained dimensions omitted. Figures 2.2 and 2.3 include examples of three formats for rectangular coordinate dimensioning.

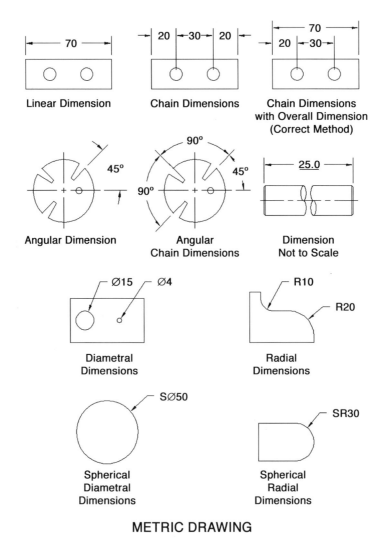

**METRIC DRAWING**

**FIGURE 2.1** Linear and angular dimensions.

The three methods shown are equivalent. There is no difference in the legal interpretation for these methods—the only difference is in their format. Figure 2.4 contains a sample drawing with Geometric Dimensions and Tolerances. Note that some dimensions with ± tolerances are also used on this drawing, but only to define the nominal size and size tolerance for features of size. This is common practice.

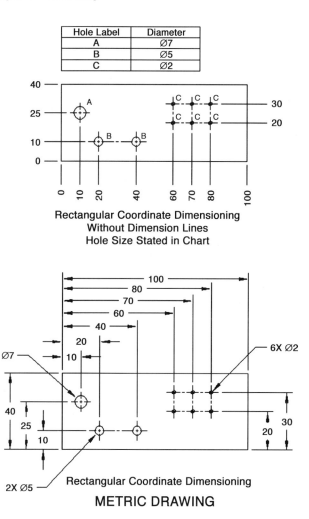

| Hole Label | Diameter |
|------------|----------|
| A | Ø7 |
| B | Ø5 |
| C | Ø2 |

Rectangular Coordinate Dimensioning
Without Dimension Lines
Hole Size Stated in Chart

Rectangular Coordinate Dimensioning
**METRIC DRAWING**

**FIGURE 2.2** Rectangular coordinate dimensioning.

The dimensioning strategy chosen for a drawing can greatly affect the tolerance between part features. Whether two features are related by a single dimension or by a series of dimensions determines the number of tolerances contributing to the variation possible between the features.

For parts depicted on traditionally dimensioned drawings, that is, drawings that have dimensions, it is important that the dimensioning matches the intended function of the part. This is true for drawings dimensioned using

| Hole | Diameter | X | Y | DEPTH |
|------|----------|-----|-----|-------|
| A1 | Ø7 | 10 | 25 | THRU |
| B1 | Ø5 | 20 | 10 | THRU |
| B2 | Ø5 | 40 | 10 | THRU |
| C1 | Ø2 | 60 | 30 | THRU |
| C2 | Ø2 | 70 | 30 | THRU |
| C3 | Ø2 | 80 | 30 | THRU |
| C4 | Ø2 | 60 | 20 | THRU |
| C5 | Ø2 | 70 | 20 | THRU |
| C6 | Ø2 | 80 | 20 | THRU |

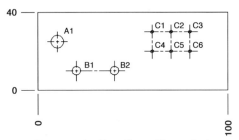

Rectangular Coordinate Dimensioning
Without Dimension Lines for Holes
Hole Size and Location
Stated in Chart

**METRIC DRAWING**

**FIGURE 2.3** Rectangular coordinate dimensioning.

± or GD&T, although GD&T is the only way to unambiguously communicate functional relationships.

Dimensions must be arranged and related in such a way to minimize tolerance accumulation between related features. Although rectangular coordinate dimensioning as shown above is convenient, easy to do in CAD, and ties in well with numerical control (NC) processing, it rarely (if ever) properly reflects the functional interrelationship between part features. All features are merely related to an arbitrarily selected origin. It is a far better approach to relate features functionally.

GD&T is primarily used with basic dimensions, which specify the exact (or nominal) location of features. Tolerances associated with the basic dimensions are found in feature control frames.

As stated earlier, some parts modeled using CAD have no explicitly stated dimensions on the face of the drawing. The dimensions are understood to exist in the CAD model data—any and every dimension is implied and must be obtained from the CAD system. For this system to work, a note must

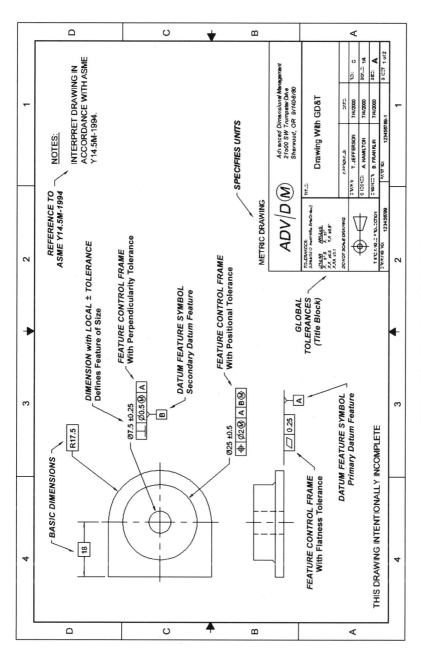

**FIGURE 2.4** Drawing with GD&T.

be added to the drawing instructing users that the dimensions are to be obtained from the CAD model data. This method also requires that GD&T is used to tolerance the features defined by the model data. The GD&T may be explicitly specified or implicitly specified.

## TYPES OF TOLERANCES

Two types of tolerances are common on mechanical drawings: plus/minus ($\pm$) tolerances and geometric tolerances. *Plus/minus tolerances* relate to linear distances or displacements and are stated in linear units (inches, millimeters, ...), or they relate to polar displacements and are stated in angular units (degrees, radians, ...). Linear tolerances are associated with linear dimensions, and angular tolerances are associated with angular dimensions. Typically, tolerances are stated in the same units as the dimension; hence a linear metric dimension has a linear metric tolerance. Tolerances may be stated specifically or generically as described below.

### Title Block or General Note Tolerances

These tolerances are specified in the title block or in the general notes and apply to the entire drawing. They may be overridden by a locally specified tolerance, which may have a larger or smaller value. Where used, the tolerance value is associated with the number of decimal places in each dimension. This is commonly found on drawings prepared to U.S. inch standards. It should be noted that many U.S. companies who have converted to the metric system have adopted this practice as well. (See Fig. 2.4.)

### Local Plus/Minus Tolerances

These are specified adjacent to each dimension and apply only to that dimension or group of dimensions. (See Fig. 2.4.)

## GEOMETRIC DIMENSIONING AND TOLERANCING (GD&T)

GD&T is a symbolic language that precisely defines the allowable variation in size, form, orientation, and location of features on a part. More importantly, GD&T precisely defines the relationship between features on a part, specifying which features are to be used to establish the origin of measurements for locating other features. Geometric tolerances are specified in feature control frames and are primarily associated with features located by basic dimensions.

It should be noted that only linear units may be specified in a feature control frame. For example, the geometric tolerances used to control an angle specify tolerance zones using linear units such as inches or millimeters, unlike ± tolerances used to control angles, which use polar units, such as degrees. Such differences are covered in depth in Advanced Dimensional Management's™ GD&T training courses and material.

GD&T is the only method for precisely defining part geometry. The geometric characteristic symbols used in feature control frames are shown in Fig. 2.5.

## Geometric Tolerances

| | GEOMETRIC TOLERANCE AND SYMBOL | |
|---|---|---|
| FORM TOLERANCES | —— | STRAIGHTNESS |
| | ▱ | FLATNESS |
| | ○ | CIRCULARITY (ROUNDNESS) |
| | ⌭ | CYLINDRICITY |
| ORIENTATION TOLERANCES | ∠ | ANGULARITY |
| | ⊥ | PERPENDICULARITY |
| | // | PARALLELISM |
| LOCATION TOLERANCES | ⊕ | POSITION |
| | ◎ | CONCENTRICITY |
| | ⌰ | SYMMETRY |
| PROFILE TOLERANCES | ⌒ | PROFILE OF A LINE |
| | ⌓ | PROFILE OF A SURFACE |
| RUNOUT TOLERANCES | ↗ | CIRCULAR RUNOUT |
| | ↗↗ | TOTAL RUNOUT |

**FIGURE 2.5** GD&T symbology.

## FEATURE CHARACTERISTICS AND ASSOCIATED TOLERANCE TYPES

This section discusses the variable geometric characteristics of part features and the associated types of tolerances. Every feature on a part is subject to variation and must be completely toleranced. This includes the geometric characteristics of the feature itself, such as its size and its form, and the relationship of the feature to the rest of the part, such as where it lies or how much it tilts relative to another feature or a *datum reference frame*. The variation that is allowed for each geometric characteristic of every feature must be fully defined. Additionally, the variation that is allowed in the relationship of every feature to the rest of the part must also be fully defined. This variation may be specified directly as a tolerance or indirectly as a subset of another tolerance.

There are four geometric characteristics that describe feature geometry and the interrelationship of part features. These are

- Size
- Form
- Orientation
- Location

Consequently, there are four types of tolerances that are possible for each feature. These are

- Size tolerances
- Form tolerances
- Orientation tolerances
- Location tolerances

Every feature on a part, however, does not necessarily possess all four characteristics.

(Note: this discussion does not address other geometric aspects of surface geometry such as surface texture.)

### Size

Size can be considered as the magnitude of the straight-line distance between two points on one or two surfaces whose surface normal vectors are colinear and point in opposite directions. Size is measured normal to each surface along the line between the points. Such points are considered to be opposed or in opposition. If every point on a nominal surface is opposed by another point on the nominal surface, the feature is said to be a *feature of size*. This matters because the ASME Y14.5M-1994 standard only discusses size as it relates to features of size, which are cylindrical surfaces, spherical surfaces, and width features that consist of two opposed parallel planes. There are other two-dimensional features that may be considered features of size—those are not addressed here.

Only features of size have *size* as defined in the ASME Y14.5M-1994 standard. Therefore, only those features that are features of size require a size tolerance. Portions of features may possess the characteristics of being a feature of size, and that portion requires a size tolerance.

A size tolerance is often specified as a ± tolerance associated with a dimension. This isn't the only way to specify a size tolerance, however, as a profile of a surface tolerance could be specified with a basic dimension to define the size limits for a feature. For example, a width feature could be specified with one planar surface as datum feature A, the opposing planar surface located a basic distance away, a flatness tolerance specified for the datum feature, and a profile of a surface tolerance specified for the other surface. In a different example a cylindrical surface could be defined with a basic dimension and toleranced using profile of a surface. Such features are not dimensioned and toleranced as traditional features of size, but their size limits and form limits have been completely defined.

Some features, such as a single planar feature, do not have size characteristics and therefore do not require a size tolerance to be completely defined.

## Form

Form can be considered as the shape of a feature. Every feature has form, regardless of whether it is nominally a flat plane, a cylinder, a width, a sphere, a cone, or a mathematically complex surface such as a paraboloid or the surface of an automobile windshield.

Consequently, every feature must have a form tolerance, either directly or indirectly specified. Examples of directly specified *form tolerances* include flatness, circularity, cylindricity, and straightness. An example of an indirectly specified form tolerance comes with rule #1, which requires perfect form at maximum material condition (MMC) when a size dimension and ± tolerance are applied to a feature of size. Another way to control form is to specify a profile of a surface tolerance to a basically defined surface. Depending on the context and datum feature references in the feature control frame, profile of a surface may control form, orientation, location, and possibly even size—however, when properly specified, it always controls form.

Such indirect methods of controlling form can be overridden by specifying a form tolerance with a smaller value. For example, consider a basically located planar surface with a profile of a surface tolerance and a flatness tolerance: if the flatness tolerance value is less than the profile tolerance value, then the flatness tolerance overrides the form control provided by the profile tolerance. The form of the surface may only vary as much as the flatness tolerance allows.

Directly or indirectly, a form tolerance must be specified for every feature of a part.

## Orientation

Orientation can be considered as the angle between features, or more precisely, *orientation* is the amount a feature may tilt relative to a datum reference frame. Aside from the primary datum feature, every feature on a part is oriented to other features. A primary datum feature is exempt because all other features are directly or indirectly oriented to it, rather than the other way around.

Consequently, every feature on a part except the primary datum feature must have an orientation tolerance, either directly or indirectly specified. An orientation tolerance must be specified for all but the main primary datum feature on parts with more than one primary datum feature. For example, on parts with more than one datum reference frame, there is usually one datum reference frame that is considered the main or global datum reference frame. It is the datum reference frame to which the majority of part features are related, and the other datum reference frames are related to it as well.

Like form, orientation may also be controlled directly or indirectly. Many drawings that use dimensions with ± tolerances for all features rely on the default angular ± tolerance in the title block to control the orientation of all features. Even some drawings that use GD&T may rely on this default angular tolerance. Such practice is problematic and should be avoided.

Other methods of indirectly specifying an orientation tolerance occur where a profile of a surface tolerance is related to a datum reference frame and where a positional tolerance is related to a datum reference frame—both of these cases control orientation and may also control size, form, and location.

## Location

Location can be considered as where a feature lies relative to another feature, or more precisely, location is where a feature lies relative to a datum reference frame.

Consequently, most features on a part must have a location tolerance. Examples where a Location tolerance is not required include: a location tolerance is not required for the planar datum features on parts with mutually perpendicular planar primary, secondary and tertiary datum features; a location tolerance is not required for the primary and secondary datum features on parts with a planar primary datum feature and a secondary datum feature of size that perpendicular to the primary datum feature. Most features, however, require a location tolerance.

Location tolerances must be directly specified, as they are not subsets of other tolerance types. For example, a positional tolerance related to a datum reference frame controls orientation as a subset of position, but only positional, runout, concentricity, symmetry and profile tolerances control location.

# 3

## Tolerance Format and Decimal Places

Figure 3.1 shows the four standard formats for linear and angular dimensions and tolerances. Formats are included for U.S. inch and metric Dimensioning and Tolerancing. The rules for angular dimensions and tolerances are the same for drawings prepared using U.S. inch and metric units.

> *Limit dimensions* do not specify a nominal value. A high (maximum) value and a low (minimum) value are specified. When a limit dimension is stated in a horizontal format, the smaller value precedes the larger value, with the values separated by a dash. When a limit dimension is stated in a vertical format, the larger value (upper limit) is placed above the smaller value (lower limit). It makes no difference whether the toleranced feature is an internal feature or an external feature.
>
> *Equal bilaterally toleranced dimensions* specify a nominal value and the amount a dimension may deviate from nominal. The tolerance values are equal in each direction.
>
> *Unequal bilaterally toleranced dimensions* specify a nominal value and the amount a dimension may deviate from nominal. The tolerance values are not equal in each direction. Neither tolerance value is zero.
>
> *Unilaterally toleranced dimensions* specify a nominal value and the amount a dimension may deviate from nominal in one direction only. The tolerance is in one direction only, either larger or smaller.

## Formatting Dimensions
## with Plus / Minus Tolerances

| Tolerancing with SI Units: (millimeters) | Tolerancing with U.S. Customary Units: (inches) |
|---|---|
| Limit Dimension<br><br>8.75<br>8.25    ★★★<br><br>Same number of decimal places in both limits. | Limit Dimension<br><br>8.75<br>8.25<br><br>Same number of decimal places in both limits. |
| Equal Bilateral Tolerancing<br><br>8.5 ±0.25<br><br>Number of decimal places may be different for dimension and tolerance. | Equal Bilateral Tolerancing<br><br>8.50 ±.25<br><br>Number of decimal places must be the same for dimension and tolerance. |
| Unequal Bilateral Tolerancing<br><br>8.5 $^{+0.25}_{-0.40}$    ★★★<br><br>Number of decimal places may be different for dimension and tolerances. Both tolerances must have the same number of decimal places. | Unequal Bilateral Tolerancing<br><br>8.50 $^{+.25}_{-.40}$<br><br>Number of decimal places must be the same for dimension and both tolerances. |
| Unilateral Tolerancing<br><br>8.5 $^{+0.25}_{0}$<br><br>Number of decimal places may be different for dimension and tolerances. The zero tolerance has no decimal places and is not preceded by a + or - sign. | Unilateral Tolerancing<br><br>8.50 $^{+.25}_{-.00}$<br><br>Number of decimal places must be the same for dimensions and both tolerances. |
| Leading zeroes for dimensions and tolerances. Trailing zeroes only used in certain applications, (marked ★★★). | No leading zeroes for dimension values. Trailing zeroes used where needed. |

Angular Dimensions and Tolerances

Equal Bilateral    23.5° ±1.0°        Unequal Bilateral    23.50° $^{+0.50°}_{-0.25°}$

Angles may be specified using decimal degrees or degrees, minutes and seconds. If decimal degrees are used, the number of decimal places must be the same for the dimension and both tolerances. Angular dimensions and tolerances follow the same rules on drawings prepared using either type of units.

The examples above show the ways to specify dimensions and +/- tolerances on drawings prepared using SI units or U.S. Customary units. The way to specify angular dimensions and tolerances is also shown.

**FIGURE 3.1**   Plus/minus dimension and tolerance format (U.S. inch and metric).

Regardless which of the above methods is selected, the nominal dimension value must be part of the tolerance range. To put it another way, the upper and lower limits must include the nominal dimension, even if it is at one extreme of the range. The tolerance range cannot ever be "off the part."

As shown in Fig. 3.1, there are four different methods to specify tolerances and ranges. Each method specifies an upper and lower limit, either directly or indirectly. Three methods include a "nominal" value and acceptable deviation limits from that nominal, while limit dimensioning gives only the range, the upper and lower limits for the dimension.

Several questions may come to mind: What is the real difference between these methods? Do any of the methods better communicate design intent? Do any of the methods alone guarantee that manufacturing will target the nominal value stated on the drawing in the manufacturing process?

The short answer is no, as all these methods specify the same legal limits. (See Fig. 3.2). Given that these methods are legally equivalent, all that you can be sure of is that manufacturing will always shoot for what is in their best interest. They must. Based on the factors discussed in Chapter 5, they will adjust the process to minimize their costs. All of the methods ultimately communicate the same information, and that information is limits. Every plus/minus tolerance communicates a high limit and a low limit. However, depending on the manufacturing process, in some cases, the nominal value may be more important than others.

For example, if a hole is to be drilled, it is common to select a standard drill size for the hole. Of course the size of the hole must be verified to work using Tolerance Analysis, but if a hole is to be drilled, it is very common to select a standard sized hole for drilling. There are several methods for tolerancing drilled holes, and each has been derived empirically, the data collected after drilling thousands of holes in sample stock. Typically, drilled hole tolerances are *unequal bilateral*. The tolerance range is biased toward larger holes, such as +0.005/−0.002 or +0.004/−0.001. In cases where a hole is drilled and the nominal hole size on the drawing is a standard drill size, it is probably likely that manufacturing will target design nominal. However, it is not necessary for them to do so. They may have some older drills on hand that because of wear and other factors may target a different value. As long as the hole is within the range specified on the drawing, it is within specification and must be accepted as a good part. There is also the possibility that the hole will be put in using a different process.

The water gets muddied a little when statistical process control (SPC) comes in to the picture. Quality control methods such as SPC, combined with drawing specifications such as critical control characteristics, sometimes attempt to link "design nominal" with "manufacturing nominal." Perhaps

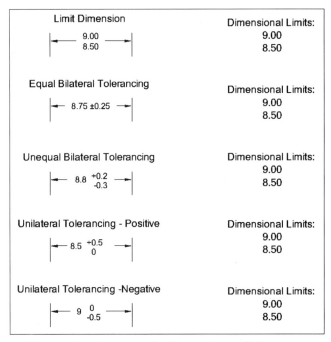

The examples above show the five ways to specify the same
dimensional limits using dimensions and +/- tolerances. The
tolerance range is the same for all five examples.

**FIGURE 3.2**    Five ways to specify limits using plus/minus dimensions and tolerances.

in an environment where these are strictly implemented there can be direct
linkage between design nominal and manufacturing nominal.

There is a theorem in statistics called the *Central Limit Theorem*, which
states that sampled values (tolerances in this case) under certain conditions
follow a normal or gaussian distribution and that it is more likely for a value
to be in the center of a range than at the extremes. Figure 3.3 shows a normal
distribution. Indeed this is the basis for the idea of statistical tolerancing as
presented in Chapter 8. The idea of a dimension "centering" about the
midpoint of its range seems to work well with equal bilaterally toleranced
dimensions but causes some grief when considered for unequal bilaterally or
unilaterally toleranced dimensions. In these cases, as in the example of the
drilled hole presented earlier, the nominal dimension value stated on the
drawing is not the midpoint of the tolerance range. This text will not attempt
to sort out these potential statistical inconsistencies.

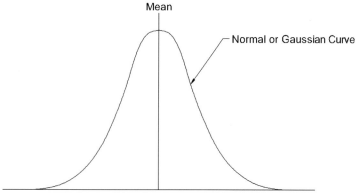

Tolerance values are more likely to be in the center of the range
near the mean than at the extremes with controlled processes.

**FIGURE 3.3** Gaussian distribution and the Central Limit Theorem.

Without SPC and the added controls it brings, we are back to the idea
that a plus/minus dimension allows anything within its tolerance range.
Worst-case tolerancing uses this as its basis and is presented in Chapter 7.

Designers may hope that stating a design nominal on the drawing will
bias manufacturing toward the stated value, and it may. However, in a legal
sense, it adds no more or less specific requirement. Using any of the above
methods merely specifies acceptable limits. All manufacturing must do is
ensure that their process lies within those limits.

# 4

Converting Plus/Minus Dimensions
and Tolerances into Equal Bilaterally
Toleranced Dimensions

The method of performing Tolerance Stackups taught in this text requires all
dimensions and tolerances to be converted into equal bilateral format.

This chapter describes the techniques for converting plus/minus dimensions and tolerances into the equal bilateral format. This is necessary for all
dimensions and tolerances, whether they are plus/minus or GD&T. Conversion of Geometric Dimensions and Tolerances into equal bilateral plus/minus
toleranced dimensions will be discussed in Chapter 9. Whether dealing with
U.S. inch or metric dimensions and tolerances, linear or angular units, the
technique for conversion is the same.

## CONVERTING LIMIT DIMENSIONS TO EQUAL
## BILATERAL FORMAT

The following example presents the procedure for converting limit dimensions into equal bilaterally toleranced dimensions.

*Example 4.1. Converting Limit Dimensions to Equal
Bilateral Format*

- Given a limit dimension,

    10.00 Upper limit (metric format)

    9.55 Lower limit (metric format)

- Subtract the lower limit from the upper limit to obtain the total tolerance.

  Total tolerance $= 10 - 9.55 = 0.45$

- Divide the total tolerance by 2 to obtain the equal bilateral tolerance value.

  Equal bilateral tolerance value $= \dfrac{0.45}{2} = 0.225$

- Add the equal bilateral tolerance value to the lower limit. This is the adjusted nominal value.

  Adjusted nominal value $= 9.55 + 0.225 = 9.775$

  (Note: The adjusted nominal value can also be obtained by subtracting the equal bilateral tolerance value from the upper limit.)

Conversion complete:

  Equal bilateral equivalent $= 9.775 \pm 0.225$

## CONVERTING UNEQUAL BILATERAL FORMAT TO EQUAL BILATERAL FORMAT

The following example presents the procedure for converting unequal bilaterally toleranced dimensions into equal bilaterally toleranced dimensions.

*Example 4.2. Converting Unequal Bilaterally Formatted Dimensions to Equal Bilateral Format*

- Given an unequal bilaterally toleranced dimension (inch format),

  $8.50 \, {}^{+\,.25}_{-\,.10}$

- Establish upper and lower limits. Add the plus tolerance to the nominal value; this is the upper limit. Subtract the minus tolerance from the nominal value; this is the lower limit.

  Upper limit $= 8.50 + .25 = 8.75$

  Lower limit $= 8.50 - .10 = 8.40$

- Subtract the lower limit from the upper limit to obtain the total tolerance. (Note: The total tolerance can also be obtained by adding the $+$ and $-$ tolerances given.)

Total tolerance derived from limits $= 8.75 - 8.40 = .35$

or

Total tolerance derived from given tolerances

$\quad = .25 + .10 = .35$

- Divide the total tolerance by 2 to obtain the equal bilateral tolerance value.

Equal bilateral tolerance value $= \dfrac{.35}{2} = .175$

- Add the equal bilateral tolerance value to the lower limit. This is the adjusted nominal value. Establish the adjusted nominal value.

$8.40 + .175 = 8.575$

(Note: The adjusted nominal value can also be obtained by subtracting the equal bilateral tolerance value from the upper limit.)

Conversion complete :

Equal bilateral equivalent $= 8.575 \pm .175$

## CONVERTING UNILATERALLY POSITIVE FORMAT TO EQUAL BILATERAL FORMAT

The following example presents the procedure for converting unilaterally positive toleranced dimensions (plus something, minus nothing) into equal bilaterally toleranced dimensions.

*Example 4.3. Converting Unilaterally Positive Formatted Dimensions to Equal Bilateral Format*

- Given a unilaterally positive toleranced dimension (inch format),

$8.50 \begin{array}{l} + .25 \\ - .00 \end{array}$

- Establish upper and lower limits. Add the plus tolerance to the nominal value; this is the upper limit. The specified nominal value is the lower limit.

Upper limit $= 8.50 + .25 = 8.75$
Lower limit $= 8.50 - .00 = 8.50$

- Subtract the lower limit from the upper limit to obtain the total tolerance. (Note: The total tolerance is equivalent to the plus tolerance.)

Total tolerance derived from limits $= 8.75 - 8.50 = .25$

or

Total tolerance derived from given tolerances

$$= .25 - .00 = .25$$

- Divide the total tolerance by 2 to obtain the equal bilateral tolerance value.

Equal bilateral tolerance value $= \dfrac{.25}{2} = .125$

- Add the equal bilateral tolerance to the lower limit. This is the adjusted nominal value. Establish the adjusted nominal value.

$8.50 + .125 = 8.625$

(Note: The adjusted nominal value can also be obtained by subtracting the equal bilateral tolerance value from the upper limit.)

Conversion complete :

Equal bilateral equivalent $= 8.625 \pm .125$

## CONVERTING UNILATERALLY NEGATIVE FORMAT TO EQUAL BILATERAL FORMAT

The following example presents the procedure for converting unilaterally negative toleranced dimensions (plus nothing, minus something) into equal bilaterally toleranced dimensions.

*Example 4.4.  Converting Unilaterally Negative Formatted Dimensions to Equal Bilateral Format*

- Given a unilaterally negative toleranced dimension (metric format),

$8.5 \; {}^{0}_{-0.25}$

- Establish upper and lower limits. The specified nominal value is the upper limit. Subtract the negative tolerance from the nominal value; this is the lower limit.

Upper limit $= 8.5 + 0 = 8.5$

Lower limit $= 8.5 - 0.25 = 8.25$

- Subtract the lower limit from the upper limit to obtain the total tolerance. (Note: The total tolerance is equivalent to the minus tolerance.)

Total tolerance derived from limits $= 8.5 - 8.25 = 0.25$

or

Total tolerance derived from given tolerances

$$= 0 + 0.25 = 0.25$$

- Divide the total tolerance by 2 to obtain the equal bilateral tolerance value.
  
  Equal bilateral tolerance value $= \dfrac{0.25}{2} = 0.125$

- Add the equal bilateral tolerance to the lower limit. This is the adjusted nominal value. Establish the adjusted nominal value.

  $8.25 + 0.125 = 8.375$

(Note: The adjusted nominal value can also be obtained by subtracting the tolerance equal bilateral value from the upper limit.)

Conversion complete:

Equal bilateral equivalent $= 8.375 \pm 0.125$

## MEAN SHIFT

As presented in Chapter 3, design nominal is not always at the midpoint or mean of the tolerance range. In converting unequal bilaterally and unilaterally toleranced dimensions to equal bilateral format, we have changed the dimension value so it is at the midpoint or *mean* of the tolerance range.

In tolerancing jargon, we have affected a *mean shift*. The new *mean* or nominal value of the dimension is different than the value specified on the drawing. The "mean" has been "shifted" to the midpoint of the tolerance range. Remember our earlier discussion that it makes no difference how a tolerance range is specified, that is, whether limits, equal bilateral, unequal bilateral, or unilateral tolerances are specified, the result is the same. All that has legally been specified are upper and lower limits for a dimension. Only with equal bilateral tolerancing is the stated dimension value on the drawing the midpoint of the range.

Where dimensions are included in the Tolerance Stackup, the mean shift is little more than a curiosity, as it has no effect on the outcome of the Tolerance Stackup. The mean shifts are accounted for in the Tolerance Stackup method. Using more advanced and streamlined methods where the dimensions are not included and only the tolerances are manipulated, the mean shift must be accounted for. This text does not address the Tolerance

Analysis methods where dimensions are not included in the Tolerance Stackup.

An easy method to determine the mean shift for a dimension and tolerance converted to equal bilateral format follows. The mean shift will be calculated for the dimensions and tolerances shown in Examples 4.1 through 4.4.

### Example 4.1a.  Mean Shift Calculation for Limit Dimensions Converted into Equal Bilateral Format

In Example 4.1, limit dimensions were converted into equal bilateral format. Limit dimensions do not state a nominal or mean, so there is no mean shift when converting limit dimensions into equal bilateral format.

### Example 4.2a.  Mean Shift Calculation for Unequal Bilateral Format Converted into Equal Bilateral Format

Given an unequal bilaterally toleranced dimension (inch format)

$$8.50 \; {}^{+\,.25}_{-\,.10}$$

that has been converted into equal bilateral format,

$$8.575 \pm .175$$

the mean shift is calculated as follows:

Converted dimension value (mean) − original dimension value (nominal) = mean shift

Mean shift = $8.575 - 8.50 = .075$

Note the sign of the mean shift is positive, indicating the mean was shifted toward the high end of the tolerance range. Note: When converting an unequal bilaterally toleranced dimension to an equal bilaterally toleranced dimension, the mean shift is always ½ the difference between the positive and negative tolerance values, and the shift is toward the larger of the two values.

### Example 4.3a. Mean Shift Calculation for Unilateral Positive Format Converted into Equal Bilateral Format

Given a unilaterally positive toleranced dimension (inch format)

$$8.50 \; {}^{+\,.25}_{-\,.00}$$

that has been converted into equal bilateral format,

$$8.625 \pm .125$$

the mean shift is calculated as follows:

> Converted dimension value(mean) − original dimension value (nominal) = mean shift
>
> Mean shift = 8.625 − 8.50 = .125

Note the sign of the mean shift is positive, indicating the mean was shifted toward the high end of the tolerance range. Note: When converting a unilaterally positive toleranced dimension to an equal bilaterally toleranced dimension, the mean shift is always ½ the positive tolerance value, and the shift is toward the high end of the tolerance range.

### Example 4.4. Mean Shift Calculation for Unilateral Negative Format Converted into Equal Bilateral Format

Given a unilaterally negative toleranced dimension (metric format)

$$8.5 \, {}^{0}_{-0.25}$$

that has been converted into equal bilateral format,

$$8.375 \pm 0.125$$

the mean shift is calculated as follows:

> Converted dimension value (mean) − original dimension value (nominal) = Mean Shift
>
> Mean shift = 8.375 − 8.50 = −0.125

Note the sign of the mean shift is negative, indicating the mean was shifted toward the low end of the tolerance range. Note: When converting a unilaterally negative toleranced dimension to an equal bilaterally toleranced dimensions, the mean shift is always ½ the negative tolerance value, and the shift is toward the low end of the tolerance range.

The *mean* is really the midpoint of the range in this context. It merely lies at the middle of the range. From a mathematical point of view, the term *mean shift* in this context is actually a misnomer, as the "mean" requires multiple values to be determined, and for a single dimension in a Tolerance Stackup, there is only one specified value or range with a corresponding midpoint. According to the *McGraw-Hill Dictionary of Scientific and Technical Terms AccessScience Webpage (2003),*

> *mean* [MATHEMATICS] A single number that typifies a set of numbers, such as the arithmetic mean, the geometric mean, or the expected value. Also known as mean value.

For the purposes of Tolerance Analysis, we are interested in the Arithmetic Mean. The Arithmetic Mean is the average value for a group of values, and is found by adding the values and dividing the sum by the number of values: According to the *McGraw-Hill Dictionary of Scientific and Technical Terms AccessScience Webpage,*

> *arithmetic mean* [MATHEMATICS] The average of a collection of numbers obtained by dividing the sum of the numbers by the quantity of numbers. Also known as average (av).

So, strictly speaking, the mean in this case is the arithmetic mean, and the arithmetic mean requires a collection (or population) of values to be of any significance. This is exactly the situation that is found in inspection and manufacturing mass-produced parts, where the same features are inspected and manufactured many times. Say 100 parts are manufactured, and it is desired to know where the mean of the process lies for the diameter of a hole, with a dimension of $\emptyset 10 \pm 0.5$ specified. The diameter of the hole on each of the 100 parts is measured and recorded. The values are plotted on a chart, the arithmetic mean is calculated using the method described above and found to be $\emptyset 10.2$. The measured mean of the manufacturing process is $\emptyset 10.2$, which shows that the result of the manufacturing process is centered 0.2 above the specified mean or nominal.

In this context, there has been a mean shift of 0.2 in the positive direction, or toward a larger value. If the mean of the manufacturing process was 9.8, there would still be a mean shift of 0.2, but it would be in the negative direction—the mean shift could be stated as $-0.2$. You see, this value really is a mean value; it really is the result of taking the sum of a number of values and dividing that sum by the number of values, and in this example the number of values was 100.

## MEAN SHIFT RECAP

Mean shift in a Tolerance Stackup with dimension values stated is of little concern. When a dimension and its tolerance are converted into equal bilateral format, the dimension value may be shifted up or down depending on how the dimension and tolerance were specified. As long as all the dimensions and tolerances are treated in the same manner and included in the Tolerance Stackup, any dimensional mean shift will be accounted for in the final result.

This material is included primarily to educate the reader. The use of the term *mean shift* is fairly common in industry, and it is important for the reader to understand what it does and does not mean.

# 5

## Variation and Sources of Variation

### WHAT IS VARIATION?

In the context of manufacturing, variation is the amount one or more measured values deviate from a specified value. It is the imperfection seen in actual as-produced parts, contrasted against the perfect models created in CAD and seen on drawings. In the context of design and specification, tolerances on the drawing set the limits for variation. In the context of Tolerance Analysis and Tolerance Stackups, variation must be considered from a specification point of view as well as a measured value point of view.

- From a design or specification point of view, *variation* is the amount a toleranced feature is allowed to deviate from its specified value. At its worst-case or most extreme value, it is the maximum amount the feature may vary from nominal. Not surprisingly, the goal of a worst-case Tolerance Stackup is to make sure the feature will satisfy its intended function at its worst-case conditions. When a tolerance is specified on a drawing, it sets the limits of variation for any given feature. Good practice dictates that the tolerances specified represent limits that are functionally acceptable when the part is manufactured or assembled.

- From a manufacturing or measurement point of view, *variation* is the amount a manufactured or assembled feature has deviated from its specified value. Measurements determine whether the variation is within the specified limits. Tolerances set the boundaries within which manufacturing must operate.

- From a Tolerance Analysis point of view, *variation* is calculated when a Tolerance Stackup is performed. It may be the variation between the features of a single part, or it may be the variation between features of different parts. The only way to verify that the variation specified for a feature, multiple features, or multiple parts in an assembly is acceptable is to perform a Tolerance Stackup.

All manufactured part features are subject to variation—there is no such thing as a perfectly produced part, and no feature has ever been created without variation from its stated dimensions. Be it a rough, flame-cut edge of structural brace on a bridge, to a highly polished surface on an interplanetary probe, the feature will be imperfect. There is always some associated error or variation for every produced feature. The amount of variation may be very large, in the case of the flame-cut structural member, or it may be very small, as in the case of the polished surface, but it is still there. No manufactured feature is perfect.

Today, most design engineers use 3D solid modeling CAD software to design their parts and assemblies. The models created by such systems are mathematically precise, and many would say they are nearly perfect in their representation of the part geometry. Unfortunately, many design engineers confuse the precision of their CAD models with the actual part.

The perfect CAD model only represents a starting point; it is similar to the dimensioned geometry on a 2D drawing in that it merely defines the *nominal* or desired dimensional state of the part. When the part is manufactured, regardless of the method, there will be variation on each produced feature, and the part will be imperfect. Specifying tolerances allows the designer or design engineer to establish the functionally allowable dimensional limits for each feature. Many times the author has approached a design engineer asking about the lack of tolerances on a drawing, only to learn that the engineer believed that the 3D model itself was all that was required, and that specifying tolerances was unnecessary because the model was perfect and the part was going to be manufactured right from the model. Of course the 3D model is only half the story, as it represents the nominal or perfect part. Tolerances must be added to communicate how much variation, and more specifically, how much of which type of variation is allowable for each feature. For example, the acceptable limits of form, orientation, and location of a surface may be required, as the surface may have to be very flat, it may be able to tilt within a few degrees from nominal, and its location may not be that important. These requirements can be communicated using several methods, but in this case it is likely that form, orientation, and profile tolerances would be used.

## SOURCES OF VARIATION

There are many factors that contribute to the variation found on a finished product; however, there are three major sources of variation that must be addressed and included in every Tolerance Stackup:

Tolerances specified on the drawing

Variation encountered in the inspection process

Variation encountered in the assembly process

Many initial attempts at a Tolerance Stackup only include the tolerances specified on the applicable part and assembly drawings. That makes sense, as one would expect specified tolerances to be included in a Tolerance Stackup. However, applicable variation from the inspection process and the assembly process must also be included. The inspection process may contribute variation where drawings are based on GD&T and datum features of size are referenced at MMC or LMC. The assembly process may contribute variation where parts are related and/or located by external features within internal features, such as fasteners within holes.

A list of general and specific sources of variation and examples follows. This list includes the sources of variation that may originate within the manufacturing process, the inspection process and the assembly process. Of all the potential sources of variation listed below, only specified tolerances, datum feature shift and assembly shift should be included in a Tolerance Stackup—the other sources of variation are merely included here for descriptive purposes.

### Manufacturing Process Limitations (Process Capability)

Manufacturing processes have a limit to their accuracy and precision. For any given process, certain tolerances are easily achievable without extra effort or care. These are well within the process capability. Closer tolerances are achievable, but at increased cost due to the extra time and labor required or because some parts are out of tolerance and thrown away. Even closer tolerances may be virtually unachievable, due to the inherent variation in the process. If a particular process is not accurate or precise enough, a different process or design should be sought.

Information describing the process capability (precision and accuracy) of shop machinery is available from its manufacturer and can be found in the User's Manual; however, this information from the User's Manual typically only applies to new equipment in a special environment. To maintain the capability of the equipment requires that many factors are in place and certain conditions are maintained. Examples include that the machinery is set level, that it is clean, that there are no obstructions to any working mechanisms, that there is no damage to any working components, that humidity and temperature levels are controlled within specified limits, and very importantly, that the machinery has been properly maintained.

To get a more accurate picture of the processing machinery's capability, the author suggests obtaining recent statistical process control (SPC) data from the machinery for the part or parts under consideration. If it is a new design, use recent SPC data from a similar part or parts. This will ensure that the data best represents the variation that will be encountered while manufacturing the part or parts.

## Tool Wear

Cutting tools, drills, dies, all wear as they age, due to friction with the workpiece (part). As tools wear they reduce in size and become dull. For example, this may cause the tool to cut a smaller hole, an out-of-round hole, or shear a surface farther from its nominal location.

## Operator Error and Operator Bias

Operator error includes aspects such as improper handling of raw materials, improper clamping of material, and improper sequence of operations, among others. In automated processes, these errors are also possible, but hopefully with less frequency and effect. Factors such as training, turnover of personnel, time of day, all may have an impact on the frequency and severity of operator error.

Operator bias includes the effects of human factors and ergonomics, such as whether the operator is left handed or right handed, taller or shorter, stronger or weaker, etc. Depending on the process, these factors may play a role in biasing the process one way or the other.

## Variations in Material

Variations in the material from the foundry, or material formed or cut by a previous process contribute to possible variation. Of primary concern in mechanical Tolerance Analysis is the variation in size or form of raw material or stock shapes, such as sheet thickness, stock size variations or the angle between the sides of an extruded structural shape. Other types of material variation may include aspects such as hardness, ductility, porosity, chemical composition, or resistivity (or conductivity) to name a few.

## Ambient Conditions

Temperature, humidity, vibration, cleanliness, and other ambient factors affect the finished product.

Operating a machine in an environment that is outside the specified temperature range can affect the process greatly. Cooling systems may not be able to maintain the required operating temperature in an environment that exceeds the specified temperature limits. Lubricant s may break down more quickly in an excessively hot environment, and lubricants may be too viscous to effectively reduce wear between machine components in an environment that is too cold.

Lack of adequate clearance around the equipment can also have an effect. For example, obstructions in the path of stock being fed in to the machine could force the operator to feed the stock at an angle, affecting the final product. Inadequate clearance for maintenance could also lead to problems, as a machine that is difficult to service is likely not serviced—the more difficult it is to service a piece of equipment, the less likely it will be serviced. Inadequate clearance for cooling could lead to higher than expected operating temperatures. Inadequate clearance for services or improper routing and alignment of services, such as cooling water or drains, can also have deleterious effects on the process.

## Difference in Processing Equipment

Parts manufactured on a piece of equipment in one plant, such as vertical gun drilling machine, may be manufactured on a completely different machine at another plant, such as a horizontal gun drilling machine. Obviously there are different factors such as gravity and the capabilities of the machinery to consider in these cases. There may also be differences between the quality of parts manufactured on the same model of equipment in the same plant, due to number of factors.

## Difference in Process

A part that is manufactured using different processes on different processing lines is likely to have very different tolerances. Each process and machine will affect the part and its tolerances in a different way. For example, if a hole is die cast in plant A, drilled in plant B, drilled using a drill bushing in plant C, and reamed in plant D, the hole will have different tolerances as a result of each operation.

## Poor Maintenance

Processing machinery may be neglected, and preventative maintenance may be lacking. Hence the precision and accuracy possible when the machine was new, or properly maintained, is lost over time. Often this is a result of the demands and pressures of productivity, which may leave little downtime for maintenance.

Remember, the more difficult it is to service a piece of equipment, the less likely it will be serviced.

## Inspection Process Variation and Shortcuts

Although this may not seem like a source of manufacturing process variation, it is perhaps one of the most likely sources of apparent variation between processes.

Parts are often inspected in a manner that doesn't quite get the correct information, but in a quick and dirty world, the method is deemed to be an adequate approximation, and the risk worth taking. For example, size tolerances apply to the entire toleranced feature (i.e., the entire feature must be within the specified size tolerances), but it is common for only a single measurement at one location on the part to be taken.

Consider a part that is manufactured in two shifts in a plant and inspected using a different inspection process for each shift. Each inspection process uses different methods to "verify" that the tolerances have been met. Yet, each process reports different results for the same measurement. The apparent measured difference between the parts manufactured during day shift and night shift doesn't exist. The error is in the shortcuts taken in the inspection process, not in the manufacturing process!

The aspect of uncertainty of measurement must also be discussed here. A measurement can not be any more precise than the device used to make the measurement. As a general rule, the precision and accuracy of all measuring devices and procedures should be tested and verified before making any measurements. The error inherent in the inspection process must be quantified. This represents the uncertainty in the process. Good practice dictates that the inspection process error should be subtracted from the limits being measured. Some forward thinking corporations require their suppliers to adhere to this practice, often to the supplier's dismay. The reader is directed to ASME Standard B89 which covers Dimensional Metrology for more information.

The various sources of inspection process variation listed above are not included in Tolerance Stackups. However, there is one inspection process variable that must be included in Tolerance Stackups where it may be a factor: datum feature shift. Datum feature shift is encountered on drawings based on GD&T, and it occurs where datum features of size are referenced at MMC or LMC in a feature control frame. Datum feature shift is discussed in great detail in Chapter 9.

## Assembly Process Variation

The assembly process can have a profound effect on assembly variation. It is very important that the designer understands the assembly process. Assuming the wrong assembly process during design can lead to serious problems. The sequence of assembly operations has a huge effect on the relationship between the features of assembled parts. For example, how parts are held, how and whether they are fixtured, which fasteners are started first, whether all fasteners are started before tightening any fasteners are factors that affect the relationship between parts of the finished assembly. Most of the Tolerance Stackups in this book assume a certain assembly process, such as starting all fasteners in a pattern simultaneously. It is critical that the Tolerance Analyst understands the assembly process and builds the Tolerance Stackup accordingly.

Assembly shift is often the largest contributor in Tolerance Stackups where parts are assembled and located by fasteners passing through holes in mating parts. Assembly shift must be included in all Tolerance Stackups where parts are located by internal features within external features, such as by fasteners passing through clearance holes or a key within a keyway in mating parts. The clearance between the mating external features and the internal features allows the parts to shift during the assembly process, hence the name *assembly shift*. Assembly shift is discussed in great detail in Chapter 9.

Many times interference is seen between mating parts at assembly, and the knee-jerk reaction is that the interfering features are out of tolerance. The individual part manufacturing processes are then modified to eliminate the interference. Unfortunately, in many cases the real problem was not solved, because the parts were not out of tolerance, they were assembled improperly, or there was a problem with the assembly process.

As well as the individual parts, the assembly process must be scrutinized any time there is a tolerance problem at assembly.

There are many other sources of variation not listed here. This is merely a sample of some of the sources of variation that are commonly encountered in industry.

# 6

## Tolerance Analysis

### WHAT IS TOLERANCE ANALYSIS?

*Tolerance Analysis* is a global term that includes two subcategories: first, it describes the methods used to determine the meaning of individual tolerancing specifications; second, it is the process of determining the cumulative variation possible between two or more features. The second part of the definition is commonly called a *Tolerance Stackup*.

Before a Tolerance Stackup can be performed, the dimensioning and tolerancing specifications applied to a drawing must be clearly understood. Each tolerancing specification must be broken down, and its meaning must be translated into a form that can be used in a Tolerance Stackup. Tolerancing specifications are complex—understanding the meaning and the ramifications of the tolerancing specifications you specify or the specifications that someone else has specified is an art. It takes training and practice to be able to fully understand tolerancing specifications. Although this is not a GD&T text, this book includes many examples and guidelines for breaking down various dimensioning and tolerancing specifications.

The second step is to perform a Tolerance Stackup. Using the Tolerance Stackup techniques presented in this text allows the Tolerance Analyst to study the cumulative effects of multiple tolerances. A distance or displacement is chosen as the subject of the study, which usually represents a nominal gap or interference. Typically the distance or gap between the features to be studied is not directly dimensioned or toleranced. This may be the distance between two parts that must not touch, the distance between a bolt head and a flange at 90°, the total variation between two parts on an assembly that must fit within a given space, or to determine if a pin will fit though a hole.

Tolerance Stackups provide a numerical answer to a question. Typical questions include

- Will these two surfaces touch in their worst case? If so, how much will they interfere?
- What is the minimum distance between the bolt head and the flange at 90°?
- What is the maximum thickness of the two parts that must fit within this groove?
- Will the pin fit within the hole?
- How large can the body of the switch be and still assemble?
- What is the worst-case largest angle possible between these surfaces?
- How do I know if the worst-case assembly will satisfy its dimensional objectives?
- Why is there interference between these existing parts? Is the interference allowed by the part tolerances and the assembly process?
- If we reduce the size of the clearance holes, will the parts still assemble?
- Will the dimensioning and tolerancing scheme used on the parts allow too much variation at assembly? Should the drawings be redimensioned and retoleranced to reduce the accumulation of tolerances?
- If we chuck the part using this diameter, how much tolerance is allowed for the smaller coaxial diameter?

Notice that some of these questions pertain to parts that may still be on the drawing board, and some pertain to parts and assemblies that already exist. Tolerance Analysis may be used to solve tolerancing problems in both situations.

Some parties make the distinction that Tolerance Analysis is the act of determining the variation between *existing* features, that the part features have already been manufactured and/or assembled, and that a Tolerance Stackup applies to new parts or part still under development. Their definition makes an unnecessary distinction and misses the point. As I stated in the first paragraph of this chapter, Tolerance Analysis is the study of individual tolerances and their meaning, and it is the study of the cumulative variation between part features. Tolerance Stackups are the means of analyzing and predicting that variation, regardless of whether the features only exist on paper or if the parts have already been manufactured. The same methods are used in both cases.

The subject of a Tolerance Analysis may be to verify a needed clearance, or it may be to verify a needed interference condition. Examples are verifying the amount of interference on a press-fit pin, the interference between the components of a switch or verifying that parts will touch in an automated welding operation.

Perhaps the most common Tolerance Analysis is to verify fit. The size, location, and orientation of every clearance hole and tapped hole that receives a fastener must be determined using Tolerance Analysis. Some of these tolerance analyses are so simple that the engineer doesn't even realize that he or she is analyzing tolerances!

When Tolerance Analysis is used to solve a given problem, the method used is commonly called a *Tolerance Stackup*. This is because as dimensions and their tolerances are added together, they "stack up" to add to the total possible variation. Dimensions and Tolerances are stacked up to form a *chain of Dimensions and Tolerances*, which can be followed head to tail from one end of the distance under consideration to the other.

## WHAT IS A TOLERANCE STACKUP?

Quite simply, a Tolerance Stackup is a decision making tool. By performing a Tolerance Stackup, information is obtained that helps to answer one or more questions about a particular design. The information obtained is numeric— the result of a Tolerance Stackup is almost always a minimum and maximum distance and typically only one of these limits is of interest.

Once a Tolerance Stackup is completed, the information obtained can be used to determine if a change must be made to the dimensions and/or tolerances of the parts being studied. The variation predicted by a Tolerance Stackup can be reduced in a number of ways. For example, if an undesired interference is found at worst case, the designer may decide to change the dimensioning and tolerancing scheme, changing one or more dimensions or decreasing the tolerance values for one or more features. The parts in an assembly may be redesigned, eliminating loose fits that lead to misalignment at final assembly. Parts may even be eliminated from an assembly by modifying the mating parts, eliminating contributing tolerances from the Tolerance Stackup.

A very effective way to reduce the variation at assembly is to assemble parts using a fixture. Assembling parts in a fixture almost always leads to less variation at final assembly. A very common technique is to use features such as holes that locate mating parts on pins in the fixture. These holes are added to the parts, and often have no function other than fixturing. The holes are typically tight fitting, having very little clearance with the locating pins. The purpose of most assembly fixtures is to reduce the variation between important features on mating parts. When a fixture is used to reduce variation, it is a good idea to use the fixturing features as datum features, relating the features being controlled to the fixturing features. This will also help reduce the variation encountered at assembly.

Although fixturing is a great idea and usually reduces both the variation at assembly and the overall cost, fixturing is often misapplied. The most common misapplication is where the fixture is an afterthought, and it is not coordinated with all parties involved. A common scenario is where assembly engineers encounter excessive variation at assembly and create an assembly or welding fixture to reduce the variation. Redlines are sent to design engineering to modify the parts, adding fixturing holes to the parts in the assembly. Design engineering changes the drawings, adding the features, but leaves the features either untoleranced or loosely toleranced because they are features for manufacturing. This is a mistake—the fixturing features are functionally very important and should be integrated into the revised design, not just dropped on the drawing.

Remember the fundamental rule in ASME Y14.5M-1994 that states that a drawing shouldn't specify manufacturing methods? The water gets a little muddy here, as these fixturing features are functional. Even though they are requested by manufacturing, they are functional because they locate one part to another at assembly. Their precision affects how precisely the parts are located by the fixture. Remember the statement in the last paragraph that fixturing features should be used as datum features? Here's the reason: The fixturing features become the principal locators for the mating parts—the features on the parts assembled in the fixture are related to the fixturing features. Fixturing features are functional if used properly.

In order to do a Tolerance Stackup between features on the parts that were assembled in a fixture, the chain of Dimensions and Tolerances must pass from one fixtured part, through its fixturing features, through the fixture, and through the fixturing features on the other part in the assembly. The fixture and its tolerances have become an integral part of the Tolerance Stackup between the parts. Thus the fixture tolerances must be included in the Tolerance Stackup if assembly fixtures are used. Fixture dimensions and tolerances are included in the Tolerance Stackup just like part dimensions and tolerances.

Fixture tolerances vary, but fixtures are usually manufactured to much tighter tolerances than the parts they assemble. Generally, fixture tolerances are often only 10% of the part tolerances. This is only a general rule: the fixture drawing should be reviewed to determine the dimensions and tolerances required for the Tolerance Stackup. The fixture manufacturer should be consulted if the fixture drawing is unavailable. If the fixture drawing and the fixture manufacturer are unavailable, the fixture tolerance values may be assumed. The same rules for assuming part tolerances must be followed in this case. See the material on assumptions near the end of Chapter 9.

In a Tolerance Stackup the chain of Dimensions and Tolerances is separated into two groups: the dimensions that are followed in one direction are labeled as positive, and the dimensions that are followed in the opposite direction are labeled as negative. Generally, the Tolerance Stackup process is as follows:

- The distance to be studied is identified and labeled.
- The positive and negative directions of the Tolerance Stackup are identified.
- A Tolerance Stackup sketch is created.
- The dimensions in the positive direction are added together.
- The dimensions in the negative direction are added together.
- The negative direction total is subtracted from the positive direction total to find the "nominal" distance.
- All applicable tolerances are added together. This is the *total possible variation.*
- Half of the total possible variation is added to the nominal distance to find the Upper Limit for the distance.
- Half of the total possible variation is subtracted from the nominal distance to find the lower limit for the distance.

This entire process is covered in detail in the next two chapters.

## WHY PERFORM A TOLERANCE STACKUP?

There are many examples where Tolerance Analysis provides valuable information whether a design will properly function. As described above, each part is comprised of toleranced (variable) features. Assemblies are comprised of variable parts, and additional variation may occur as part of the assembly process. As more toleranced dimensions stack up, more and more variation is possible. Obviously, the greatest variation is possible in complex assemblies of many parts.

A Tolerance Stackup allows us to determine the maximum possible variation between two features on a single part or more commonly between components in an assembly. There are several reasons it is important to know how much variation is possible.

A Tolerance Stackup allows the designer to

- Optimize the tolerances of parts and assemblies in a new design.
- Balance accuracy, precision, and cost with manufacturing process capability.
- Determine the part tolerances required to satisfy a final assembly condition.
- Determine the allowable part tolerances if the assembly tolerance is known.
- Determine if the parts will work at their worst-case condition or with the maximum statistical variation.
- Determine if the specified part tolerances yield an acceptable amount of variation between assembled components.
- Troubleshoot malfunctioning existing parts or assemblies.
- Determine the effect changing a tolerance value will have on assembly function.
- Explore design alternatives using different or modified parts.

It is very important to understand that the there are four main factors that determine which dimensions and tolerances are included in a Tolerance Stackup:

- The geometry of parts and assemblies that contribute to the distance being studied in the Tolerance Stackup
- The dimensioning and tolerancing schemes on the drawings of the parts and assemblies in the Tolerance Stackup
- The assembly process, how the parts are assembled
- The direction of the Tolerance Stackup and the direction of the dimensions and tolerances

The geometry of the parts and assemblies being studied plays a huge role in determining which features affect the distance being studied. How parts mate at assembly, which surfaces touch, the angle of the interface(s), and which features locate the parts relative to one another also play a role. This is a result of the physical shape of the parts, the physical relationship of the part features to one another, and the physical relationship between the parts in the final assembly. Again, these are all physical functions, directly attributable to the geometry of the parts and assemblies being studied.

The dimensioning and tolerancing schemes used on the part and assembly drawings also play a huge role in determining which dimensions and tolerances must be included in the chain of Dimensions and Tolerances.

Drawings of parts that have been functionally dimensioned and toleranced will add fewer dimensions and tolerances (less variation) to the chain of Dimensions and Tolerances, because the designer looked ahead and tried to minimize the accumulation of the tolerances for an important feature relationship. Drawings of parts that have been dimensioned and toleranced using plus/minus typically will add more dimensions and tolerances to the chain of Dimensions and Tolerances, because the plus/minus system is imprecise, inaccurate and incapable of communicating the information required to reduce the accumulation of the tolerances for an important feature relationship. That said, parts dimensioned and toleranced using GD&T may also be imprecise, inaccurate and also may add extra tolerances to the chain of Dimensions and Tolerances, but to a lesser degree than with plus/minus. If a part is dimensioned and toleranced poorly, it is likely that more dimensions and tolerances will be "chained" and that there will be more than one dimension and tolerance between important features.

Many times I have been asked to a perform a Tolerance Stackup for a client using drawings that are dimensioned and toleranced in a manner that was easy to do with their CAD system but didn't reflect the functional requirements of the part at all. An example of this can be seen in back in Fig. 2.2 where all the dimensions and tolerances are related to a corner of the part. This is very easy to do using most CAD systems, and in many cases it makes it easy for an NC programmer to enter the data into a CNC Manufacturing program, but it absolutely makes no sense when considering the functional requirements of the part. Perhaps surprisingly, it is also difficult to establish a clearly defined origin for all the part features from such a dimensioning and tolerancing scheme. Again, this is a result of the ambiguity of the plus/minus system.

Unfortunately, this sort of dimensioning and tolerancing scheme is very common. As mentioned earlier, the dimensioning and tolerancing scheme on the drawings can have a huge effect on a Tolerance Stackup. In fact, a Tolerance Stackup done to find the variation possible between two features on poorly dimensioned and toleranced drawings, and the same Tolerance Stackup done on the same parts with their drawings revised with functional dimensioning and tolerancing will be very different. There will be far more dimensions and tolerances in the Tolerance Stackup for the poorly dimensioned and toleranced parts, and there will likely be far more assumptions required. The functionally dimensioned and toleranced drawings will yield a more compact Tolerance Stackup with far less ambiguity, therefore requiring far fewer assumptions.

The assembly process also plays a large role in which dimensions and tolerances are included in the chain of Dimensions and Tolerances. Understanding if parts will be assembled by hand, if they will be assembled on an assembly line, or if they will be assembled using a fixture is very important. The assembly process can add or remove variation in several ways. Later we will learn about assembly shift, which is a function of clearance between mating locating features. In some cases, such as where large clearance holes locate one part to another and no fixturing is used, the assembly process may add more variation than the sum of the tolerances on the parts. In other cases, parts may be assembled using a fixture that locates the parts using fixturing features that have nothing to do with the final application of the part. In these cases the chain of Dimensions and Tolerances must pass through the fixturing features to accurately represent the potential tolerance accumulation. Fixture tolerances should also be included in the chain of Dimensions and Tolerances in these cases. The effect of the assembly process will be discussed in detail in later chapters.

The direction of a linear Tolerance Stackup is always along a straight line. The methods taught in this text are for linear, one-dimensional Tolerance Stackups. Once the direction is chosen, all the dimensions and tolerances that affect the distance being studied are included in the Tolerance Stackup. Dimensions and tolerances on surfaces at an angle to the Tolerance Stackup direction may need to be projected into the direction of the Tolerance Stackup using trigonometry. Dimensions and tolerances that are perpendicular to the Tolerance Stackup direction typically have no effect on the Tolerance Stackup and are usually not included in the chain of Dimensions and Tolerances. This topic is addressed in greater detail later in the text.

## METHODS AND TYPES OF TOLERANCE ANALYSIS

There are two methods of performing a Tolerance Analysis: manually modeled and computer modeled. Manually modeled analyses are done by hand, using pen and paper, or spreadsheet programs. Manual analyses are limited to linear (one-dimensional) variation. Several linear analyses may be combined to determine two- or three-dimensional variation, but great care must be taken to ensure redundant items are not included in the analyses. Three-dimensional analyses are best suited to computer-modeling tools. Computer modeled analyses are performed by computer statistical simulation programs. Programs are available for one-, two-, and three-dimensional analyses.

There are two major types of Tolerance Analysis: worst-case (arithmetic) and statistical. Worst-case tolerance analyses represent the largest (worst-case) possible variation. For a Tolerance Stackup with many dimensions and tolerances, statistical tolerance analyses may be more appropriate. Statistical tolerance analyses use one of several techniques for determining the likely maximum variation, which is usually less than the worst-case result. The most common technique is the root-sum-square (RSS) method. Statistical tolerancing is based on a number of factors that will be discussed in Chapter 8. Computer programs and spreadsheets make statistical tolerance analyses easier to perform, as the math can be built into the program.

Tolerance Stackups may be done on any toleranced part, or any assembly of toleranced parts. A Tolerance Stackup cannot be done on a part or assembly that is not toleranced. Tolerances are required on each contributing feature of each part affecting the dimension to be studied. If the tolerance values are assumed or it is decided to use the manufacturing process capability (be very careful here), the result of the Tolerance Stackup will only be a guess. The more uncertain you are about the accuracy of the data entered into the Tolerance Stackup, the less certain you can be about the output, which is true of any mathematical exercise. The only way to be sure of the results of a Tolerance Stackup is to use tolerances that are clearly specified on the drawing or in a related document.

To recap, Tolerance Stackups are performed to determine the variation of a single untoleranced dimension or distance. An example of an untoleranced distance on a single part can be seen in Fig. 6.1, and untoleranced distances on several assemblies can be seen in Fig. 6.2.

What are the Minimum and Maximum Distance for Gap A-B?

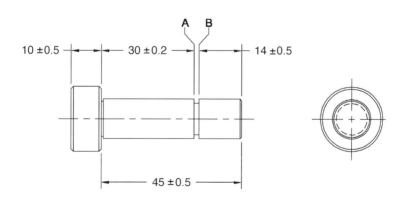

Pin with Groove

**FIGURE 6.1** Single part with missing dimensions.

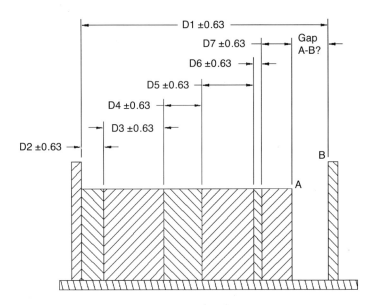

Simple Assembly with Missing Dimension

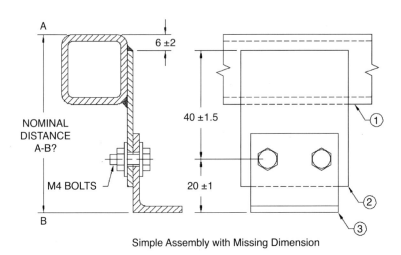

Simple Assembly with Missing Dimension

**FIGURE 6.2** Assemblies with missing dimensions.

# 7

## Worst-case Tolerance Stackups

Worst-case Tolerance Stackups determine the absolute maximum variation possible for a selected distance or gap. This distance is usually not dimensioned (it may have a reference dimension) and is not directly toleranced. If it were directly toleranced, no study would be necessary, as the limits would already be defined. When performing a Tolerance Stackup, other dimensions and tolerances are added and subtracted to obtain the total variation of the distance being considered. This method assumes that all dimensions in the Tolerance Stackup may be at their worst-case maximum or minimum, regardless of the improbability.

*Tolerance Stackups* as defined in this text follow a chain of Dimensions and Tolerances. The dimensions and tolerances in a Tolerance Stackup are called a *chain of Dimensions and Tolerances* because the Dimensions and Tolerances that make up the Tolerance Stackup are arranged like the links in a chain, and followed head-to-tail from one end of the distance being studied (call it point $A$) to the other (call it point $B$.)

A step-by-step explanation of how to perform worst-case (arithmetic) Tolerance Stackups follows.

### WORST-CASE TOLERANCE STACKUP WITH DIMENSIONS

1. Select the distance (gap or interference) whose variation is to be determined. Label one end of the distance $A$ and the other end $B$ (see Fig. 7.1).

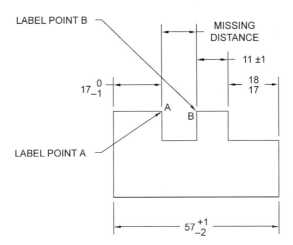

**FIGURE 7.1**   Worst-case chain of Dimensions and Tolerances #1.

2. Determine if a one-, two-, or three-dimensional analysis is required.

   a. If a two-dimensional analysis is required, determine if both directions can be resolved into one dimension using trigonometry. If not, a linear Tolerance Stackup is not appropriate, and a computer program should be used for the Tolerance Analysis.

   b. If a three-dimensional analysis is required, a linear Tolerance Stackup is probably not appropriate, and a computer program should be used for the Tolerance Analysis.

3. Determine a positive direction and a negative direction.

   a. If there is a dimension that spans distance *A–B*, label it as a positive dimension by placing a " + " sign adjacent to the dimension value (see Fig. 7.2). It should also be assigned a direction by placing a dimension origin symbol at the end where the dimension starts and an arrowhead at the other end where the dimension terminates. All dimensions in the chain of Dimensions and Tolerances that are followed in this direction shall be labeled as positive dimensions. All dimensions that are followed in the opposite direction shall be labeled as negative dimensions.

   b. If no dimension spans distance *A–B*, start at point *A*. If the direction of the dimension from point *A* goes toward, through or terminates at point *B*, then label it positive using a " + "

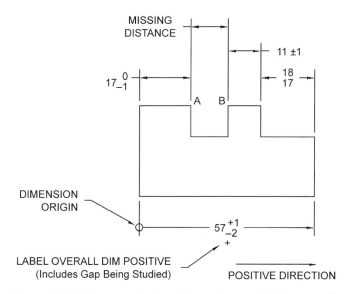

**FIGURE 7.2** Worst-case chain of Dimensions and Tolerances #2.

sign, a dimension origin symbol, and arrowhead as described in item 3.a above. If its direction is away from point *B*, label it negative (see Fig. 7.3). Identify the chain of Dimensions and Tolerances from point *A* to point *B*, and label all dimensions in the same direction positive or negative, as indicated by the sign of the first dimension. Remember, if there is a dimension passing through distance *A–B* it should be labeled as a positive dimension.

c. Follow the chain of Dimensions and Tolerances from point *A* to point *B*. You should be able to follow a continuous path from the start to the end of each dimension in the chain from point *A* to point *B* (see Fig. 7.4).

In this example, the first dimension starts at point *A* and ends at the left edge of the part. The second dimension starts where the first dimension ends, and ends at the right edge of the part. The third dimension starts where the second dimension ends. The fourth dimension starts where the third dimension ends, and ends at point *B*.

If the dimensions are not properly labeled, the nominal distance will be negative after the negative total is subtracted from the positive total. If this happens, check the + or − labels assigned to the dimensions, making sure that the sum of

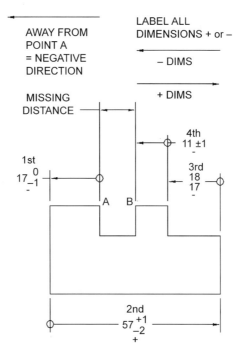

**FIGURE 7.3**   Worst-case chain of Dimensions and Tolerances #3.

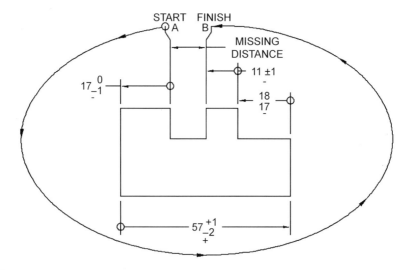

**FIGURE 7.4**   Worst-case chain of Dimensions and Tolerances #4. Follow the chain
of dimensions and tolerances from point *A* to point *B* to make sure there are no
breaks or discontinuities in the chain.

the positively labeled dimensions is larger than the sum of the negatively labeled dimensions. Remember that the total value of the positive dimensions must include distance *A–B*.

4. Convert all dimensions and tolerances to equal bilateral format (See Fig. 7.5). Instructions for how to do this are included in Chapter 4.

5. Now all the dimensions and tolerances must be put into a chart and totaled for reporting purposes. Place each positive dimension value in the positive column on a separate line. Place each negative dimension value in the negative column on a separate line. (See Fig. 7.6.)

6. Place the tolerance value for each dimension in the tolerance column adjacent to each dimension. This value is half the total variation allowed by the tolerance (see Fig. 7.6).

7. Add the entries in each column, entering the results at the bottom of the chart (see Fig. 7.6).

8. Subtract the negative total from the positive total. This gives the nominal dimension or distance. (See Fig. 7.7.) In cases where all the dimensions and tolerances in the chain were not originally in equal to bilateral format this value will probably be different than the

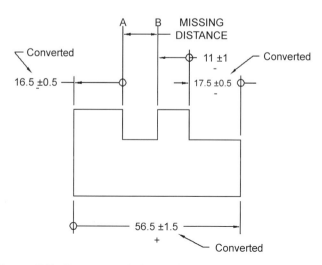

**FIGURE 7.5** Worst-case chain of Dimensions and Tolerances #5.

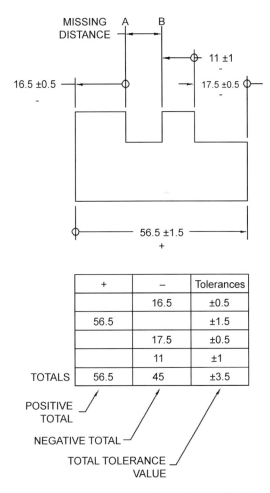

**FIGURE 7.6**  Worst-case chain of Dimensions and Tolerances #6.

distance that is measured directly from a drawing or CAD model. [Note: This value should be positive. If it is negative, some dimensions were given the wrong sign or distance *A–B* was included in the negative total value. Remember that the total value of the positive dimensions must be greater than the total value of the negative dimensions. (See step 3.)]

9. Apply the total tolerance. Adding and subtracting the tolerance from the nominal dimension gives the maximum and minimum distance values (see Fig. 7.7).

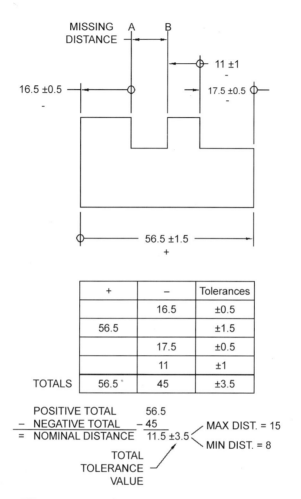

**FIGURE 7.7** Worst-case chain of Dimensions and Tolerances #7.

In this example the chain of Dimensions and Tolerances started at the at the left surface of the distance being studied at point *A* and proceeded counterclockwise around the part until point *B* was reached. The chain of Dimensions and Tolerances could also have started at the right side of the distance being studied and proceeded clockwise. In most cases it doesn't matter whether the chain of Dimensions and Tolerances goes clockwise or counterclockwise, to the left or to the right, or up or down—all that matters is that all the dimensions and tolerances that contribute to the Tolerance Stackup are included in the chain of Dimensions and Tolerances, and that the direction of the dimensions are properly included as being in the positive or negative direction.


## ASSEMBLY SHIFT

Assembly shift represents the amount that parts can move during assembly due to the clearance between a hole and a fastener, a hole and a shaft, a width and a slot (like a key and keyway) or between any external feature within an internal feature. To put it another way, Assembly shift accounts for the freedom parts have to move from their nominal locations due to the clearance between mating internal and external features at assembly.

An internal cylindrical feature (such as a hole) may shift about an external cylindrical feature (such as a bolt or a pin) in all directions normal to the axes of the features. Consider the example of a flat washer with a bolt passing through its center hole shown in Fig. 7.8. The Washer is free to shift or move in any direction perpendicular to the bolt. All it takes is a slight force to nudge the washer one way or the other about the fastener.

Parts are routinely subjected to forces during assembly. If there is clearance between mating parts with holes and fasteners, assembly forces may push the parts until the holes and fasteners are in contact. Torquing one set of bolts may rotate a part about the bolts, making it difficult or impossible to engage a second set of fasteners or other mating features. Care must be taken to accurately reflect the assembly sequence when performing a Tolerance Stackup. Gravity is an example of a force that is always present (at least here on earth). It is common for parts assembled vertically to always be biased downwards, the force of gravity pulling the parts and fasteners down against the holes. The effect of gravity and assembly sequence are discussed in greater detail in Chapter 18.

Assembly shift is often overlooked in Tolerance Stackups. Many Tolerance Stackups include the tolerances found on the drawing(s), as that variation is clearly specified. The tolerances on the drawing typically represent the variation allowed by and attributed to the manufacturing process. The tolerance values must be functional, that is, they must still allow the part to function as intended, but the tolerances specified on the drawing are typically understood to represent the variational limits allowed for the manufacturing process.

Assembly shift is different from tolerances in that it is not specified. Indeed it is often not even considered until there is a problem at assembly or until a Tolerance Stackup is performed. Assembly shift is merely a result of clearances between mating parts at assembly. It is a measure of how much parts can move relative to one another about their locating features.

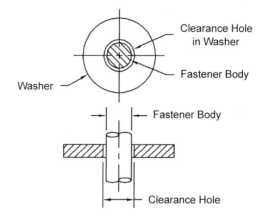

Nominal Washer & Fastener

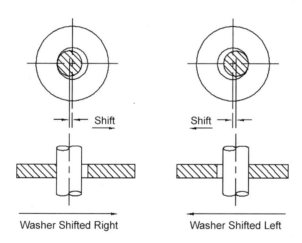

Washer Shifted Right          Washer Shifted Left

Washer Shifted about Fastener

**FIGURE 7.8** Shift about a fastener.

Consider an 8-mm fastener passing through a 10-mm hole – there is 2-mm clearance, and the part can shift 2 mm total, or ±1 mm in any direction normal to its axis.

Assembly shift is greatest (the most shift is possible) when the hole and the fastener are at their least material conditions (**LMC**), which are the largest hole and the smallest fastener. The difference between the two represents the worst-case assembly shift. When considering assembly shift of a clearance hole about a bolt or screw, the major diameter of the fastener must be used in the Tolerance Stackup, because the outermost surface of the fastener contacts the surface of the clearance hole. A common shortcut, however, is to use the nominal size of the fastener. For example, for a metric M8 bolt, 8 mm would be used in the calculations. For an inch series .250–20 UNC bolt, .250 in. would be used in the calculations.

If more accurate results are required, use the minimum value for the fastener size, which can be obtained from a fastener drawing, a catalog, commercial or military specifications, or a source such as the Machinery's Handbook. Using the nominal diameter of a fastener is a liberal approach, as fastener major diameters are usually slightly smaller than the nominal size. Using a smaller value for the fastener diameter yields a larger assembly shift value when subtracted from the maximum hole diameter. However, in many applications using this technique is adequate, and the shortcut is taken. The Tolerance Analyst must be aware of the implications of this shortcut and determine if the risk is acceptable. Tolerance Stackups in this text have been solved using both methods. Where applicable, a notation was added to each Tolerance Stackup stating that the nominal diameter was used instead of the major diameter to calculate assembly shift.

Given the hole and fastener combination in Fig. 7.9, it is apparent that the maximum assembly shift is possible when the hole is manufactured at its largest (LMC) size of 10.6 mm. The worst-case assembly shift is determined by subtracting the smallest possible fastener diameter from the largest possible hole diameter.

Assembly shift is added to the Tolerance Stackup as a line item without a sign or direction, as it allows the parts to shift in both the positive and negative directions, similar to an equal bilateral tolerance. Different than a tolerance, assembly shift is a function of mating parts at assembly, and is not associated with a dimension value, nor is it shown with a dimension origin symbol.

For floating-fastener applications where both parts have clearance holes, assembly shift is added to the Tolerance Stackup twice, each line representing the amount the holes in each part may shift about the fasteners. The amount each part may shift about the fastener is independent of the mating parts and must be calculated separately.

For fixed-fastener applications where one part has tapped or press-fit holes or studs and the other part has clearance holes, assembly shift only needs to be calculated for the part with clearance holes. In fixed-fastener cases assembly shift is added to the Tolerance Stackup once representing the amount the clearance holes may shift about the fastener.

Assembly shift is typically not calculated for fasteners within threaded holes because fasteners are commonly assumed to be "fixed" within the threaded holes, as threads are assumed to be self-centering. From that line of reasoning, the threaded holes cannot shift about the threaded fasteners at

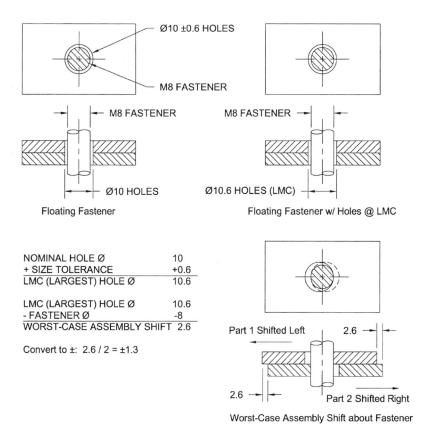

**FIGURE 7.9** Worst-case assembly shift. The Worst-Case Assembly Shift applies to each part. Each part may shift +/-1.3mm relative to the fastener, leading to Total Assembly Shift of 2 *±1.3 = ±2.6.

assembly. This is an oversimplification, as there is always some clearance between internal and external threads, and assembly forces do bias the threads at assembly. However, due to complexity of the geometry, it is very difficult, if not impossible, to quantify the amount of assembly shift of a threaded fastener within a threaded hole. A simplified approach could be to compare the difference between the pitch diameters of mating male and female threads, which is sometimes called the *allowance*, and use that value for their assembly shift. Whether this assembly shift would ever be seen is debatable, but in some critical cases it may be a good idea to account for the possibility.

The idea that threads mate along the pitch diameter or pitch cylinder is also an oversimplification. How actual imperfect threads mate is geometrically complex, and the author tends to think of it in a statistical context. There are potentially a great number of points in contact between the imperfect helical surfaces of the threads contributing to their mating relationship. When the effects of these points of contact are considered as a group all along the thread, it is probable that the threads approximate being centered.

Additionally, the final contact of the head of the fastener against the mating surface most likely biases or tilts the fastener to one side of the threaded hole. There is always some amount of angular variation between the mating surface and the axis of a threaded hole and between the underside of the head of the fastener and the axis of its screw thread. However, for all intents and purposes this is a minor consideration except in the most extreme circumstances, as it occurs last in the assembly process and does not affect the fit until disassembly.

Remember, the shortcuts for threaded features described above must be used with discretion in critical applications.

Interestingly, assembly shift can also be used to reduce the total variation in certain cases. It is common in many industries to use slots and oversized holes in mating parts for the sole purpose of manual adjustment at assembly. The adjustment's sole purpose is to counteract the accumulated tolerances and allow the assembler to optimize the relationship between functionally related features. The author has designed many assemblies where mating parts had horizontal slots in one part and vertical slots in the other part. This allows for a great amount of adjustment, and it negates much if not all of the tolerance accumulation in some circumstances. If a Tolerance Stackup was done on such an assembly, the assembly shift (the clearance between the slots and the fastener in the direction of the Tolerance Stackup) would be subtracted from the total tolerance. Great care must be exercised in such situations, as there is a crucial fact that must be understood: the assembly procedure must be absolutely understood to use this technique. It should be formally stated, preferably in writing, that the assemblers are to manually adjust the parts at final assembly and that the assemblers will use the large assembly shift of the slots to their advantage and find the optimal location for the parts. A problem with this approach is by definition it is subjective and subject to human error. What is "optimal" or "aligned" to one assembler may not be the same for another assembler.

Many low-volume assembly methods rely upon their skilled assemblers to make adjustments at final assembly. The assemblers are trusted to make good decisions and locate the parts correctly. There are three options how to handle the assembly shift in such situations.

1.  The assembly shift may be included in the Tolerance Stackup, even though the assemblers will likely counteract its effects and use it to their advantage. This is probably not a good approach, as it is likely overly pessimistic.

2. The assembly shift may be eliminated from the Tolerance Stackup. That is, it may not be necessary to include the applicable assembly shift values in the Tolerance Stackup. This is a moderate approach, as it recognizes that the assembly shift will not add to the total tolerance predicted by the Tolerance Stackup.

3. The assembly shift may be subtracted from the total tolerance predicted by the Tolerance Stackup. This reflects the idea that the assemblers will use the assembly shift to their full advantage at assembly. However, if there is any doubt that the assemblers will use the assembly shift to their full advantage and adjust the parts to their absolute optimal location every time, it may not be a good idea to subtract the assembly shift from the total tolerance predicted by the Tolerance Stackup. This may be an overly optimistic approach.

Many high-volume assembly methods do not allow their assemblers to make adjustments at final assembly. Even though the assemblers may possess excellent skills and they may have a good understanding of the products being assembled, they may not have the time or the authority to make the necessary adjustments. Many industries that use assembly lines speed the process up as much as possible to maximize throughput and increase productivity. Such an environment allows parts to end up where they may at assembly, allowing assembly shift to fully manifest itself in the process. Most automated or semiautomated assembly line methods have no means for adjustment at final assembly. In these cases the assembly shift should be included in the Tolerance Stackup.

There is one other important point regarding adjustment at assembly: the parts must be able to be properly adjusted at assembly if the assembly shift is to be eliminated or subtracted from the total tolerance. Some parts may be too heavy, too large, too small, too awkward, or difficult to access or see the critical dimension to allow for proper adjustment at assembly. In these cases the assembly shift should be included in the Tolerance Stackup.

This text assumes that there is no adjustment at assembly and that any and all possible assembly shift will show up at final assembly. Given that premise, each occurrence of assembly shift must be included in the Tolerance Stackup.

## Rules for Assembly Shift

• Assembly shift is the amount that parts can move at assembly due to the clearance between an internal feature such as a hole and an external feature such as a fastener.

• In floating-fastener cases assembly shift is added to the Tolerance Stackup twice, each line representing the amount the clearance holes in each part can shift about the fastener. The amount each part may shift about the fastener is independent of the mating parts and must be calculated separately.

- In fixed-fastener cases assembly shift is added to the Tolerance Stackup once, representing the amount the clearance holes can shift about the fastener.

- Assembly shift is typically not calculated for fasteners within a threaded hole because fasteners are commonly assumed to self-center within the threaded holes.

- In cases where the results of the Tolerance Stackup are very critical and the tolerances are tight, it may be necessary to calculate or estimate the amount that a threaded fastener may move within a threaded hole.

- In cases where oversized holes or slots are used to allow for adjustment at assembly, the assembly shift may be eliminated or even subtracted from the total tolerance. This must be done with utmost caution, as the Tolerance analyst must be absolutely certain that the assembly process will allow time for adjustment, that the assemblers understand the purpose of this extra adjustment, and the parts can be adjusted at assembly, i.e., they are not too heavy or awkward to properly be adjusted to an optimal position.

## THE ROLE OF ASSUMPTIONS IN TOLERANCE STACKUPS

Assumptions play a very important role in Tolerance Stackups. It is very common to find that all the required information is not available when performing a Tolerance Stackup. There are a number of reasons, including incompletely dimensioned and toleranced drawings, purposefully incomplete supplier drawings, mistakenly incomplete supplier drawings, drawings dimensioned using ± instead of GD&T, and incompletely documented parts taken from catalog data sheets to name a few.

Many drawings, especially older drawings, are incompletely dimensioned and toleranced. A good example is a part with several nominally coaxial diameters, such the pin in Fig. 7.10.

The features are drawn or modeled coaxial, but their coaxial relationship is not toleranced, only their sizes are toleranced. Everyone dealing with such a part must guess how accurately they must be positioned. A common stance is to fall back upon the process: "Hey, these are turned on a lathe in the same operation—they'll be coaxial." Ask the question "How much variation can there be in their coaxiality?" or "How far apart can their axes be?" The

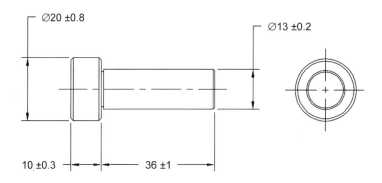

COAXIALITY BETWEEN THESE DIAMETERS IS NOT SPECIFIED.

**FIGURE 7.10** Coaxial pin without GD&T.

answer will almost always fall back on the process capability; say a few thousandths of an inch or the metric equivalent. This is, of course, a guess, and does not establish legal limits of acceptability in the same sense as a stated tolerance. If a guess is deemed acceptable and used in a Tolerance Stackup, it must be so stated in the Tolerance Stackup report.

Many supplier or vendor drawings are incomplete, either purposefully for proprietary reasons or as a result of oversight or tradition. A phone call to the supplier requesting either the actual tolerance or an estimate must be made. Again, the source of the information and the fact that it is an assumption should be stated in the report.

Aside from the size dimension and tolerance on a feature of size, all plus/minus dimensioned and toleranced components require assumptions to make them work in Tolerance Stackup. As most engineers and shop personnel have been using ± dimensions and tolerances for a long time, they are likely unaware of these assumptions in interpretation. It is beyond the scope of this text to explain the difference between ± tolerancing and GD&T. Suffice to say that if it is critical that parts function and fit at assembly, GD&T is the only way to ensure this will happen.

Catalog parts present a special problem, as it is very common to have no tolerances available on a catalog data sheet. More commonly a detail is included showing the required mating part geometry. Unfortunately, these are typically inadequate from a Tolerance Analysis point of view, as without the tolerances for the catalog component it is unclear how their numbers were derived. Using the dimensioning and tolerancing data verbatim from the catalog may lead to an undesirable situation in many circumstances. It is a good idea in such cases to call the manufacturer and ask for the required information.

Many parts such as bearings, bushings, press-fit inserts, any parts that are based on some sort of fit, are usually toleranced such that the dimensions and tolerances specified in the catalog are necessary for performance. Using different dimensions and tolerances than the ones specified in the catalog can lead to diminished performance. Care must be taken when dealing with such components, as the catalog data usually does not allow for mating features to be misaligned, which almost always happens when mating features are subject to positional or location tolerances.

Parts more likely to be a problem and require altered tolerances are switches, covers and connectors, and other parts where the cutouts are shown in a catalog data sheet. Typically these cutout details do not take location or positional tolerances into account.

As shown in Chapter 18, fixed- and floating-fastener calculations are used to calculate how much holes must be oversized to account for the size of the mating fastener or pin and the positional tolerance on the holes. It is common for holes to be improperly sized on drawings prepared using the plus/minus system—the clearance holes are too small to account for their potential variation in orientation and location. Most drawings prepared using $\pm$ use rectangular coordinate dimensions and tolerances, and the holes have locational $\pm$ tolerances in two directions, $X$ and $Y$. For the holes to be properly sized, the fixed- and floating-fastener formulas in Chapter 18 must be used, and the diagonal distance or hypotenuse of the rectangular tolerance zone must be calculated and used in the formulas. This is rarely done. Many companies use a standard clearance hole chart that sizes all clearance holes 1/32 in. or 1/16 in. larger than the fastener. This method is inadequate and leads to interference at assembly. If it is necessary that a clearance hole does not interfere with a mating fastener, then the formulas presented in Chapter 18 must be used.

In all fairness, many manufacturers that do not completely tolerance their component drawings simply don't know the value of the missing tolerance. The best they can do is tell you the tolerance their process can produce capably. They may not even know this information, and it may require that they do some research to get it. The time it takes for them to get back with the information may not work with your schedule, and an assumption may be needed to meet a deadline. It is common to insert values into the Tolerance Stackup based on educated guesses, label them as such, and replace the guesses in a revised Tolerance Stackup with the actual values when available.

## FRAMING THE PROBLEM REQUIRES ASSUMPTIONS: IDEALIZATION

The Tolerance Analysis techniques presented in this text are for solving one-dimensional, linear Tolerance Stackups. These techniques work well for solving many of the geometric problems encountered on all sorts of parts and assemblies. However, all of these parts and assemblies are three-dimensional, and it is likely that the geometric problems to be solved are also three-dimensional. How is a three-dimensional problem solved using one-dimensional techniques? The problem is idealized. The problem is framed in a way that projects the potential variation along the direction of the Tolerance Stackup. The Tolerance Analyst must be confident that the considered tolerances adequately represent all of the tolerances that may contribute to the Tolerance Stackup. As will be seen throughout this text, many problems are framed and solved in several ways to make sure the chain of Dimensions and Tolerances includes the required contributors. For example, several problems are solved as if all the tolerances only act in a straight-line, and then the problems are solved as if some of the tolerances allow features to tilt or rotate, adding a geometric effect to the analysis.

Tolerance Stackups are performed with these considerations:

- All parts are considered in a static state. The Tolerance Stackup allows parts to shift or rotate relative to one another during assembly, but the study is performed in a static condition.

  ○ This is typically a worst-case static condition, reflecting worst-case misalignment, minimum clearance, or maximum interference. If desired, statistics may be used to reduce the predicted worst-case total variation.

  ○ If more than one position or orientation of a part must be studied, as in the case of a linkage or a mechanism, then a Tolerance Stackup should be done for the considered feature at each important position or orientation.

- Tolerance Stackups are performed at a specified temperature. Unless specified otherwise, Tolerance Stackups are performed at ambient temperature, the temperature at which the parts are assembled and/ or inspected.

  ○ If a study is needed to account for differential thermal expansion, then the study should be done at the operating temperature. It may be common in some industries to perform Tolerance Stackups at a number of temperatures to account for various stages of cooling or heating during operation. It must be understood that where parts are assembled at one temperature and operate at a different temperature, it is important to study both conditions, as the parts must be assembled before they can operate.

Tolerance Stackups are most accurate when done on parts and assemblies at the temperature at which they were inspected, as that is likely the only verifiable geometric data obtained for the part geometry. Many more assumptions are required for Tolerance Stackups done at reduced or elevated temperatures, as it is likely that the changes in part geometry due to thermal expansion are predicted (e.g., by FEA) and not empirical.

## WORST-CASE TOLERANCE STACKUP EXAMPLES

The Tolerance Stackup examples that follow increase in complexity from finding a minimum and maximum distance on a very basic part to finding a minimum and maximum distance on a complex bolted assembly with assembly shift. All of the examples are based on parts dimensioned and toleranced using the plus/minus ($\pm$) system. Although $\pm$ is fallible and not the best way to dimension and tolerance parts, these early examples are intended to be simple and versatile in their application. By starting with $\pm$, the material is applicable to a broader swath of industry, including those companies that have not yet adopted GD&T as part of their standard engineering practice. These same examples are included in chapter 8 on statistical Tolerance Stackups, so the reader can compare the results.

   The reporting methods of the following examples also increase in complexity from very simple to more complete. Ultimately the reader will be taught to use a formal Tolerance Stackup report form such as the one available from Advanced Dimensional Management™. These problems, however, are based on simpler reporting formats, as it is important at this early stage to keep the topic of Tolerance Analysis as uncluttered as possible. More formal and complete reporting tools and practices are covered in depth in later chapters.

   The first example shows how to determine the minimum and maximum distance between two surfaces on a single part. The second example is similar to the one presented earlier in this chapter but with slightly different dimensions and tolerances. The third example is a very simple assembly. The fourth example is an assembly of parts that are assembled in the vertical position. The force of gravity affects the parts, and assembly shift must be included in the chain of Dimensions and Tolerances. The fifth example is a complex weldment. The sixth example is geometrically similar to the weldment in example five, but the parts are bolted together, so assembly shift must be included in the chain of Dimensions and Tolerances.

## Example 7.1

In this example, a pin is the subject of the study. (See Fig. 7.11). The goal of this Tolerance Stackup is to determine the minimum and maximum width of the groove in the pin. For some reason, the groove was not directly dimensioned and toleranced on the drawing, but the width of the groove is important. If the part had been dimensioned and toleranced functionally, the width of the groove would have been directly dimensioned and toleranced. In that case a Tolerance Stackup would not be required, as the minimum and maximum values for the groove width could be easily calculated right from the drawing. Unfortunately, drawings are not always dimensioned and toleranced functionally, and this example shows how tolerances may accumulate when the dimensioning and tolerancing scheme is not optimized. The Tolerance Stackup results are shown in Fig. 7.12.

The Tolerance Stackup sketch in Fig. 7.12 shows the chain of Dimensions and Tolerances for this problem. One end of the groove is labeled point A and the other end labeled point B. The 45 ± 0.5 dimension spans the distance being studied, so it is labeled as being in the positive direction. All of the dimensions are already presented in equal bilateral format, so conversion is not required. Now the remaining dimensions and tolerances in the chain of Dimensions and Tolerances must be identified.

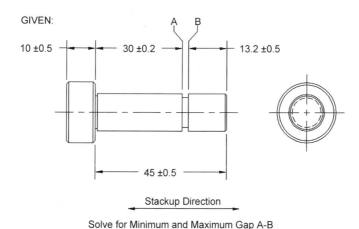

GIVEN:

10 ±0.5   30 ±0.2   A  B   13.2 ±0.5

45 ±0.5

Stackup Direction

Solve for Minimum and Maximum Gap A-B

**FIGURE 7.11**   Pin with groove.

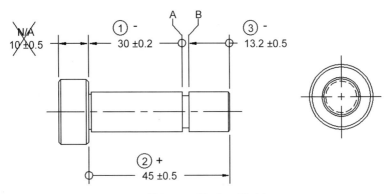

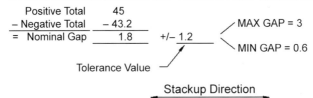

Tolerance Stackup Sketch

| + | − | Tolerances | Description |
|---|---|---|---|
|  | 30 | ±0.2 | GROOVE - HEAD |
| 45 |  | ±0.5 | OVERALL LENGTH |
|  | 13.2 | ±0.5 | TIP - GROOVE |
| 45 | 43.2 | ±1.2 | Totals |

```
   Positive Total      45
 – Negative Total    – 43.2                    MAX GAP = 3
 =  Nominal Gap        1.8      +/– 1.2
                                               MIN GAP = 0.6
          Tolerance Value
```

Stackup Direction

Solve for Minimum and Maximum Gap A-B

**FIGURE 7.12**   Pin with groove solved.

The first dimension in the chain of Dimensions and Tolerances is $30 \pm 0.2$, which starts at the side of the groove at point A and terminates at the head of the pin. This dimension is labeled as item 1. The next dimension is $45 \pm 0.5$, which starts at the head of the pin and terminates at the end of the pin. This dimension is labeled as item 2. The last dimension is $13.2 \pm 0.5$, which starts at the end of the pin and terminates at the other side of the groove at point B. This dimension is labeled as item 3.

As stated earlier, the $45 \pm 0.5$ dimension spans the groove width, so it is labeled as being in the positive direction, which is left to right in this example. Following the chain of Dimensions and Tolerances from point A to point B, we see that the other two dimensions are in the opposite direction, and they are labeled as being in the negative direction, which is right to left in this example. A dimension origin symbol is placed at the start side of each dimension and an arrowhead is placed at the terminating end of each dimension. It is a good idea to visually follow the chain of Dimensions and Tolerances from point A to point B to make sure nothing was missed.

Only the most essential information is included in the Tolerance Stackup report in Fig. 7.12. The positive dimension value is entered in the positive ( + ) direction dimension column, and the other two dimension values are entered in the negative (−) direction dimension column. The equal bilateral tolerance for each dimension is entered in the Tolerances column on the same line as the dimension. Each dimension is described in the Description column.

The dimension values in each column are totaled, and the negative total is subtracted from the positive total, which gives the nominal distance being studied. The tolerance values are totaled, and this total is subtracted from and added to the nominal distance to determine the minimum and maximum distances respectively. The minimum and maximum distances are reported.

## Example 7.2

In this example, a part like the one presented at the beginning of this chapter is the subject of the study. (See Fig. 7.13.) The goal of this Tolerance Stackup is to determine the minimum and maximum distance between two parallel surfaces on the part. The distance being studied is different than in the material presented earlier in the chapter. This distance was not directly dimensioned and toleranced on the drawing. If this distance had been directly dimensioned and toleranced a Tolerance Stackup would not be required, as the minimum and maximum values could be easily calculated right from the drawing.

The Tolerance Stackup results are shown in Fig. 7.14. The Tolerance Stackup sketch in Fig. 7.14 shows the chain of Dimensions and Tolerances for this problem. One end of the distance is labeled point A and the other end labeled point B. The 57 + 3/−1 dimension spans the distance being studied, so it is labeled as being in the positive direction. None of the dimensions and tolerances are presented in equal bilateral format, so conversion is required. Once the conversion is complete, the equal bilateral formatted data can be used in the Tolerance Stackup. Now the remaining dimensions and tolerances in the chain of Dimensions and Tolerances must be identified.

The first dimension in the chain of Dimensions and Tolerances is 12 + 1/ −0.85, which starts at point A and terminates at the surface to the left. This dimension is labeled as item 1. The next dimension is 17 + 1/−0, which starts

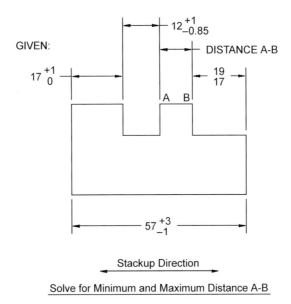

**FIGURE 7.13**  Simple part.

at the end of the previous dimension and terminates at the left side of the part. This dimension is labeled as item 2. The next dimension is 57 $+3/-1$, which starts at the left side of the part and terminates at the right side of the part. This dimension is labeled as item 3. The last dimension is the limit dimension 19/17, which starts at the right side of the part and terminates at point B. This dimension is labeled as item 4.

As stated earlier, the 57 $+3/-1$ (converted to 58 $\pm$ 2) dimension spans the distance A–B, so it is labeled as being in the positive direction, which is left-to-right in this example. Following the chain of Dimensions and Tolerances from point A to point B, we see that the other three dimensions are in the opposite direction, and they are labeled as being in the negative direction, which is right-to-left in this example. A dimension origin symbol is placed at the start side of each dimension, and an arrowhead is placed at the terminating end of each dimension. It is a good idea to visually follow the chain of Dimensions and Tolerances from point A to point B to make sure nothing was missed.

Only the most essential information is included in the Tolerance Stackup report in Fig. 7.14. The positive dimension value is entered in the positive ( + ) direction dimension column, and the negative dimension values

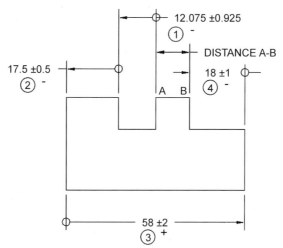

Tolerance Stackup Sketch

| + | − | Tolerances | Description |
|---|---|---|---|
| | 12.075 | ±0.925 | DIM 1 |
| | 17.5 | ±0.5 | DIM 2 |
| 58 | | ±2 | DIM 3 |
| | 18 | ±1 | DIM 4 |
| 58 | 47.575 | ±4.425 | Totals |

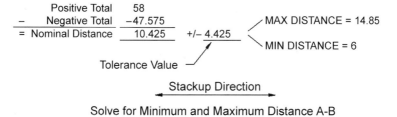

FIGURE 7.14 Simple part solved.

are entered in the negative (−) direction dimension column. The equal
bilateral tolerance for each dimension is entered in the Tolerances column
on the same line as the dimension. Each dimension is described in the Description column. In this case, each dimension is described by its item number.

    The dimension values in each column are totaled, and the negative total
is subtracted from the positive total, which gives the nominal distance being
studied. The tolerance values are totaled, and this total is subtracted from and
added to the nominal distance to determine the minimum and maximum
distances, respectively. The minimum and maximum distances are reported.

## Example 7.3

In this example, a simple assembly is studied. (See Fig. 7.15.) The goal of this
Tolerance Stackup is to determine the minimum and maximum distance
between opposing surfaces on two parts in the assembly. This distance was
not directly dimensioned and toleranced on the assembly drawing. If this

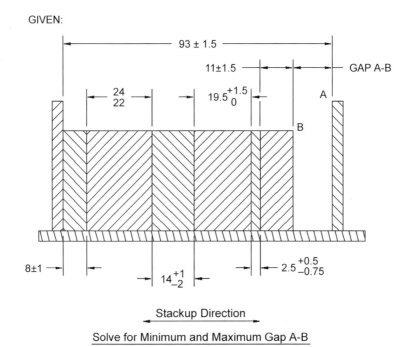

**FIGURE 7.15** Simple assembly.

distance had been directly dimensioned and toleranced a Tolerance Stackup would not be required, as the minimum and maximum values could be easily calculated right from the drawing.

The Tolerance Stackup results are shown in Fig. 7.16. The Tolerance Stackup sketch in Fig. 7.16 shows the chain of Dimensions and Tolerances for this problem. One end of the distance is labeled point A and the other end labeled point B. The distance is labeled Gap A-B. The 93 ± 1.5 dimension spans the distance being studied, so it is labeled as being in the positive direction. Some of the dimensions and tolerances are not presented in equal bilateral format, so conversion is required. Once the conversion is complete, the equal bilateral formatted data can be used in the Tolerance Stackup. Now the remaining dimensions and tolerances in the chain of Dimensions and Tolerances must be identified.

The first dimension in the chain of Dimensions and Tolerances is 93 ± 1.5, which starts at point A and terminates at the left inside surface. This dimension is labeled as item 1. The next dimension is 8 ± 1, which is the thickness of the leftmost part. This dimension is labeled as item 2. The next dimension is the limit dimension 24/22, which is the thickness of the adjacent part to the right. This dimension is labeled as item 3. The next dimension is 14 + 1/−2, which is the thickness of the next part to the right. This dimension is labeled as item 4. The next dimension is 19.5 + 1.5/−0, which is the thickness of the next part to the right. This dimension is labeled as item 5. The next dimension is 2.5 + 0.5/−0.75, which is the thickness of the next part to the right. This dimension is labeled as item 6. The last dimension is 11 ± 1.5, which terminates at point B. It is the thickness of the last part to the right. This dimension is labeled as item 7.

As stated earlier, the 93 ± 1.5 dimension spans Gap A-B, so it is labeled as being in the positive direction, which is right to left in this example. Following the chain of Dimensions and Tolerances from point A to point B, we see that all the other dimensions are in the opposite direction, and they are labeled as being in the negative direction, which is left to right in this example. A dimension origin symbol is placed at the start side of each dimension and an arrowhead is placed at the terminating end of each dimension. It is a good idea to visually follow the chain of Dimensions and Tolerances from point A to point B to make sure nothing was missed.

Only the most essential information is included in the Tolerance Stackup report in Fig. 7.16. The positive dimension value is entered in the positive (+) direction dimension column, and the negative dimension values are entered in the negative (−) direction dimension column. The Equal-bilateral tolerance for each dimension is entered in the Tolerances column on the same line as the dimension. Each dimension is described in the Description column. In this case, each dimension is described by its item number.

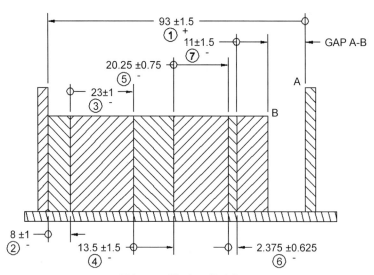

Tolerance Stackup Sketch

| + | − | Tolerances | Description |
|---|---|---|---|
| 93 | | ±1.5 | DIM 1 |
| | 8 | ±1 | DIM 2 |
| | 23 | ±1 | DIM 3 |
| | 13.5 | ±1.5 | DIM 4 |
| | 20.25 | ±0.75 | DIM 5 |
| | 2.375 | ±0.625 | DIM 6 |
| | 11 | ±1.5 | DIM 7 |
| 93 | 78.125 | ±7.875 | Totals |

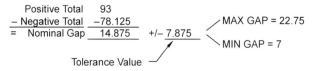

```
    Positive Total      93
  − Negative Total    −78.125                              MAX GAP = 22.75
  =   Nominal Gap      14.875      +/− 7.875
                                                           MIN GAP = 7
                      Tolerance Value
```

Stackup Direction

Solve for Minimum and Maximum Gap A-B

FIGURE 7.16   Simple assembly solved.

The dimension values in each column are totaled, and the negative total is subtracted from the positive total, which gives the nominal distance being studied. The tolerance values are totaled, and this total is subtracted from and added to the "nominal" distance to determine the minimum and maximum distances, respectively. The minimum and maximum distances are reported.

## Example 7.4

In this example, an assembly with parts assembled in the vertical direction is studied. (See Fig. 7.17.) The assembly will be greatly affected by the force of gravity, which will most likely pull the bracket (part number 3) down against the fasteners. The fasteners will in turn be pulled down against the holes in the hanger (part number 2). This will add assembly shift to the chain of Dimensions and Tolerances twice, once for the holes in the hanger and once for the holes in the bracket. It is assumed that the frame (part number 1) and the hanger are fixed in space. Individual part drawings for items 2 and 3 are shown in Fig. 7.18.

The goal of this Tolerance Stackup is to determine the maximum distance between the upper surface of the frame and the lower surface of the bracket. This distance was not directly dimensioned and toleranced on the assembly drawing. If this distance had been directly dimensioned and toleranced, a Tolerance Stackup would not be required, as the maximum value could be easily calculated right from the drawing.

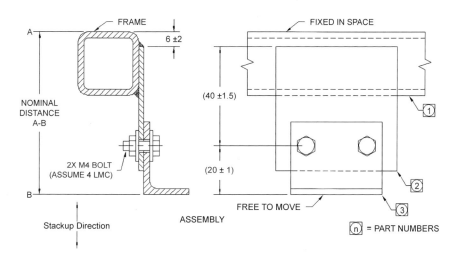

**FIGURE 7.17**   Hanger assembly (with gravity and assembly shift).

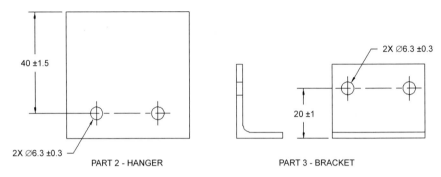

**FIGURE 7.18**   Parts for hanger assembly.

The Tolerance Stackup results are shown in Figs. 7.19 and 7.20.

The Tolerance Stackup sketch in Fig. 7.19 shows the chain of Dimensions and Tolerances for this problem. The upper end of the distance is labeled point A and the lower end labeled point B. The distance is labeled as A-B. No dimension spans the distance being studied, all three dimensions in the chain act in the same direction, and will therefore be labeled as being in the positive

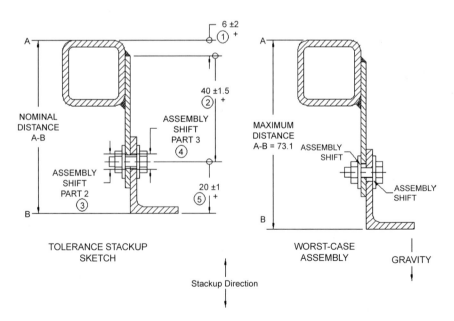

**FIGURE 7.19**   Worst-case hanger assembly and Tolerance Stackup sketch.

| + | − | Tolerances | Description |
|---|---|---|---|
| 6 | | ±2 | DIM 1: PART 1 - PART 2 |
| 40 | | ±1.5 | DIM 2: PART 2 EDGE - HOLES |
| | | ±1.3 | DIM 3: ASSY SHIFT PART 2: 6.3(H) + 0.3(ST) − 4(F) = 2.6 / 2 = ±1.3 |
| | | ±1.3 | DIM 4: ASSY SHIFT PART 3: 6.3(H) + 0.3(ST) − 4(F) = 2.6 / 2 = ±1.3 |
| 20 | | ±1 | DIM 5: PART 3 HOLES - FLANGE |
| 66 | 0 | ±7.1 | Totals |

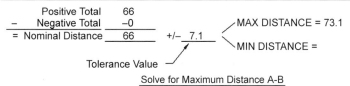

```
        Positive Total    66
  −     Negative Total   −0                      ⟋ MAX DISTANCE = 73.1
  = Nominal Distance      66      +/−  7.1   ⟨
                                            ⟍ MIN DISTANCE =
            Tolerance Value  ⟋
              Solve for Maximum Distance A-B
```

**FIGURE 7.20**   Tolerance Stackup report solved.

direction. All of the dimensions are already presented in equal bilateral format, so conversion is not required. Now the dimensions and tolerances in the chain of Dimensions and Tolerances must be identified.

The first dimension in the chain of Dimensions and Tolerances is $6 \pm 2$, which starts at point A and terminates at the top of the hanger. This dimension is labeled as item 1. The next dimension is $40 \pm 1.5$, which is the distance from the top of the hanger to the centerline of the holes. This dimension is labeled as item 2. The third item in the chain of Dimensions and Tolerances is assembly shift. This is the assembly shift of the fasteners within the holes in the hanger. The fourth item in the chain of Dimensions and Tolerances is also assembly shift. This is the assembly shift of the holes in the bracket about the fasteners. The last dimension is $20 \pm 1$, which is the distance from the centerline of the holes in the bracket to the bottom of the bracket. This dimension is labeled as item 5.

Notice that the two occurrences of assembly shift are numbered as items 3 and 4. They are numbered as they are encountered in the chain of Dimensions and Tolerances.

As stated earlier, all three dimensions act in the same direction to increase the total distance between points A and B: so all the dimensions are labeled as being in the positive direction, which is top to bottom in this example. A dimension origin symbol is placed at the start of each dimension and an arrowhead is placed at the terminating end of each dimension. It is a good idea to visually follow the chain of Dimensions and Tolerances from point A to point B to make sure nothing was missed.

Only the most essential information is included in the Tolerance Stackup report in Fig. 7.20. The positive dimension values are entered in the positive ( + ) direction dimension column. The equal bilateral tolerance for each dimension is entered in the Tolerances column on the same line as the dimension. Each dimension is described in the Description column. In this case, each dimension is described by its item number and by a short text description. Both occurrences of assembly shift are entered into the Tolerance Stackup report on the appropriate lines. Assembly shift is entered as an equal bilateral tolerance value per the previous section—there are no dimension values associated with the assembly shift values.

The dimension values in each column are totaled, and the negative total is subtracted from the positive total, which gives the nominal distance being studied. In this example, there are no dimensions in the negative direction, so the sum of the positive direction dimensions is used as the nominal distance. The tolerance values are totaled, and this total is added to the nominal distance to determine the maximum distance. In this example, only the maximum distance is reported, as that was the purpose of this Tolerance Stackup. The minimum value could have been reported as well, but this example shows that sometimes only one extreme is of interest.

## Example 7.5

In this example, an inseparable assembly (or weldment) is studied. (See Fig. 7.21). The goal of this Tolerance Stackup is to determine the minimum and maximum distance between parts 5 and 6 in the assembly. This distance was not directly dimensioned and toleranced on the assembly drawing. If this distance had been directly dimensioned and toleranced a Tolerance Stackup would not be required, as the minimum and maximum values could be easily calculated right from the drawing.

The Tolerance Stackup results are shown in Figs. 7.22 and 7.23.

The Tolerance Stackup sketch in Fig. 7.22 shows the chain of Dimensions and Tolerances for this problem. One end of the distance is labeled point A and the other end labeled point B. The distance is labeled as *Gap A-B*. The 32/30 limit dimension spans the distance being studied, so it is labeled as being in the positive direction. Some of the dimensions and tolerances are not presented in equal bilateral format; so conversion is required. Once the conversion is complete, the equal bilateral formatted data can be used in the Tolerance Stackup. Now the remaining dimensions and tolerances in the chain of Dimensions and Tolerances must be identified.

The first dimension in the chain of Dimensions and Tolerances is the limit dimension 11.6/11.4, which starts at point A and is the distance the pin (part number 5) protrudes from part number 4. This dimension is labeled as

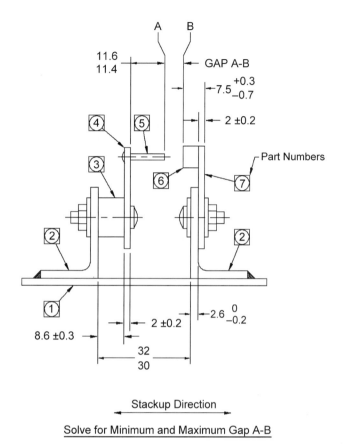

**Figure 7.21** Complex welded assembly.

item 1. The next dimension is 2 ± 0.2, which is the thickness of part number 4. This dimension is labeled as item 2. The next dimension is 8.6 ± 0.3, which is the thickness of the spacer (part number 3). This dimension is labeled as item 3. The next dimension is the limit dimension 32/30, which is the distance between the flange faces of the left and right brackets. This dimension is labeled as item 4. The next dimension is 2.6 + 0/−0.2, which is the thickness of the flange on the right bracket. This dimension is labeled as item 5. The next dimension is 2 ± 0.2, which is the thickness of part number 7. This dimension is labeled as item 6. The last dimension is 7.5 + 0.3/−0.7, which terminates at point B. It is the thickness of parts 6 and 7 combined. This dimension is labeled as item 7.

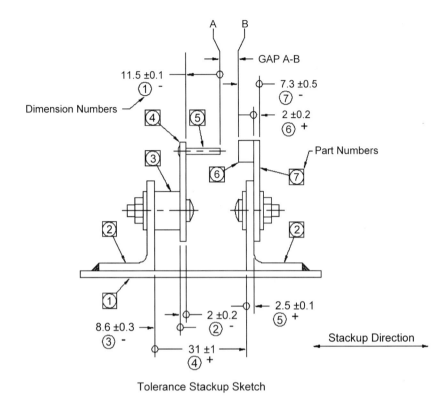

Tolerance Stackup Sketch

**FIGURE 7.22** Complex welded assembly (Tolerance Stackup sketch).

As stated earlier, the 32/30 limit dimension spans Gap A-B, so it is labeled as being in the positive direction, which is left-to-right in this example. Following the chain of Dimensions and Tolerances from point A to point B, we see that some of the remaining dimensions are in the positive direction and some are in the negative direction, which is right to left in this example. The directions of the remaining dimensions are labeled accordingly. A dimension origin symbol is placed at the start side of each dimension and an arrowhead is placed at the terminating end of each dimension. It is a good idea to visually follow the chain of Dimensions and Tolerances from point A to point B to make sure nothing was missed.

For this problem a little more information is included in the Tolerance Stackup report in Fig. 7.23. There are new columns for the dimension number ("Dim No.") and the part number ("Part No.") This makes the report a bit more complete, and makes it easier to cross-reference with the Tolerance

Worst-Case Tolerance Stackup

| Dim No | Part No | + | − | +/− | Description |
|---|---|---|---|---|---|
| 1 | 5 | | 11.5 | +/− 0.1 | Pin Length |
| 2 | 4 | | 2 | +/− 0.2 | LH Plate Thickness |
| 3 | 3 | | 8.6 | +/− 0.3 | Standoff Thickness |
| 4 | 2 | 31 | | +/− 1 | Flange to Flange Dist Between LH & RH Item 2 |
| 5 | 2 | 2.5 | | +/− 0.1 | RH Angle Brkt Web Thickness |
| 6 | 7 | 2 | | +/− 0.2 | RH Plate Thickness |
| 7 | 6 & 7 | | 7.3 | +/− 0.5 | Thickness of RH Plate and Boss |
| | | 35.5 | 29.4 | +/− 2.4 | Totals |

| | | |
|---|---|---|
| Positive Total | 35.5 | |
| − Negative Total | - 29.4 | |
| = Nominal Gap | 6.1 +/− 2.4 | Tolerance |

| | | |
|---|---|---|
| Max Gap | 8.5 | |
| Min Gap | 3.7 | Clearance |

**FIGURE 7.23** Tolerance Stackup report solved.

Stackup sketch. The positive dimension values are entered in the positive (+) direction dimension column, and the negative dimension values are entered in the negative (−) direction dimension column. The equal bilateral tolerance for each dimension is entered in the ± column on the same line as the dimension. Each dimension is described in the description column.

The dimension values in each column are totaled, and the negative total is subtracted from the positive total, which gives the nominal distance being studied. The tolerance values are totaled, and this total is subtracted from and added to the nominal distance to determine the minimum and maximum distances, respectively. The minimum and maximum distances are reported.

### Example 7.6

In this example, a complex assembly is studied. (See Fig. 7.24.) The assembly is very similar to the weldment in the previous example, except the brackets are bolted to the base plate instead of being welded. This will add a great deal of potential variation, as assembly shift will be added to the chain of Dimensions and Tolerances four times: twice for the holes in the base plate and once for the holes in each bracket. The base plate and the bracket are detailed in Fig. 7.25.

The goal of this Tolerance Stackup is the same as in Example 7.5: determine the minimum and maximum distance between parts 5 and 6 in the assembly. This distance was not directly dimensioned and toleranced on the

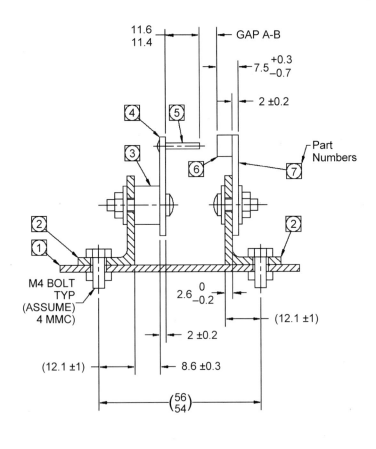

Stackup Direction

Solve for Minimum and Maximum Gap A-B

FIGURE 7.24   Complex bolted assembly.

assembly drawing. If this distance had been directly dimensioned and toleranced a Tolerance Stackup would not be required, as the minimum and maximum values could be easily calculated right from the drawing.

The Tolerance Stackup results are shown in Figs. 7.26 and 7.27.

The Tolerance Stackup sketch in Fig. 7.26 shows the chain of Dimensions and Tolerances for this problem. One end of the distance is labeled point A and the other end labeled point B. The distance is labeled as Gap A-B. The

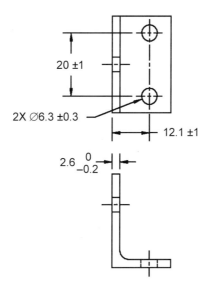

PART 2 - ANGLE BRKT

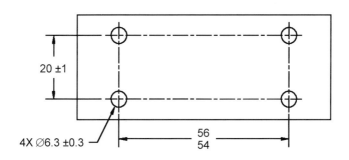

PART 1 - BASE PLATE

**FIGURE 7.25** Parts for complex bolted assembly.

56/54 limit dimension (converted to 55 ± 1) spans the distance being studied, so it is labeled as being in the positive direction. Some of the dimensions and tolerances are not presented in equal bilateral format, so conversion is required. Once the conversion is complete, the equal bilateral formatted data can be used in the Tolerance Stackup. Now the remaining dimensions and tolerances in the chain of Dimensions and Tolerances must be identified.

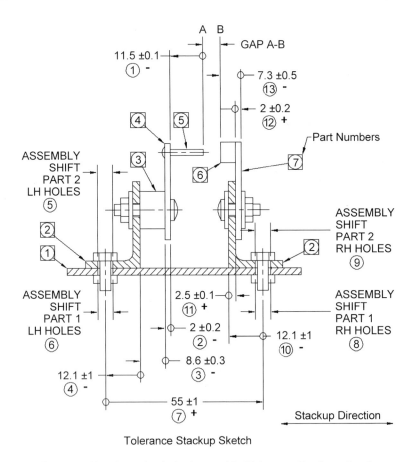

A   B

GAP A-B

11.5 ±0.1
①  -

7.3 ±0.5
⑬  -

2 ±0.2
⑫  +

④      ⑤                                    ─Part Numbers

ASSEMBLY
SHIFT
PART 2        ③
LH HOLES              ⑥                          ⑦
⑤

ASSEMBLY
SHIFT
PART 2
②       RH HOLES
②                                      ⑨

ASSEMBLY                    2.5 ±0.1                        ASSEMBLY
SHIFT                        ⑪  +                           SHIFT
PART 1                                                      PART 1
LH HOLES               2 ±0.2                          RH HOLES
⑥                      ②  -         12.1 ±1              ⑧
                                    ⑩  -

12.1 ±1                    8.6 ±0.3
④  -                       ③  -

55 ±1
⑦  +                         Stackup Direction

Tolerance Stackup Sketch

**Figure 7.26**   Complex bolted assembly Tolerance Stackup sketch.

   The first dimension in the chain of Dimensions and Tolerances is the
11.6/11.4 limit dimension, which starts at point A and is the distance the pin
(part number 5) protrudes from part number 4. This dimension is labeled as
item 1. The next dimension is $2 \pm 0.2$, which is the thickness of part number 4.
This dimension is labeled as item 2. The next dimension is $8.6 \pm 0.3$, which is
the thickness of the spacer (part number 3). This dimension is labeled as item
3. The next dimension is $12.1 \pm 1$, which is the distance between the flange face
and the center of the holes in the left bracket. This dimension is labeled as item
4. The next item in the chain of Dimensions and Tolerances is assembly shift.
This is the assembly shift of the holes in the left bracket about the fasteners.
This assembly shift is labeled as item 5. The next item in the chain of

Worst-Case Tolerance Stackup

| Dim No | Part No | + | – | +/– | Description |
|---|---|---|---|---|---|
| 1 | 5 | | 11.5 | +/–0.1 | Pin Length |
| 2 | 4 | | 2 | +/–0.2 | LH Plate Thickness |
| 3 | 3 | | 8.6 | +/–0.3 | Standoff Thickness |
| 4 | 2 | | 12.1 | +/–1 | CL Hole - Edge on LH Angle Brkt |
| 5 | 2 | | | +/–1.3 | Assy Shift in LH Angle Brkt Holes @ LMC: 6.6–4 = 2.6 / 2 = +/–1.3 |
| 6 | 1 | | | +/–1.3 | Assy Shift in Base Plate LH Holes @ LMC: 6.6–4 = 2.6 / 2 = +/–1.3 |
| 7 | 1 | 55 | | +/–1 | CL - CL Holes Dim on Base Plate |
| 8 | 1 | | | +/–1.3 | Assy Shift in Base Plate RH Holes @ LMC: 6.6–4 = 2.6 / 2 = +/–1.3 |
| 9 | 2 | | | +/–1.3 | Assy Shift in RH Angle Brkt Holes @ LMC: 6.6–4 = 2.6 / 2 = +/–1.3 |
| 10 | 2 | | 12.1 | +/–1 | CL Hole - Edge on RH Angle Brkt |
| 11 | 2 | 2.5 | | +/–0.1 | RH Angle Brkt Flange Thickness |
| 12 | 7 | 2 | | +/–0.2 | RH Plate Thickness |
| 13 | 6 & 7 | | 7.3 | +/–0.5 | Thickness of RH Plate & Boss |
| Totals | | 59.5 | 53.6 | +/–9.6 | Worst Case Tolerance |

| | |
|---|---|
| Positive Total | 59.5 |
| Negative Total | –53.6 |
| Nominal Gap | 5.9 +/– 9.6 |

| | | |
|---|---|---|
| Max Gap | 15.5 | Clearance |
| Min Gap | –3.7 | Inteference!!! |

**FIGURE 7.27** Complex bolted assembly - spreadsheet with solution.

Dimensions and Tolerances is also assembly shift. This is the assembly shift of the fasteners within the left pair of holes in the base plate. This assembly shift is labeled as item 6. The next dimension is the 56/54 limit dimension, which is the distance between the left pair of holes and the right pair of holes in the base plate. This dimension is labeled as item 7. The next item in the chain of Dimensions and Tolerances is assembly shift. This is the assembly shift of the fasteners within the right pair of holes in the base plate. This assembly shift is labeled as item 8. The next item in the chain of Dimensions and Tolerances is also assembly shift. This is the assembly shift of the holes in the right bracket about the fasteners. This assembly shift is labeled as item 9. The next dimension is 12.1 ± 1, which is the distance between the flange face and the center of the holes in the right bracket. This dimension is labeled as item 10. The next dimension is 2.6 +0/−0.2, which is the thickness of the flange on the right bracket. This dimension is labeled as item 11. The next dimension is 2 ± 0.2, which is the thickness of part number 7. This dimension is labeled as item 12. The last dimension is 7.5 +0.3/−0.7, which terminates at point B. It is the thickness of parts 6 and 7 combined. This dimension is labeled as item 13.

As stated earlier, the 56/54 limit dimension (converted to 55 ± 1) spans Gap A-B, so it is labeled as being in the positive direction, which is left to right in this example. Following the chain of Dimensions and Tolerances from point A to point B, we see that some of the remaining dimensions are in the positive direction and some are in the negative direction, which is right to left in this example. The directions of the remaining dimensions are labeled accordingly. A dimension origin symbol is placed at the start side of each dimension and an arrowhead is placed at the terminating end of each dimension. It is a good idea to visually follow the chain of Dimensions and Tolerances from point A to point B to make sure nothing was missed.

The Tolerance Stackup report for this problem is shown in Fig. 7.27. This report is formatted similar to the report the previous example. The

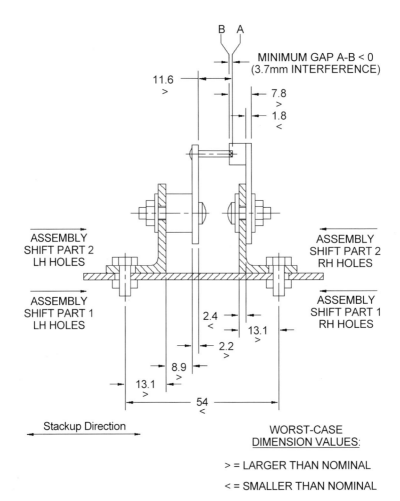

**FIGURE 7.28**   Complex bolted assembly solved (interference).

positive dimension values are entered in the positive ( + ) direction dimension column, and the negative dimension values are entered in the negative (−) direction dimension column. The equal bilateral tolerance for each dimension is entered in the ± column on the same line as the dimension. Each dimension is described in the description column. All four occurrences of assembly shift are entered into the Tolerance Stackup report on the appropriate lines. Assembly shift is entered as an equal bilateral tolerance value per the previous section—there are no dimension values associated with the assembly shift values.

The dimension values in each column are totaled, and the negative total is subtracted from the positive total, which gives the nominal distance being studied. The tolerance values are totaled, and this total is subtracted from and added to the nominal distance to determine the minimum and maximum distances, respectively. The minimum and maximum distances are reported. Notice that in this example the maximum distance is 15.5-mm clearance, and the minimum distance is 3.7-mm interference. Changing the geometry of the assembly by bolting the parts together instead of welding added quite a bit of variation to the Tolerance Stackup. Figure 7.28 shows the complex bolted assembly with the worst-case predicted interference.

## TOLERANCE STACKUPS AND ASSEMBLIES

### Moving Across an Interface from One Part to the Other in a Tolerance Stackup

Most Tolerance Stackups are done on assemblies to find a distance between features on distinct parts. The chain of Dimensions and Tolerances starts at one end of the distance being studied, makes its way from part to part, and ends at the other end of the distance being studied. This section discusses how to move from one part to another in the Tolerance Stackup.

There are two common types of interfaces encountered in Tolerance Stackups: mating planar surfaces, and clearance holes in mating parts or clearance holes and tapped holes that share common fasteners. This section presents general guidelines for traversing an interface from one part to another in the chain of Dimensions and Tolerances. These are only guidelines—there are cases where these guidelines must be modified.

The guidelines are based on the following assumptions: the mating features in the interface are part of the Tolerance Stackup, their dimensions and tolerances contribute to the Tolerance Stackup, and they are not directly part of the distance being studied. It is also assumed that the dimension and tolerance values are in the same direction as the Tolerance Stackup direction. If they are not, the dimensions and/or tolerance values must be trigonometrically projected into Tolerance Stackup direction as required. Lastly it is assumed that the dimensions and tolerances are in equal bilateral format. If they are not, they must be converted to equal bilateral format.

The first set of guidelines addresses traversing a planar interface (two nominally flat mating surfaces) between mating parts. The second set of guidelines addresses traversing a feature-of-size interface, such as coaxial clearance holes in mating parts, or coaxial clearance and threaded holes, with common fasteners. The fixed- and floating-fastener situations described in Chapter 18 are examples of a feature-of-size interface.

### Planar Interface: Traversing a Planar Interface from One Part to Another in the Tolerance Stackup

1.  For ± dimensions and tolerances:

    a.  The dimension to the interfacial surface on the first part is included in the Tolerance Stackup.

    b.  The ± location tolerance associated with the dimension is included in the Tolerance Stackup.

    c.  Now the Tolerance Stackup moves from the interfacial surface on the first part to the mating surface on the second part.

    d.  Steps 1.a and 1.b are repeated in reverse order for the second part.

2.  For GD&T:

    a.  If the planar feature is a referenced datum feature:

        i.   The basic dimension to the datum feature is included in the Tolerance Stackup.

        ii.  If there is a profile tolerance specified for the datum feature, lines for profile tolerance and datum feature shift are added to the Tolerance Stackup report.

             (1) The values for profile and datum feature shift are entered if the location of the datum feature contributes to the Tolerance Stackup. (The value for datum feature shift may be zero.) (See Chapters 9, 13, and 14.)

             (2) The values for profile and datum feature shift are set to zero and the lines are marked "N/A" if the location of the datum feature does not contribute to the Tolerance Stackup.

(3) If the location of the surface does not affect the Tolerance Stackup, but the profile tolerance controls the form of the feature, the profile tolerance may be included in the Tolerance Stackup as described in Chapter 20.

iii. If the datum feature has a form tolerance, the form tolerance is typically not included in the chain of Dimensions and Tolerances. However, the form tolerance may be included in the Tolerance Stackup per the guidance in Chapter 20 if desired.

iv. Special cases may require using an orientation tolerance or a lower segment composite profile tolerance in the Tolerance Stackup. These are uncommon applications and must be carefully addressed on a case-by-case basis. For more information see Chapter 9.

v. Now the Tolerance Stackup moves from the interfacial surface on the first part to the mating surface on the second part.

vi. Steps 2.a.i–2.a.iv are repeated in reverse order for the second part.

b. If the planar feature is not a datum feature:

i. The basic dimension from the datum reference frame related to the feature is included in the Tolerance Stackup.

ii. Lines for profile and datum feature shift are added to the Tolerance Stackup report. The values for profile and datum feature shift are entered. (The value for datum feature shift may be zero.) (See Chapters 9, 13, and 14.)

iii. Special cases may require using an orientation tolerance or a lower segment composite profile tolerance in the Tolerance Stackup. These are uncommon applications and must be carefully addressed on a case-by-case basis. For more information see Chapter 9.

iv. Now the Tolerance Stackup moves from the interfacial surface on the first part to the mating surface on the second part.

v. Steps 2.b.i–2.b.iii are repeated in reverse order for the second part.

Generally speaking, these guidelines also apply to interfacial surfaces that are complex-curved or warped, such as parabolic surfaces. Great care must be taken when dealing with such surfaces in a Tolerance Stackup, as these surfaces may share some properties with planar surfaces, and they may share some properties with features of size. This depends greatly the shape of the mating surfaces (how close they are to being nominally flat, or how close they are to mimicking the geometry of a feature of size), how they contribute to the Tolerance Stackup, and the direction of the Tolerance Stackup.

**Feature-of-Size Interface: Traversing a Feature-of-Size Interface (Mating Clearance and/or Threaded Holes with Common Fasteners) from One Part to Another in the Tolerance Stackup**

1. For ± dimensions and tolerances:

   a. The dimension to the Feature of Size on the first part is included in the chain of Dimensions and Tolerances.
   (This is the dimension in the direction of the Tolerance Stackup where rectangular or polar coordinate dimensioning is used.)

   b. The ± location tolerance associated with the dimension is included in the Tolerance Stackup.

   c. If the features are clearance holes, assembly shift is calculated and added to the chain of Dimensions and Tolerances for the holes in the first part. If the features are threaded or press-fit holes assembly shift is not added.

   d. Now the Tolerance Stackup moves from the interfacial feature on the first part to the mating feature on the second part.

   e. Steps 1.a–1.c are repeated in reverse order for the second part.

2. GD&T:

   a. If the feature of size (hole, pin, etc.) is a referenced datum feature,

      i. The basic dimension to the datum feature is included in the Tolerance Stackup.

      ii. If a positional or orientation tolerance is specified for the datum feature, lines for the positional/orientation tolerance, bonus tolerance, and datum feature shift are added to the Tolerance Stackup report.

      (1)   The values for position/orientation, bonus toler-
            ance, and datum feature shift are entered if the
            location of the datum feature contributes to the
            Tolerance Stackup. (The values for bonus toler-
            ance and datum feature shift may be zero.) (See
            Chapter 9.)

      (2)   The values for position/orientation, bonus toler-
            ance, and datum feature shift are set to zero and
            the lines are marked "N/A" if the location of the
            datum feature does not contribute to the Toler-
            ance Stackup. (This is common where the datum
            feature of size is the primary or secondary datum
            feature in a referenced feature control frame.)
            (See Chapters 9, 13, and 14.)

   iii.  Assembly shift is calculated and added to the chain
      of Dimensions and Tolerances for the datum feature
      of size in the first part. Assembly shift is typically not
      added if the datum features of size are threaded
      holes.

   iv.  Now the Tolerance Stackup moves from the datum
      feature on the first part to the datum feature on the
      second part.

   v.  Steps 2.a.i–2.a.iii are repeated in reverse order for the
      second part.

b.   If the feature of size (hole, pin, etc.) is a not a datum feature:

   i.  The basic dimension from the datum reference frame
      related to the feature is included in the Tolerance
      Stackup.

   ii.  Lines for positional tolerance, bonus tolerance, and
      datum feature shift are added to the Tolerance Stackup
      report. The values for position, bonus tolerance, and
      datum feature shift are entered. (The values for bonus
      tolerance and datum feature shift may be zero.) (See
      Chapters 9, 13, and 14.)

   iii.  Assembly shift is calculated and added to the chain of
      Dimensions and Tolerances for the holes in the first
      part. Assembly shift is typically not added if the
      features of size are threaded holes.

iv.  Now the Tolerance Stackup moves from the interfacial feature on the first part to the mating feature on the second part.

v.  Steps 2.b.1–2.b.iii are repeated in reverse order for the second part.

3.  Special cases may require using an orientation tolerance or a lower segment composite position tolerance in the Tolerance Stackup. This is not common and must be addressed on a case-by-case basis. For more information see Chapter 9.

These guidelines address the most common situations encountered. There are many special cases and circumstances that affect whether and how dimensions and tolerances should be included in the chain of Dimensions and Tolerances. Unfortunately, no set of rules or guidelines is universally applicable. The Tolerance Analyst must carefully consider the problem and determine whether and how each dimension and tolerance may affect the Tolerance Stackup result.

## THE TERM *CHAIN OF DIMENSIONS AND TOLERANCES*

Describing the dimensions and tolerances that contribute to a Tolerance Stackup as a *chain of Dimensions and Tolerances* is unique to this text. It is a technically accurate description, as the dimensions in the Tolerance Stackup lay head to tail, and can be visualized and followed like the links in a chain. An older and less accurate practice was to describe Tolerance Stackups as "loops", inferring that a loop could be followed from point A to point B. Sometimes, as is seen in Examples 7.1 to 7.3, the chain of Dimensions and Tolerances can be followed along a circular course, which could be considered as a loop. A *loop* implies a circular or elliptical course, starting at one end and looping around to the other end. However, not all Tolerance Stackups follow a circular or elliptical course. In Example 7.4 the contributing dimensions in the Tolerance Stackup are all in the same direction. The chain of Dimensions and Tolerances follows a straight line, which is most certainly not a loop. The chain of Dimensions and Tolerances in Examples 7.5 and 7.6 generally follow a counterclockwise course, but change direction several times along the way. These are not loops either. So, in the interest of technical accuracy, the term *Chain of Dimensions and Tolerances* will be used throughout this text.

# 8

## Statistical Tolerance Stackups

Statistical Tolerance Stackups determine the *probable* or *likely* maximum variation possible for a selected dimension. Similar to worst-case Tolerance Stackups, all tolerances are added to obtain the total variation. This method, however, more realistically assumes that it is highly improbable that all the dimensions in the Tolerance Stackup will be at their worst-case low limit or high limit at the same time. Remember, the worst-case Tolerance Stackup result requires some dimensions to be at their low limit and others to be at their high limit. So the direction of the deviation as well as the amount of deviation must be just so to achieve a worst-case condition.

It is more likely that the actual variations will be different than what is predicted by the worst-case model. The sum of the dimensions and tolerances will likely approximate a normal distribution. Most or all of the dimensions will likely be closer to their nominal value than either extreme. Also, some of the dimensions that the worst-case model required to be at their upper limit may actually be closer to their lower limit, and vice versa. The combination of these factors leads to the idea of a statistical Tolerance Stackup.

A question arises as to when it is appropriate to use a statistical versus a worst-case Tolerance Stackup. The answer to this question depends on a number of factors, including the number of tolerances in the Tolerance Stackup, the quantity of parts to be manufactured, manufacturing process controls, design sensitivity, past company practices, and willingness to accept risk, to name a few. A simple rule of thumb is as the number of tolerances in a Tolerance Stackup increases, the benefits and validity of using a statistical analysis increases. There are various rules in industry that state for more than 3, 4, 6, 10, etc., dimensions that a statistical analysis is the right choice.

This author does not adhere to the idea of an arbitrary number of dimensions being an automatic reason to switch from a worst-case to a statistical approach. No doubt, as the number of tolerances in a Tolerance Stackup increases, a statistical solution may not only be a good idea, but may more accurately represent the variation that will be seen at assembly. The number of tolerances alone, however, is insufficient reason to select a statistical approach. All factors, especially those factors relating to manufacturing and process controls, must be considered, and must be weighed against the risk of an overly conservative or overly liberal result.

Statistical tolerance analyses are based on several conditions being in place. These include

- The manufacturing processes for the parts must be controlled processes. This requires, among other things, that manufacturing nominal is the same as design nominal.
- Processes must be centered and output normal or gaussian distributions (see Fig. 8.1). This presents a problem where unequal bilateral or unilateral tolerances have been specified.
- Parts must be randomly selected for assembly.
- The design must be able to tolerate the possibility that some small percentage of the as-produced parts or assemblies exceed the calculated statistical result.
- The enterprise must be willing to tolerate the possibility that some parts or assemblies will be rejected due to exceeding the calculated statistical result.

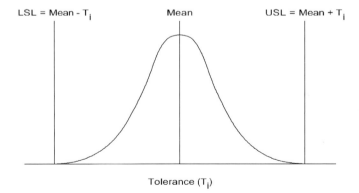

$LSL = Mean - T_i$  Mean  $USL = Mean + T_i$

Tolerance $(T_i)$

**FIGURE 8.1**  Gaussian distribution.

As presented in Chapter 3, *design nominal* and *manufacturing nominal* are rarely if ever the same. This presents a problem when considering the above assumptions. The requirements for RSS statistical Tolerance Analysis are that manufacturing processes shall be centered and output normal distributions. This correlates to the $C_p$ and $C_{pk}$ values encountered in statistical process control (SPC), which address process spread and process centering. It is beyond the scope of this text to address the statistical implications regarding these apparent discontinuities in great detail. Suffice to say that the difference between the design nominal and manufacturing nominal and the fact that some processes aren't as "controlled" as they should be lead to multiplying the statistical result by a coefficient greater than one. This practice also addresses the fact that most of the conditions listed in the previous paragraph are not always 100% applicable to every dimension and tolerance. Multiplying the RSS result by a coefficient greater than 1 gives the *adjusted statistical result*.

There are several statistical methods available for Tolerance Analysis. Root-sum-square (RSS) and Monte Carlo simulations are the two most common. Root-sum-square is commonly used on manually modeled and spreadsheet-based statistical Tolerance Stackups.

Monte Carlo simulation is typically used with computer-based Tolerance Analysis simulation software. Simply put, Monte Carlo simulations take all the variables in a Tolerance Stackup, give them a random value within their range, derive a result, iterate this process thousands of times, and average the result. This is a purely statistical approach. These are very powerful tools, and are great for solving three-dimensional Tolerance Stackups, as these tools allow the Tolerance Analyst to look at many combinations of translational and rotational variation. These tools are also fairly expensive and complex to learn and use. It is a good idea to dedicate staff to learn how to use the tool, as modeling can be complicated.

This text uses the root-sum-square method. The root-sum-square method takes each tolerance value, squares it, adds the squared tolerance values, and takes the square root of the result. Hence the name *root-sum-square*. The formula can be seen in Fig. 8.2. This result is the statistical tolerance. There are several variations on this method, using combinations of worst-case and statistical tolerancing, or adjusting the result by multiplying it by a value $>1$. As mentioned earlier, the statistical tolerance is often multiplied by an adjustment factor such as 1.5. This is not a text on statistics

$$\text{RSS Tolerance} = \sqrt{T_1^2 + T_2^2 + T_3^2 \ldots + T_n^2}$$

**FIGURE 8.2**   Root sum formula for statistical Tolerancing. Where: $T_n =$ Tolerances in the Tolerance Stackup.

and does not attempt to cover this important area in great detail. The reader is directed to more definitive sources for background, justification and derivation of formulas and methods, such as the *Dimensioning and Tolerancing Handbook**, which contains several chapters devoted to the study of various statistical tolerancing techniques.

I have been asked many times what the RSS Tolerance Stackup result represents in terms of sigma (σ) or standard deviations. Students want to know if an RSS Tolerance Stackup represents a ±1σ, ±3σ, ±6σ, etc., distribution. It is generally assumed that if all the individual tolerances entered into the Tolerance Stackup are produced by processes controlled to ±3σ, then the RSS Tolerance Stackup result also represents ±3σ. To put it another way, it is generally assumed that the level of process controls of the inputs represents the level of process controls of the output. (See Fig. 8.3). Likewise if all the component tolerances are assumed to be ±1σ, ±2σ, or ±6σ, then the RSS Tolerance Stackup result represents ±1σ, ±2σ, or ±6σ, respectively. The better you know your processes, the more accurate the statistical Tolerance Stackup result. It is very important to learn about the manufacturing processes where possible and to obtain reliable data from statistically controlled processes.

In the real world, however, it is likely that the processes used to manufacture all the part features in a Tolerance Stackup and their associated tolerances are not controlled to the same level. That is, the tolerances in a Tolerance Stackup are probably manufactured using a few ±2σ processes, a few ±3σ processes, a few ±4σ processes, etc., and perhaps even some processes where the level of control is unknown. So, in many environments it is likely that the tolerances in a Tolerance Stackup represent a mixed bag of process capabilities. This is especially true where other factors enter into the equation, such as datum feature shift or assembly shift. Assembly shift is particularly problematic, as it is a function of the assembly process, and unless the assembly process is monitored and measured like every other process, it is likely not controlled. Indeed assembly shift often manifests itself in its worst-case form. See Chapters 7 and 9 for more coverage on assembly shift. Again, this is all the more reason to use an adjustment factor when interpreting statistical results.

A step-by-step explanation of how to perform statistical Tolerance Stackups follows. This is exactly the same process as presented in Chapter 7, except for a few additional steps in which the tolerance values are squared and the square root of their sum is taken and multiplied by an adjustment factor as described above. These methods can be easily performed simultaneously using a spreadsheet.

---

*Dimensioning and Tolerancing Handbook. Drake, Paul J., Jr.; McGraw-Hill, New York, 1999.

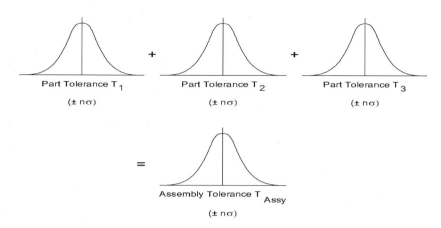

**FIGURE 8.3**  Part sigmas = assembly sigma. It is generally understood that the RSS Statistical Tolerance Stackup Result represents the same level of process controls as the Tolerances that make up the Tolerance Stackup. So, for example, if all the tolerances in the Tolerance Stackup were produced using processes controlled to $\pm3\sigma$, the RSS result would also represents $\pm3\sigma$.

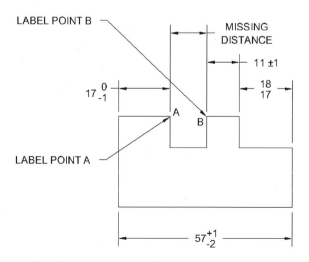

**FIGURE 8.4**  Statistical chain of Dimensions and Tolerances #1.

## STATISTICAL TOLERANCE STACKUP WITH DIMENSIONS (DIFFERENCES FROM WORST-CASE HIGHLIGHTED IN ITALICS)

1. Select the distance (gap or interference) whose variation is to be determined. Label one end of the distance $A$ and the other end $B$ (see Fig. 8.4).

2. Determine if a one-, two-, or three-dimensional analysis is required.

   a. If a two-dimensional analysis is required, determine if both directions can be resolved into one-dimension using trigonometry. If not, a linear Tolerance Stackup is not appropriate, and a computer program should be used for the Tolerance Analysis.

   b. If a three-dimensional analysis is required, a linear Tolerance Stackup is probably not appropriate, and a computer program should be used for the Tolerance Analysis.

3. Determine a positive direction and a negative direction.

   a. If there is a dimension that spans distance $A$–$B$, label it as a positive dimension by placing a "+" sign adjacent to the dimension value (see Fig. 8.5). It should also be assigned a direction by placing a dimension origin symbol at the end where the dimension starts and an arrowhead at the other end

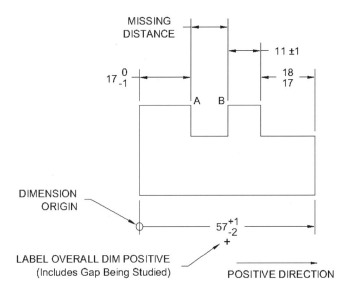

**FIGURE 8.5** Statistical chain of Dimensions and Tolerances #2.

where the dimension terminates. All dimensions in the chain of Dimensions and Tolerances that are followed in this direction shall be labeled as positive dimensions. All dimensions that are followed in the opposite direction shall be labeled as negative dimensions.

b. If no dimension spans distance *A–B*, start at point *A*. If the direction of the dimension from point *A* goes toward, through, or terminates at point *B*, then label it positive using a " + " sign, a dimension origin symbol, and arrowhead as described in 3.a. If its direction is away from point *B*, label it negative (see Fig. 8.6). Identify the chain of Dimensions and Tolerances from point *A* to point *B*, and label all dimensions in the same direction positive or negative, as indicated by the sign of the first dimension. Remember, if there is a dimension passing through distance *A–B*, it should be labeled as a positive dimension.

c. Follow the chain of Dimensions and Tolerances from point *A* to point *B*. You should be able to follow a continuous path

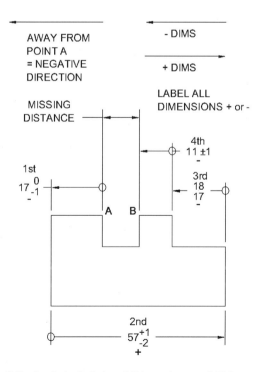

**FIGURE 8.6** Statistical chain of Dimensions and Tolerances #3.

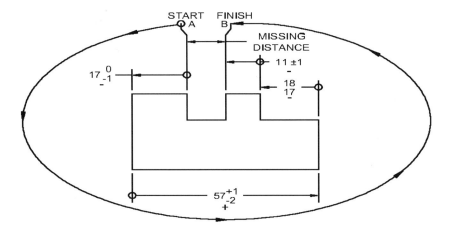

**FIGURE 8.7** Statistical chain of Dimensions and Tolerances #4. Follow the chain of Dimensions and Tolerances from point *A* to point *B* to make sure there are no breaks or discontinuities in the chain.

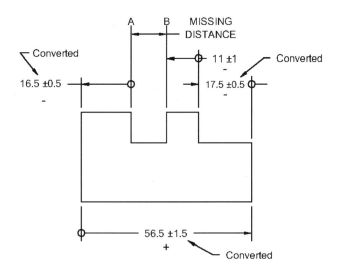

**FIGURE 8.8** Statistical chain of Dimensions and Tolerances #5. Convert all Dimensions and Tolerances to equal bilateral format.

from the start to the end of each dimension in the chain from point *A* to point *B* (see Fig. 8.7).

In this example, the first dimension starts at point *A* and ends at the left edge of the part. The second dimension starts where the first dimension ends and ends at the right edge of the part. The third dimension starts where the second dimension ends. The fourth dimension starts where the third dimension ends and ends at point *B*.

If the dimensions are not properly labeled, the nominal distance will be negative after the negative total is subtracted from the positive total. If this happens, check the + or − labels assigned to the dimensions, making sure that the sum of the positively labeled dimensions is larger than the sum of the negatively labeled dimensions. Remember that the total value of the positive dimensions must include distance *A*–*B*.

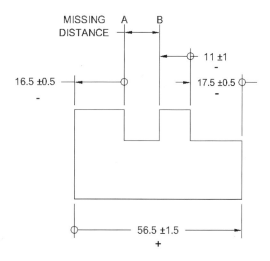

| + | − | Tolerances | Squared Tolerances |
|---|---|---|---|
|   | 16.5 |   |   |
| 56.5 |   |   |   |
|   | 17.5 |   |   |
|   | 11 |   |   |
|   |   |   |   |
|   |   |   |   |

**FIGURE 8.9**  Statistical chain of Dimensions and Tolerances #6.

4. Convert all dimensions and tolerances to equal bilateral format (± the same value). Instructions for how to do this are included in Chapter 4 (see Fig. 8.8).

5. Now all the dimensions and tolerances must be put into a chart and totaled for reporting purposes. Place each positive dimension value in the positive column on a separate line. Place each negative dimension value in the negative column on a separate line (see Fig. 8.9).

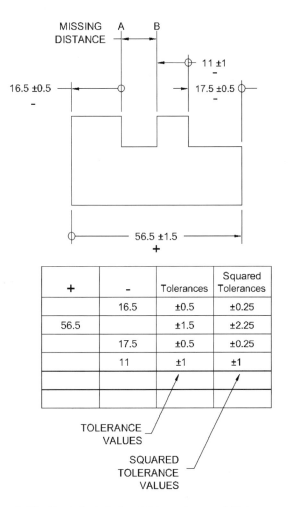

| + | – | | Tolerances | Squared Tolerances |
|---|---|---|---|---|
| | | 16.5 | ±0.5 | ±0.25 |
| 56.5 | | | ±1.5 | ±2.25 |
| | | 17.5 | ±0.5 | ±0.25 |
| | | 11 | ±1 | ±1 |
| | | | | |
| | | | | |

TOLERANCE VALUES

SQUARED TOLERANCE VALUES

**FIGURE 8.10**   Statistical chain of Dimensions and Tolerances #7.

6. Place the tolerance value for each dimension in the tolerance column adjacent to each dimension. This value is half the total variation allowed by the tolerance.
7. *Take each tolerance value and square it. Place this value in the Statistical Tolerance column next to each tolerance* (see Fig. 8.10).
8. Add the entries in each column, entering the results at the bottom of the chart (see Fig. 8.11).
9. *Take the square root of the sum of statistical tolerances (RSS).*

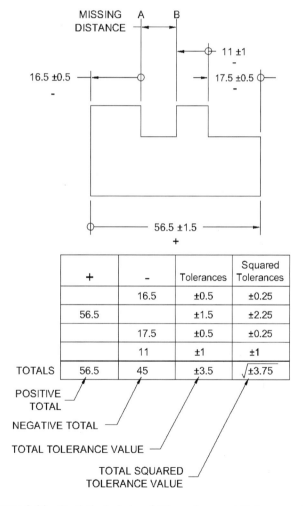

**FIGURE 8.11** Statistical chain of Dimensions and Tolerances #8.

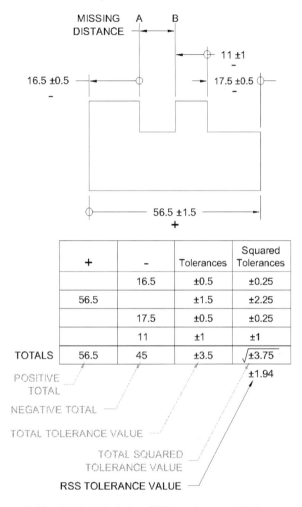

**FIGURE 8.12**   Statistical chain of Dimensions and Tolerances #9.

*Enter this result at the bottom of the chart. This is the RSS Tolerance Value* (see Fig. 8.12).

10.  Subtract the negative total from the positive total. This gives the nominal dimension or distance (see Fig. 8.13).

11.  *Apply the total statistical tolerance. Adding and subtracting the statistical tolerance from the nominal dimension gives the likely or probable maximum and minimum distance values* (see Fig. 8.13).

12.  *If it is desired to take a slightly more conservative approach, multiply the RSS tolerance by an adjustment factor ( such as 1.5 in this ex-*

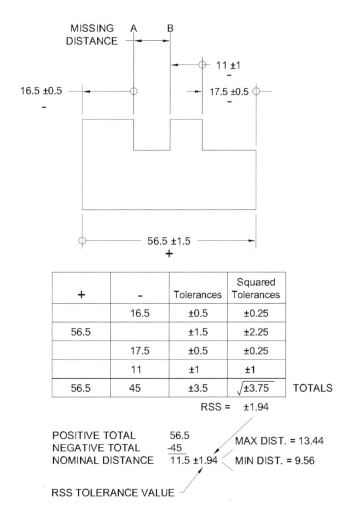

**FIGURE 8.13** Statistical chain of Dimensions and Tolerances #10.

*ample), perform that step here, substituting the larger Adjusted RSS value for the RSS value* (see Fig. 8.14).

Far and away the easiest method to solve linear Tolerance Stackup problems is to use a custom report format designed for a spreadsheet program such as Microsoft Excel or Lotus 1-2-3. The additional mathematical steps can be built into the spreadsheet; once the data is entered, the worst-case, statistical, and adjusted statistical solutions are calculated and displayed simultaneously, making it easy to compare the results of both methods. The

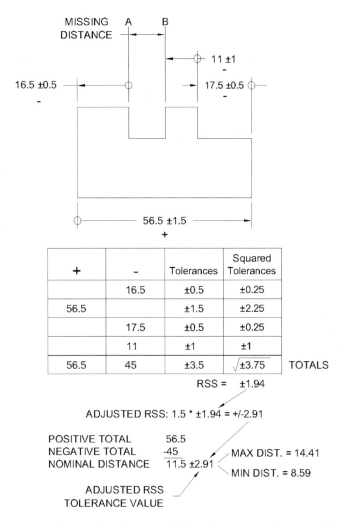

**FIGURE 8.14**   Statistical chain of Dimensions and Tolerances #11.

math and the format are inseparable, making for an easy-to-use and easy-to-understand problem-solving and reporting tool.

It cannot be overstated how important it is to use a clear and easy-to-read reporting format, as communication of the Tolerance Stackup results is almost always required. Using such a tool as described above consistently makes it easier for everyone involved in a project to understand the information quickly. It saves time and money.

| Program: | Electronics Packaging Program AV-11 | | | | | | | Stack Information: | |
|---|---|---|---|---|---|---|---|---|---|

| Product: | Part Number | Rev | Description | | | | | Stack No: | AV-11-010a |
|---|---|---|---|---|---|---|---|---|---|
| | 12345678-001 | A | Ground Plate Enclosure Assembly Option 1 w 8 Holes as Datum Feature B | | | | | Date: | 07/04/02 |
| Problem: | Edges of Ground Plate must not Touch Walls of Enclosure | | | | | | | Revision: | A |
| Objective: | Option 1: Determine if Ground Plate Contacts Enclosure Walls | | | | | | | Direction: | Along Plane of Ground Plate (Y Axis) |
| | | | | | | | | Author | BR Fischer |

| Description of Component / Assy | Part Number | Rev | Item | Description | + Dims | - Dims | Tol | Percent Contrib | Dim. / Tol Source & Calcs |
|---|---|---|---|---|---|---|---|---|---|
| Enclosure | 12345678-002 | A | 1 | Profile: Edge Along Pt A | | | +/- 0.5000 | 19% | Profile 1, A, Bm |
| | | | 2 | Datum Feature Shift ($DF_{B@LMC} - DFS_B$) / 2 | | | +/- 0.2900 | 11% | = (3.422 - (3.242 - 0.4)) / 2   (Shift within Minor Dia) |
| | | | 3 | Dim: Edge of Enclosure - Datum B | 8.5000 | | +/- 0.0000 | 0% | 8.5 Basic on Dwg |
| | | | 4 | Position: $DF_B$ M4 Holes | | | +/- 0.2000 | 8% | Position dia 0.4 @ MMC A |
| | | | 5 | Bonus Tolerance | | | +/- 0.0000 | 0% | N/A - Threads |
| | | | 6 | Datum Feature Shift ($DF_{B@LMC} - DFS_B$) / 2 | | | +/- 0.0000 | 0% | N/A - $DF_A$ not a Feature of Size |
| | | | 7 | Assembly Shift (Mounting Holes$_{MC}$ - F$_{MC}$) / 2 | | | +/- 0.6650 | 25% | = ((5 + 0.15) - 3.82) / 2 |
| Ground Plate | 12345678-004 | A | 8 | Position: $DF_B$ Dia 5 +/- 0.1 Holes | | | +/- 0.2250 | 9% | Position dia 0.45 @ MMC A |
| | | | 9 | Bonus Tolerance | | | +/- 0.1000 | 4% | = (0.1 + 0.1) / 2 |
| | | | 10 | Datum Feature Shift ($DF_{B@LMC} - DFS_B$) / 2 | | | +/- 0.0000 | 0% | N/A - $DF_A$ not a Feature of Size |
| | | | 11 | Dim: Datum B - Edge of Ground Plate | | 6.0000 | +/- 0.0000 | 0% | 6 Basic on Dwg |
| | | | 12 | Profile: Edge Along Pt B | | | +/- 0.5000 | 19% | Profile 1, A, Bm |
| | | | 13 | Datum Feature Shift ($DF_{B@LMC} - DFS_B$) / 2 | | | +/- 0.0150 | 6% | = ((5 + 0.15) - (5 - 0.15)) / 2 |

| Dimension Totals | 8.5000 | 6.0000 |
|---|---|---|
| Nominal Distance: Pos Dims - Neg Dims = | 2.5000 | |

**RESULTS:**

| | Nom | Tol | Min | Max |
|---|---|---|---|---|
| Arithmetic Stack (Worst Case) | 2.5000 | +/- 2.6300 | -0.1300 | 5.1300 |
| Statistical Stack (RSS) | 2.5000 | +/- 1.0721 | 1.4279 | 3.5721 |
| Adjusted Statistical 1.5*RSS | 2.5000 | +/- 1.6082 | 0.8918 | 4.1082 |

Notes:
- M4 Screw Dimensions: Major Dia 4 / 3.82   - M4 Tapped Hole Dimensions: Minor Dia: 3.422 / 3.242
- Used min and max screw thread minor dia in Datum Feature Shift Calculations on line 2.
- Used smallest screw major dia in Assembly Shift Calculations on line 7.

Assumptions:
- Assume threads are self centering. Do not include bonus tolerance on line 5.

Suggested Action:
- May want to use two holes as locators instead of all eight. See Stack Opt-2.

FIGURE 8.15   Sample Tolerance Stackup report form.

Preformatted Tolerance Stackup spreadsheet tools for solving worst-case and statistical linear Tolerance Stackups are available from Dimensioning and Tolerancing firms such as Advanced Dimensional Management™. An example is shown in Fig. 8.15.

The Tolerance Stackup report form shown in Fig. 15 contains the following information:

- Product name
- Model number (if applicable)
- Date
- Revision of study
- Direction of study
- Problem statement
- Objective of study
- Component or assy name
- Item number (for reference)
- Item description
- Tolerance value ( + and − )
- Tolerance source and calculations
- Worst-case results
- Statistical results
- Adjusted statistical results
- Notes at bottom if needed

Use of a consistent, standard format is very important. It makes learning to perform complex Tolerance Stackups easier as it facilitates a consistent approach to solving Tolerance Analysis problems. Data is gathered, calculations are made, and data is entered in the format the same way every time. It also makes it easier for your customers, those people who must interpret the data you provide and understand the work you have done.

Note that all assumptions have been noted, and sources of data are listed where applicable. It is common to have to search for information when performing Tolerance Stackups. Sometimes assumptions and supplier data is the only source for the required information. Make sure to include the sources for your information.

## STATISTICAL TOLERANCE STACKUP EXAMPLES

The Tolerance Stackup examples that follow are the same as Examples 7.1–7.6 that were solved worst-case in Chapter 7. The reader can compare the statistical results with the worst-case results, which are also presented here. This gives the reader the opportunity to see the effect that statistical techniques and manipulation have on the results of the Tolerance Stackups. As in the previous chapter, all of the examples are based on parts dimensioned and toleranced using the plus/minus ($\pm$) system.

As presented earlier in this chapter, solving Tolerance Stackups for the root-sum-square (RSS) and adjusted RSS results is easy and only requires a few more steps than solving for the worst-case result. Because the majority of the steps are the same as discussed in the worst-case chapter, only the additional steps required to obtain the statistical results will be discussed here.

## Example 8.1

In this example, a pin is the subject of the study. (See Fig. 8.16.) The goal of this Tolerance Stackup is to determine the minimum and maximum width of the groove in the pin. The Tolerance Stackup results are shown in Fig. 8.17.

The Tolerance Stackup sketch in Fig. 8.17 shows the chain of Dimensions and Tolerances for this problem, which is the same as when solving the problem worst-case.

The Tolerance Stackup report in Fig. 8.17 includes a new column for the squared tolerances—the other columns are the same as in the worst-case example in Chapter 7. Several additional lines are added below the chart to report the RSS tolerance value and the adjusted RSS tolerance value.

The dimension values in each column are totaled, and the negative total is subtracted from the positive total, which gives the nominal distance being studied. Each tolerance value is squared, and the squared values are totaled.

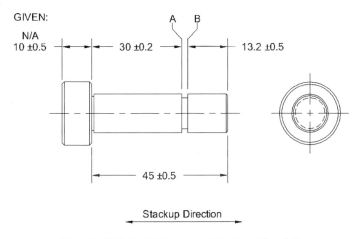

**FIGURE 8.16**  Pin with groove.

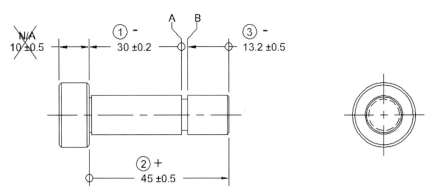

Tolerance Stackup Sketch

| + | - | Tolerances | Squared Tolerances | Description |
|---|---|---|---|---|
|  | 30 | ±0.2 | ±0.04 | GROOVE - HEAD |
| 45 |  | ±0.5 | ±0.25 | OVERALL LENGTH |
|  | 13.2 | ±0.5 | ±0.25 | TIP - GROOVE |
| 45 | 43.2 | ±1.2 | √±0.54 | Totals |

±0.74    RSS Tolerance Value

*1.5

±1.1    Adjusted RSS Tolerance Value

Positive Total    45
- Negative Total  -43.2                        MAX GAP = 2.9
= Nominal Gap     1.8      +/-  1.1
                                               MIN GAP = 0.7
Adjusted RSS Tolerance Value

Note: Because there are only three dimensions in this stackup, it is recommended that the statistical method not be used. Use the worst-case method shown in Problem 1 in the previous section.

The statistical solution is shown here for comparison only.

## Stackup Direction

## Solve for Statistical Minimum and Maximum Gap A-B

**FIGURE 8.17** Pin with groove solved statistically.

The square root is taken of the sum of the squared tolerances, which is the RSS tolerance. For these examples, the adjusted RSS tolerance will be used, which in this case means the RSS tolerance is multiplied by 1.5. The adjusted RSS tolerance is subtracted from and added to the nominal distance to determine the statistical minimum and maximum distances, respectively. The minimum and maximum distances are reported.

The worst-case tolerance is ±1.2. The adjusted RSS tolerance is ±1.1. The worst-case minimum groove width was reported as 0.6, and the maximum was 3.0 in Chapter 7. The adjusted RSS minimum groove width is 0.7, and the maximum is 2.9. Because there are only three dimensions and tolerances in this Tolerance Stackup, it is probably a better idea to use the worst-case results.

## Example 8.2

In this example, a part like the one presented at the beginning of this chapter is the subject of the study. (See Fig. 8.18.) The goal of this Tolerance Stackup is to determine the minimum and maximum distance between two parallel surfaces on the part. The distance being studied is different than in the material presented earlier in the chapter.

The Tolerance Stackup sketch in Fig. 8.19 shows the chain of Dimensions and Tolerances for this problem, which is the same as when solving the problem worst-case.

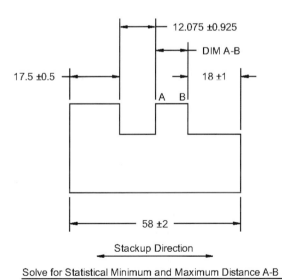

12.075 ±0.925

DIM A-B

17.5 ±0.5

18 ±1

A    B

58 ±2

Stackup Direction

Solve for Statistical Minimum and Maximum Distance A-B

FIGURE **8.18**  Simple part.

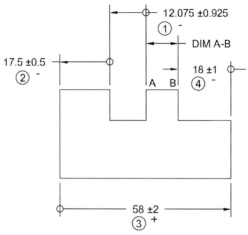

Tolerance Stackup Sketch

| + | - | Tolerances | Squared Tolerances | Description |
|---|---|---|---|---|
|  | 12.075 | ±0.925 | ±0.856 | DIM 1 |
|  | 17.5 | ±0.5 | ±0.25 | DIM 2 |
| 58 |  | ±2 | ±4 | DIM 3 |
|  | 18 | ±1 | ±1 | DIM 4 |
| 58 | 47.575 | ±4.425 | ±6.11 | Totals |

|  |  |
|---|---|
| ±2.47 | RSS Tolerance Value |
| *1.5 |  |
| ±3.71 | Adjusted RSS Tolerance Value |

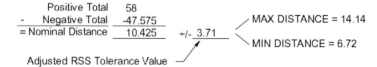

Positive Total       58
-    Negative Total   -47.575                        MAX DISTANCE = 14.14
= Nominal Distance    10.425      +/- 3.71
                                                      MIN DISTANCE = 6.72
Adjusted RSS Tolerance Value

Stackup Direction

Solve for Statistical Minimum and Maximum Distance A-B

FIGURE 8.19   Simple part solved statistically.

The Tolerance Stackup report in Fig. 8.19 includes a new column for the squared tolerances—the other columns are the same as in the worst-case example in Chapter 7. Several additional lines are added below the chart to report the RSS tolerance value and the adjusted RSS tolerance value.

The dimension values in each column are totaled, and the negative total is subtracted from the positive total, which gives the nominal distance being studied. Each tolerance value is squared, and the squared values are totaled. The square root is taken of the sum of the squared tolerances, which is the RSS tolerance. For these examples, the adjusted RSS tolerance will be used, which in this case means the RSS tolerance is multiplied by 1.5. The adjusted RSS tolerance is subtracted from and added to the nominal distance to determine the statistical minimum and maximum distances, respectively. The minimum and maximum distances are reported.

The worst-case tolerance is ±4.425. The adjusted RSS tolerance is ±3.71. The worst-case minimum distance was reported as 6, and the maximum was 14.85 in Chapter 7. The adjusted RSS minimum distance is 6.72, and the maximum is 14.14.

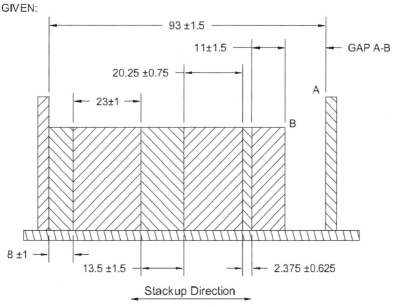

FIGURE 8.20 Simple assembly.

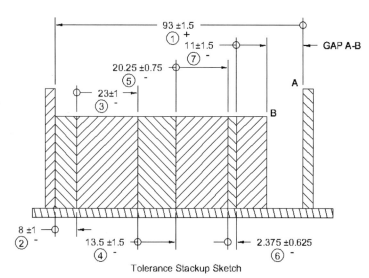

Tolerance Stackup Sketch

| + | - | Tolerances | Squared Tolerances | Description |
|---|---|---|---|---|
| 93 | | ±1.5 | ±2.25 | DIM 1 |
| | 8 | ±1 | ±1 | DIM 2 |
| | 23 | ±1 | ±1 | DIM 3 |
| | 13.5 | ±1.5 | ±2.25 | DIM 4 |
| | 20.25 | ±0.75 | ±0.5625 | DIM 5 |
| | 2.375 | ±0.625 | ±0.391 | DIM 6 |
| | 11 | ±1.5 | ±2.25 | DIM 7 |
| 93 | 78.125 | ±7.875 | $\sqrt{\pm 9.70}$ | Totals |

| | |
|---|---|
| ±3.115 | RSS Tolerance Value |
| *1.5 | |
| ±4.67 | Adjusted RSS Tolerance Value |

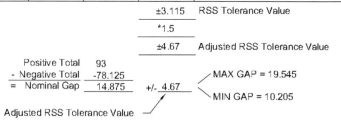

Stackup Direction

Solve for Statistical Minimum and Maximum Gap A-B

FIGURE 8.21   Simple assembly solved statistically.

## Example 8.3

In this example, a simple assembly is studied. (see Fig. 8.20.) The goal of this Tolerance Stackup is to determine the minimum and maximum distance between opposing surfaces on two parts in the assembly.

The Tolerance Stackup sketch in Fig. 8.21 shows the chain of Dimensions and Tolerances for this problem, which is the same as when solving the problem worst-case.

The Tolerance Stackup report in Fig. 8.21 includes a new column for the squared tolerances—the other columns are the same as in the worst-case example in Chapter 7. Several additional lines are added below the chart to report the RSS tolerance value and the adjusted RSS tolerance value.

The dimension values in each column are totaled, and the negative total is subtracted from the positive total, which gives the nominal distance being studied. Each tolerance value is squared, and the squared values are totaled. The square root is taken of the sum of the squared tolerances, which is the RSS tolerance. For these examples, the adjusted RSS tolerance will be used, which in this case means the RSS tolerance is multiplied by 1.5. The adjusted RSS tolerance is subtracted from and added to the nominal distance to determine the statistical minimum and maximum distances, respectively. The minimum and maximum distances are reported.

The worst-case tolerance is ±7.875. The adjusted RSS tolerance is ±4.67. The worst-case minimum gap was reported as 7, and the maximum was 22.75 in Chapter 7. The adjusted RSS minimum gap is 10.205, and the maximum is 19.545.

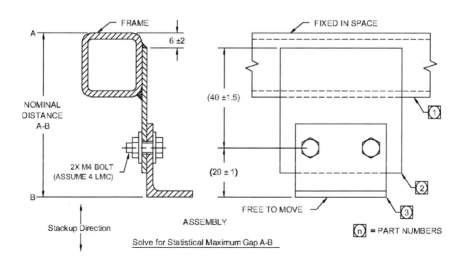

**FIGURE 8.22** Hanger assembly (with gravity and assembly shift).

*Example 8.4*

In this example, an assembly with parts assembled in the vertical direction is studied. (See Fig. 8.22.) The goal of this Tolerance Stackup is to determine the maximum distance between the upper surface of the frame and the lower surface of the bracket.

The assembly will be greatly affected by the force of gravity, which will most likely pull the bracket (part number 3) down against the fasteners. The fasteners will in turn be pulled down against the holes in the hanger (part number 2). It is assumed that the frame (part number 1) and the hanger are fixed in space. Individual part drawings for items 2 and 3 are shown in Fig. 8.23.

The Tolerance Stackup results are shown in Figs. 8.24 and 8.25. The Tolerance Stackup sketch in Fig. 8.24 shows the chain of Dimensions and Tolerances for this problem, which is the same as when solving the problem worst-case. The Tolerance Stackup report in Fig. 8.25 includes a new column for the squared tolerances—the other columns are the same as in the worst-case example in Chapter 7. Several additional lines are added below the chart to report the RSS tolerance value and the adjusted RSS tolerance value.

The dimension values in the positive column are totaled, which is the nominal distance being studied. Each tolerance value is squared, and the squared values are totaled. The square root is taken of the sum of the squared tolerances, which is the RSS tolerance. For these examples, the adjusted RSS tolerance will be used, which in this case means the RSS tolerance is multiplied by 1.5. The adjusted RSS tolerance is added to the nominal distance to determine the statistical maximum distance. The maximum distance is reported.

The worst-case tolerance is $\pm 7.1$. The adjusted RSS tolerance is $\pm 4.89$. The worst-case maximum distance was reported as 73.1 in Chapter 7. The adjusted RSS maximum distance is 70.89.

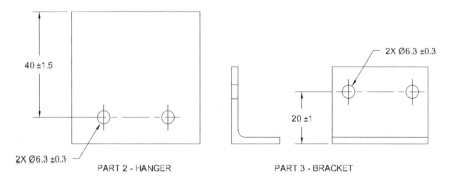

PART 2 - HANGER                    PART 3 - BRACKET

**FIGURE 8.23** Parts for hanger assembly.

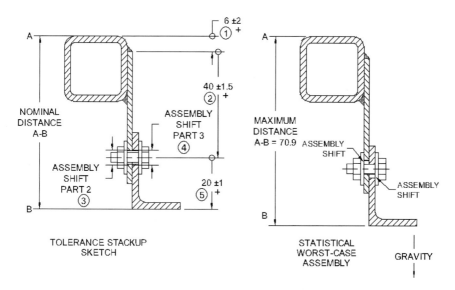

**FIGURE 8.24** Statistical worst-case hanger assembly and Tolerance Stackup sketch.

| + | – | Tolerances | Squared Tolerances | Description |
|---|---|---|---|---|
| 6 | | ±2 | ±4 | DIM 1: PART 1 - PART 2 |
| 40 | | ±1.5 | ±2.25 | DIM 2: PART 2 EDGE - HOLES |
| | | ±1.3 | ±1.69 | DIM 3: ASSY SHIFT PART 2: 6.3(H) + 0.3(ST) - 4(F) = 2.6 / 2 = ±1.3 |
| | | ±1.3 | ±1.69 | DIM 4: ASSY SHIFT PART 3: 6.3(H) + 0.3(ST) - 4(F) = 2.6 / 2 = ±1.3 |
| 20 | | ±1 | ±1 | DIM 5: PART 3 HOLES - FLANGE |
| 66 | 0 | ±7.1 | $\sqrt{}$ ±10.63 | Totals |

±3.26   RSS Tolerance Value

\*1.5

±4.89   Adjusted RSS Tolerance Value

Positive Total      66
–   Negative Total      -0
= Nominal Distance    66        +/- 4.89

Adjusted RSS Tolerance Value

MAX DISTANCE = 70.89        Stackup Direction

MIN DISTANCE =

Solve for Statistical Maximum Gap A-B

**FIGURE 8.25** Tolerance Stackup report solved statistically.

## Example 8.5

In this example, an inseparable assembly (or weldment) is studied. (see Fig. 8.26). Notice the dimensions and tolerances have already been converted to equal-bilateral format. The goal of this Tolerance Stackup is to determine the minimum and maximum distance between parts 5 and 6 in the assembly.

The Tolerance Stackup results are shown in Figs. 8.27 and 8.28. The Tolerance Stackup sketch in Fig. 8.27 shows the chain of Dimensions and Tolerances for this problem, which is the same as when solving the problem worst-case. The Tolerance Stackup report in Fig. 8.28 includes a new column for the squared tolerances—the other columns are the same as in the worst-case example in Chapter 7. Several additional lines are added below the chart to report the RSS tolerance value and the adjusted RSS tolerance value.

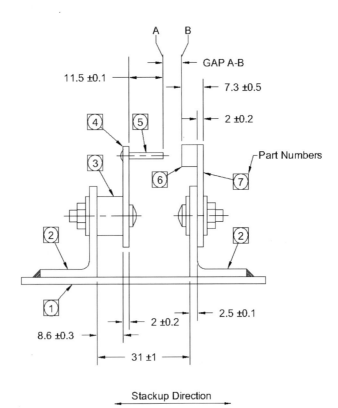

**FIGURE 8.26** Complex welded assembly.

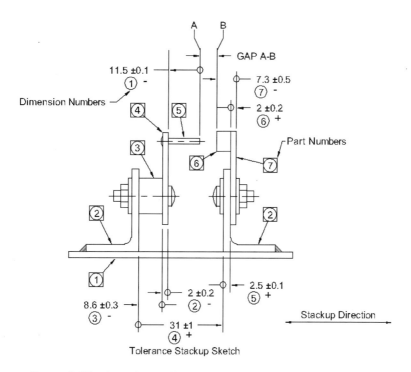

**FIGURE 8.27** Complex welded assembly (Tolerance Stackup sketch).

Statistical Tolerance Stackup

| Dim No | Part No | + | - | +/- | Squared Tolerances | Description |
|--------|---------|---|-----|-------|------------|-------------|
| 1 | 5 | | 11.5 | ± 0.1 | ± 0.01 | Pin Length |
| 2 | 4 | | 2 | ± 0.2 | ± 0.04 | LH Plate Thickness |
| 3 | 3 | | 8.6 | ± 0.3 | ± 0.09 | Standoff Thickness |
| 4 | 2 | 31 | | ± 1 | ± 1 | Flange to Flange Dist Between LH & RH Item 2 |
| 5 | 2 | 2.5 | | ± 0.1 | ± 0.01 | RH Angle Brkt Web Thickness |
| 6 | 7 | 2 | | ± 0.2 | ± 0.04 | RH Plate Thickness |
| 7 | 6 & 7 | | 7.3 | ± 0.5 | ± 0.25 | Thickness of RH Plate and Boss |
| Totals | | 35.5 | 29.4 | ± 2.4 | √± 1.44 | Totals |

+/- 1.20   RSS Tolerance

+/- ⬚1.8   Adjusted RSS Tolerance ( RSS * 1.5 )

| Positive Total | 35.5 | |
|---|---|---|
| - Negative Total | - 29.4 | |
| = Nominal Gap | 6.1 ± 1.8 | Adjusted RSS Tolerance |

| Max Gap | 7.9 |
|---|---|
| Min Gap | 4.3 |

Solve for Statistical Minimum and Maximum Gap A-B

**FIGURE 8.28** Tolerance Stackup report solved statistically.

The dimension values in each column are totaled, and the negative total is subtracted from the positive total, which gives the nominal distance being studied. Each tolerance value is squared, and the squared values are totaled. The square root is taken of the sum of the squared tolerances, which is the RSS tolerance. For these examples, the adjusted RSS tolerance will be used, which in this case means the RSS tolerance is multiplied by 1.5. The adjusted RSS tolerance is subtracted from and added to the nominal distance to determine the statistical minimum and maximum distances, respectively. The minimum and maximum distances are reported.

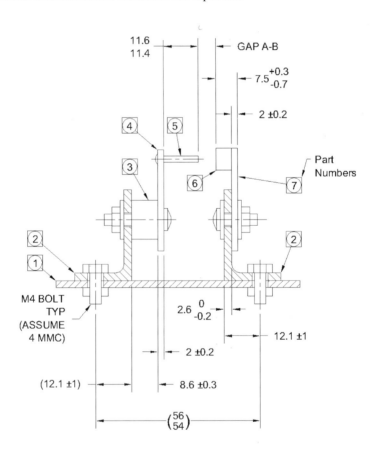

Stackup Direction

Solve for Minimum and Maximum Gap A-D

**FIGURE 8.29** Complex bolted assembly.

The worst-case tolerance is ±2.4. The adjusted RSS tolerance is ±1.8. The worst-case minimum gap was reported as 3.7, and the maximum was 8.5 in Chapter 7. The adjusted RSS minimum gap is 4.3, and the maximum is 7.9.

### Example 8.6

In this example, a complex assembly is studied. (See Fig. 8.29.) The assembly is very similar to the weldment in the previous example, except the brackets

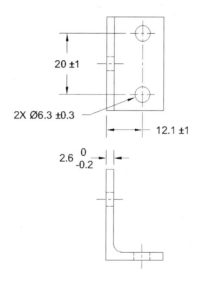

PART 2 - ANGLE BRKT

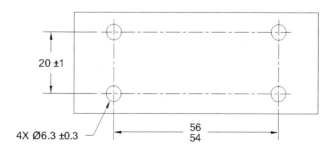

PART 1 - BASE PLATE

**FIGURE 8.30** Parts for complex bolted assembly.

are bolted to the base plate instead of being welded. The goal of this Tolerance Stackup is to determine the minimum and maximum distance between parts 5 and 6 in the assembly. The base plate and the bracket are detailed in Fig. 8.30.

The Tolerance Stackup results are shown in Figs. 8.31 and 8.32. The Tolerance Stackup sketch in Fig. 8.31 shows the chain of Dimensions and Tolerances for this problem, which is the same as when solving the problem worst-case. The Tolerance Stackup report in Fig. 8.32 includes a new column for the squared tolerances—the other columns are the same as in the worst-case example in Chapter 7. Several additional lines are added below the chart to report the RSS tolerance value and the adjusted RSS tolerance value.

The dimension values in each column are totaled, and the negative total is subtracted from the positive total, which gives the nominal distance being

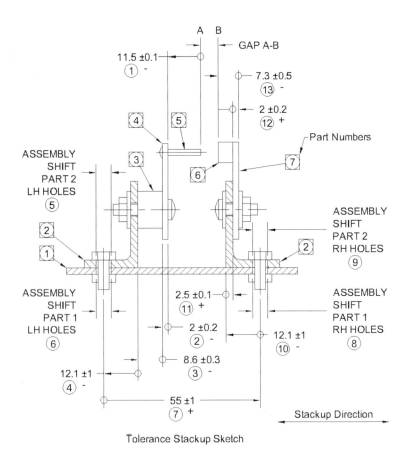

Tolerance Stackup Sketch

**FIGURE 8.31**   Complex bolted assembly solved statistically.

Statistical Tolerance Stackup

| Dim No | Part No | + | - | +/- | Squared Tolerances | Description |
|---|---|---|---|---|---|---|
| 1 | 5 | | 11.5 | +/- 0.1 | +/- 0.01 | Pin Length |
| 2 | 4 | | 2 | +/- 0.2 | +/- 0.04 | LH Plate Thickness |
| 3 | 3 | | 8.6 | +/- 0.3 | +/- 0.09 | Standoff Thickness |
| 4 | 2 | | 12.1 | +/- 1 | +/- 1 | CL Hole - Edge on LH Angle Brkt |
| 5 | 2 | | | +/- 1.3 | +/- 1.69 | Assy Shift in LH Angle Brkt Holes @ LMC: 6.6 - 4 = 2.6 / 2 = +/-1.3 |
| 6 | 1 | | | +/- 1.3 | +/- 1.69 | Assy Shift in Base Plate LH Holes @ LMC: 6.6 - 4 = 2.6 / 2 = +/-1.3 |
| 7 | 1 | 55 | | +/- 1 | +/- 1 | CL - CL Holes Dim on Base Plate |
| 8 | 1 | | | +/- 1.3 | +/- 1.69 | Assy Shift in Base Plate RH Holes @ LMC: 6.6 - 4 = 2.6 / 2 = +/-1.3 |
| 9 | 2 | | | +/- 1.3 | +/- 1.69 | Assy Shift in RH Angle Brkt Holes @ LMC: 6.6 - 4 = 2.6 / 2 = +/-1.3 |
| 10 | 2 | | 12.1 | +/- 1 | +/- 1 | CL Hole - Edge on RH Angle Brkt |
| 11 | 2 | 2.5 | | +/- 0.1 | +/- 0.01 | RH Angle Brkt Web Thickness |
| 12 | 7 | 2 | | +/- 0.2 | +/- 0.04 | Thickness of RH Plate |
| 13 | 6 & 7 | | 7.3 | +/- 0.5 | +/- 0.25 | Thickness of RH Plate & Boss |
| Totals | | 59.5 | 53.6 | +/- 9.6 | √+/- 3.19 | |
| | | | | | +/- 3.19 | RSS Tolerance |
| | | | | | +/- 4.79 | Adjusted RSS Tolerance ( RSS * 1.5) |

| | | |
|---|---|---|
| Positive Total | 59.5 | |
| Negative Total | -53.6 | |
| Nominal Gap | 5.9 +/- 4.79 | Adjusted RSS Tolerance |

| | | |
|---|---|---|
| Max Gap | 10.7 | |
| Min Gap | 1.1 | Clearance! |

**FIGURE 8.32**  Complex bolted assembly spreadsheet with statistical solution.

studied. Each tolerance value is squared, and the squared values are totaled. The square root is taken of the sum of the squared tolerances, which is the RSS tolerance. For these examples, the adjusted RSS tolerance will be used, which in this case means the RSS tolerance is multiplied by 1.5. The adjusted RSS tolerance is subtracted from and added to the nominal distance to determine the statistical minimum and maximum distances, respectively. The minimum and maximum distances are reported.

The worst-case tolerance is ±9.6. The Adjusted RSS tolerance is ±4.79. The worst-case minimum gap was reported as -3.7, which indicates interference, and the maximum was a 15.5 clearance in Chapter 7. The adjusted RSS minimum gap is a 1.1 clearance, and the maximum is a 10.7 clearance. In this example, using the adjusted RSS tolerance changes the result from 3.7-mm interference to 1.1-mm clearance. Given the number of contributors in the chain of Dimensions and Tolerances, it is probably a good idea to use the statistical result. Figure 8.33 shows the complex bolted assembly with statistically predicted clearance.

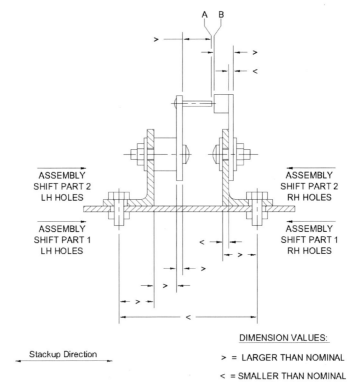

ASSEMBLY
SHIFT PART 2
LH HOLES

ASSEMBLY
SHIFT PART 2
RH HOLES

ASSEMBLY
SHIFT PART 1
LH HOLES

ASSEMBLY
SHIFT PART 1
RH HOLES

DIMENSION VALUES:

Stackup Direction

> = LARGER THAN NOMINAL

< = SMALLER THAN NOMINAL

Note: In this example, an adjusted RSS statistical tolerance was used.

**FIGURE 8.33** Complex bolted assembly solved statistically (clearance). All positive dimensions are biased toward their largest value. The values of the tolerances are within their ranges, but are not at their extremes, due to statistical manipulation. The holes in the Angle Brackets (Part 2) shift inward against the fasteners, and the fasteners shift inward against the holes in the Base Plate. This leads to the smallest statistical gap, which in this case is 1.1mm clearance!

# 9

# Geometric Dimensioning and Tolerancing (GD&T)

The most important difference between drawings created using GD&T and $\pm$ dimensioning and tolerancing is that GD&T creates coordinate systems based on datum reference frames, and all features on a part are unambiguously related to these coordinate systems. Properly applied, GD&T specifies which part features are to be used as datum features, creating the basis for each coordinate system. The rest of the features on the part are related to these coordinate systems through geometric tolerances in feature control frames.

Tolerance Stackups done on parts and assemblies that have been properly dimensioned and toleranced with GD&T are easier and more straightforward than with parts defined by $\pm$ dimensions and tolerances. This is because Tolerance Stackups performed on parts with GD&T require far fewer assumptions regarding how to interpret the tolerance specifications.

Mating parts should have coordinated datum reference frames, with the interfacing surfaces specified as primary datum features. From these surfaces, two coordinate systems are established, one on each part. Relating features on each part to these datum reference frames minimizes variation between related features on each part.

On simple parts, there may be only one datum reference frame, and all part features are related to it or the datum features themselves. More complex parts may have many datum reference frames due to geometry or functional necessity. Each datum reference frame on a part must be related to the other datum reference frames on the part, either directly or indirectly. For example, take the part in Fig. 9.1. There are four datum reference frames on this part. If

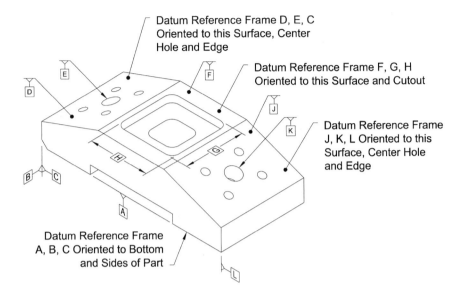

**FIGURE 9.1** Complex part with 4 datum reference frames.

the tolerance between features related to different datum reference frames needs to be studied in a Tolerance Stackup, the accumulated variation between the features and their datum reference frames must be studied.

## CONVERTING GD&T INTO EQUAL BILATERAL ± TOLERANCES

The previous examples depicted parts and assemblies that were toleranced with traditional ± tolerances for size and location. These tolerances were then converted into equal bilateral ± tolerances as required. Using the techniques presented here, parts and assemblies dimensioned with GD&T must also be converted to equal bilateral ± tolerances before a Tolerance Analysis can be completed.

Plus/minus dimensions and tolerances are still used with drawings based on GD&T, but their use should be limited to defining features of size and the depth or length of features such as holes and pins. For many reasons, ± dimensions and tolerances should not be used to locate features.

Several types of geometric tolerances and the conversion procedure will be discussed in this section. The discussion is a simplification of what actually must be considered in comparing geometrically dimensioned and toleranced parts with parts dimensioned and toleranced using the ± system.

## PROFILE TOLERANCES

Profile tolerances can be readily translated into ± tolerances. Profile tolerances specify a total width tolerance zone that follows the shape of a nominal surface (or true profile). The tolerance zone may be equal bilaterally, unequal bilaterally, or unilaterally displaced about the nominal surface, much like a ± tolerance. For typical single segment profile tolerances as shown in Fig. 9.2, the total profile tolerance zone is equally and bilaterally displaced about the nominal surface, and the equivalent ± tolerance is half the profile tolerance value. This is a simplification, as a profile tolerance is usually associated with a datum reference frame, which affects the location and orientation of the tolerance zone.

Profile tolerances are far superior to traditional ± tolerances in several ways. First, as previously mentioned, they are usually related to a datum reference frame, which precisely locates the tolerance zone relative to other

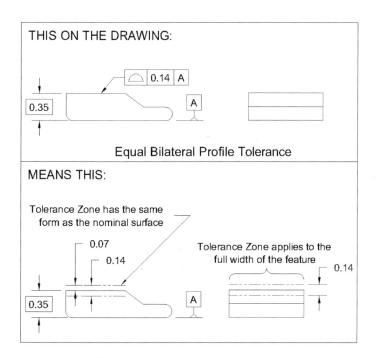

**FIGURE 9.2**  Equal bilateral profile.

toleranced features. Second, they may be applied unambiguously to any
surface, regardless of its shape, location and orientation; ± tolerances fall far
short of this universal applicability, and are only clear in their intent for
defining size limits for a single feature of size.

As described in Chapter 3 and shown in Fig. 3.1, ± tolerances may
expressed as a limit dimension, equal bilateral, unequal bilateral, unilateral
positive, or unilateral negative tolerances. Aside from the limit dimension
method, profile offers tolerancing methods analogous to ±, with the added
benefits described above.

### The Equal Bilateral Profile Tolerance Shown in Fig. 9.2
### May Be Converted as Follows:

- Given an equal bilaterally displaced profile Tolerance,

  Profile 0.14      profile Tolerance = total Tolerance

- Divide the profile Tolerance by 2 to obtain the ± equal bilateral
  Tolerance value

  $$\text{Equal bilateral Tolerance value} = \frac{0.14}{2} = 0.07$$

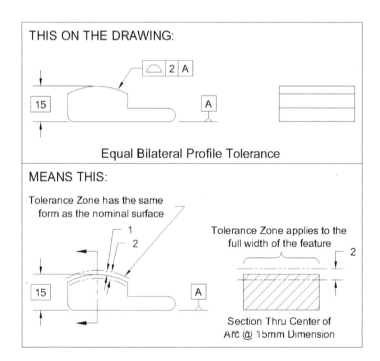

**FIGURE 9.3**   Equal bilateral profile curved surface.

- Take the value of the basic dimension locating the surface as nominal

  $\boxed{0.35}$

  Nominal dimension value = 0.35

Conversion complete:

Equal bilateral equivalent = 0.35 ± 0.07

Note: The dimension value and the tolerance value will be placed on separate lines in the Tolerance Stackup report.

The equal bilateral profile Tolerance shown in Fig. 9.3 may be converted as follows:

- Given an equal bilaterally displaced profile Tolerance on a curved surface (considered at tangential point at top),

  Profile 2      profile Tolerance = total Tolerance

- Divide the profile tolerance by 2 to obtain the ± equal bilateral Tolerance value

  Equal bilateral Tolerance value $= \dfrac{2}{2} = 1$

- Take the value of the basic dimension locating the surface as nominal

  $\boxed{15}$

  Nominal dimension value = 15

Conversion complete:

Equal bilateral equivalent = 15 ± 1

Note: The dimension value and the tolerance value will be placed on separate lines in the Tolerance Stackup report. The methods described in these examples work well for plane surfaces that are perpendicular to the direction of the Tolerance Stackup. For curved or sloped surfaces, additional steps may be necessary to convert the tolerances.

A Profile tolerance applied to a curved or sloped surface creates a three-dimensional tolerance zone offset from the surface. Any point on the surface must be located within the tolerance zone along a line perpendicular to the nominal surface. Consequently, care must be taken when including this type of surface in a Tolerance Stackup. If the surface or the contributing portion of the surface is not perpendicular to the direction of the Tolerance Stackup, trigonometry will likely be necessary to resolve the tolerance value into the direction of the Tolerance Stackup.

If the portion of the surface being considered is not directly located by a basic dimension parallel to the direction of the Tolerance Stackup, the dimension must be determined by other means. If the drawing was created using CAD, the easiest method is to measure the distance to the surface using the CAD program. If the surface is a plane at an angle to the direction of the study, trigonometry may be used to locate a point on the surface. For irregularly curved surfaces it may be impossible to locate a point on the surface without the aid of a CAD program. This calculated or measured dimension value will be used as the nominal dimension in the Tolerance Stackup.

Often, when performing Tolerance Stackups on complex, irregularly curved surfaces and parts, the Tolerance Stackup is done without dimensions. Only the variation is analyzed (tolerances only), and the nominal dimension or gap is taken directly from the CAD model. The mean shift described in Chapter 4 must be calculated and included for all the dimensions and tolerances in the Tolerance Stackup that were converted into equal bilateral format.

## UNEQUAL BILATERAL PROFILE TOLERANCES

Unequal bilateral profile Tolerances specify unequal variation in each direction from nominal. The variation is not centered about the nominal value. These are similar to unequal bilateral ± Tolerances.

### The Unequal Bilateral Profile Tolerance Shown in Fig. 9.4 May Be Converted as Follows:

- Given an unequal bilaterally displaced profile Tolerance of 3,

    Profile 3    profile Tolerance = total Tolerance

- Convert the unequal profile Tolerance into the unequal ± Tolerance value. The total Tolerance zone is 3 mm wide. The zone extends 2 mm outside the nominal surface and 1 mm inside the nominal surface. These values can be directly converted into + and − Tolerances.

    2 mm outward = +2 Tolerance value

    1 mm inward = −1 Tolerance value

    $^{+2}_{-1}$ Unequal ± Tolerance Value

- Take the value of the basic dimension locating the surface as nominal

    | 12.5 |

    Nominal dimension value = 12.5

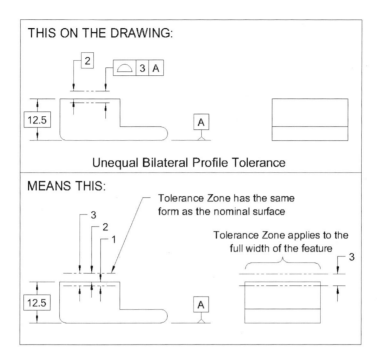

**FIGURE 9.4** Unequal bilateral profile.

- Establish upper and lower limits. Add the plus Tolerance to the nominal value; this is the upper limit. Subtract the minus Tolerance from the nominal value; this is the lower limit.

    Upper limit $= 12.5 + 2 = 14.5$
    Lower limit $= 12.5 - 1 = 11.5$

- Divide the total Tolerance by 2 to obtain the equal bilateral Tolerance value.

    Equal bilateral Tolerance value $= 3 = \pm 1.5$

- Add the equal bilateral Tolerance value to the lower limit. This is the adjusted nominal value. Establish the adjusted nominal value.

    $11.5 + 1.5 = 13$

(Note: The adjusted nominal value can also be obtained by subtracting the equal bilateral Tolerance value from the upper limit.)

Conversion complete:

    Equal bilateral equivalent $= 13 \pm 1.5$

Note: The dimension value and the tolerance value will be placed on separate lines in the Tolerance Stackup report.

## UNILATERAL PROFILE TOLERANCES

Unilateral profile Tolerances specify variation only in one direction from nominal, either into or out from the part. The nominal value represents one end of the Tolerance range. These are similar to unilateral ± Tolerances, in that the Tolerance specifications may be considered as follows:

- Unilateral outward (or positive): Similar to plus (+) something / minus (−) nothing
- Unilateral inward (or negative): Similar to plus (+) nothing / minus (−) something

### The Unilateral Profile Tolerance Shown in Fig. 9.5 May Be Converted as Follows:

- Given a unilaterally displaced profile Tolerance of 0.1,
  Profile 0.1      profile Tolerance = total Tolerance

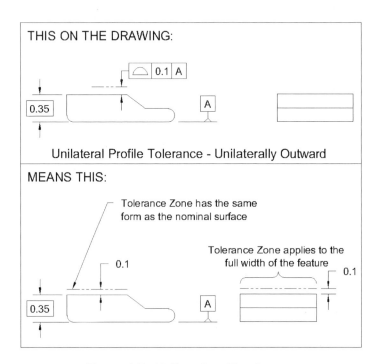

**FIGURE 9.5**   Unilateral profile tolerance.

- Convert the unilateral profile Tolerance into the unilateral ± Tolerance value. The total Tolerance zone is 0.1 mm wide. The zone extends from 0.1 mm outside the surface to the nominal surface, or 0 mm inside the surface. These values can be directly converted into + and − Tolerances.

  0.1 mm outward = +0.1 Tolerance value

  0 mm inward = −0 Tolerance value

  $+0.1 \atop -0$ Unilateral ± Tolerance value

- Take the value of the basic dimension locating the surface as nominal

  | 0.35 |

  Nominal dimension value = 0.35

- Establish upper and lower limits. Add the plus Tolerance to the nominal value; this is the upper limit. Subtract the minus Tolerance from the nominal value; this is the lower limit.

  Upper limit = 0.35 + 0.1 = 0.45

  Lower limit = 0.35 − 0 = 0.35

- Divide the total Tolerance by 2 to obtain the equal bilateral Tolerance value.

  Equal bilateral Tolerance value = 0.1 = ±0.05

- Add the equal bilateral Tolerance value to the lower limit. This is the adjusted nominal value. Establish the adjusted nominal value.

  0.35 + 0.05 = 0.4

- (Note: The adjusted nominal value can also be obtained by subtracting the equal bilateral Tolerance value from the upper limit.)

Conversion complete:

  Equal bilateral equivalent = 0.4 ± 0.05

Note: The dimension value and the tolerance value will be placed on separate lines in the Tolerance Stackup report. The same procedure may be performed to convert a unilaterally inward profile Tolerance ( + nothing; − something) to an equal bilateral ± Tolerance.

## COMPOSITE PROFILE TOLERANCES

Composite profile tolerances are specified in composite (multiple segment) feature control frames. The tolerance zones defined in the upper and lower segments of a composite profile feature control frame are described in Fig. 9.6. The profile tolerance specified in the uppermost segment of the feature control frame represents the total allowable variation in location of the feature(s) to a datum reference frame. Typically the tolerance defined in the uppermost segment is used in Tolerance Stackups. The profile tolerance defined in the uppermost segment of a composite feature control frame is exactly the same as a single segment profile tolerance feature control frame with the same contents.

The tolerance zone(s) defined in the uppermost segment of a composite profile feature control frame are basically located to a datum reference frame, so the feature(s) are also located to the datum reference frame.

The tolerance zones defined in the lower segments of a composite feature control frame are not basically located to a datum reference frame—they may only be basically oriented to a datum reference frame and are basically located to each other in the case of a pattern.

In some cases, the profile tolerance specified in a lower segment may be of more importance, but it is less common. The profile tolerance specified in a lower segment may be used in a Tolerance Stackup if the Tolerance Stackup is between features in a pattern. For example, consider the two side surfaces of the groove in Fig. 9.7. Both surfaces are toleranced using the same composite profile feature control frame. Using this technique makes the surfaces into a pattern—they are grouped together by virtue of the tolerancing scheme.

The profile tolerance specified in a lower segment may also be used with more advanced tolerancing techniques, such as where "simultaneous requirements" is explicitly stated beneath two or more composite feature control frames related to the same datum reference frame. This technique also combines the features toleranced by the composite feature control frames into a single pattern.

Even though there is a difference between the upper segment tolerance and the lower segment tolerance, both are included and formatted in the Tolerance Stackup report the same way as a single segment profile tolerance. Two lines are entered into the Tolerance Stackup report; the profile tolerance is entered on the first line and datum feature shift is entered on the second line. The technique for including profile tolerance information in a Tolerance Stackup sketch and a Tolerance Stackup report is presented in Chapters 13 and 14.

As a general rule, if the location of the feature (or pattern of features) affects the distance being studied, the upper segment profile tolerance should

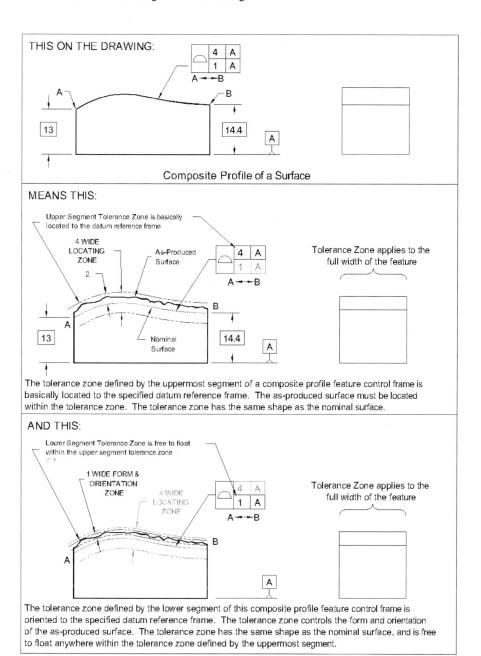

**FIGURE 9.6**   Composite profile.

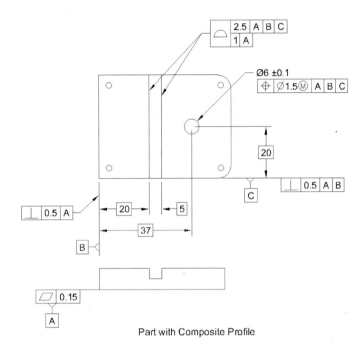

Part with Composite Profile

A Groove is cut into this part. Two leaders are directed from the Composite Profile Feature Control Frame to the walls of the Groove. Using this method makes the two Groove Walls into a pattern.

The Upper Segment Profile tolerance locates the Groove Walls to the Datum Reference Frame. The Lower Segment Profile tolerance orients the Groove Walls to Datum A, and locates the Groove Walls to each other, which determines the minimum and maximum width of the Groove.

**FIGURE 9.7**   Composite profile tolerancing sample part.

be used in the Tolerance Stackup. The upper segment profile tolerance is also used in the Tolerance Stackup when the feature(s) toleranced with the composite feature control frame and a feature toleranced with a different feature control frame on the same part are included in the chain of Dimensions and Tolerances.

As a general rule, the profile tolerance defined in the lower segment of a composite feature control frame is used in the Tolerance Stackup when the chain of Dimensions and Tolerances only includes the toleranced features within a pattern and does not include any other features on the same part. If the distance being studied is only affected by the feature-to-feature relationship within the pattern of features, the lower segment profile tolerance should be used in the Tolerance Stackup.

Several examples that show the uppermost segment and the lower segments of composite profile feature control frames used in Tolerance Stackups follow.

## Example 9.1

Refer back to the part shown in Fig. 9.7. In this example the goal of the Tolerance Stackup is to determine how close to the left edge of the part the groove may be. The groove walls are toleranced with composite profile, and the left edge of the part is datum feature B. The profile tolerance in the upper segment of the composite feature control frame is used in the Tolerance Stackup because the upper segment tolerance locates the groove walls to the

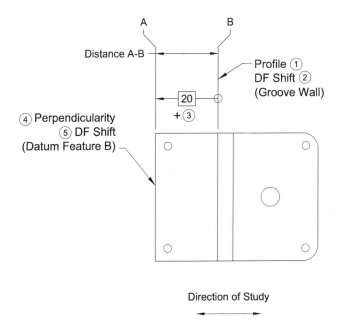

Tolerance Stackup Between Left Groove Wall and
Left Edge of Part (Datum Feature B):
Upper Segment Tolerance is Used for Line Item 1

Composite Profile Tolerancing:

**FIGURE 9.8** Composite profile Tolerance Stackup sketch.

Tolerance Stack                                                                    Release 1.2a

| Program: | Tolerance Analysis and Stackup Manual | | | | | Stack Information: | | | |
|---|---|---|---|---|---|---|---|---|---|

Stack No: Figure 9-9
Date: 07/04/02
Revision: A

**Product:**
Part Number: -
Rev: A
Description: Part with Groove

Direction: Horizontal
Author: BR Fischer

**Problem:** It is Important to Know the Minimum Distance Between the Groove Wall and the Left Edge of the Part

**Objective:** Determine the Minimum Distance Between the Groove Wall and the Left Edge of the Part

| Description of Component / Assy | Part Number | Rev | Item | Description | + Dims | - Dims | Tol | Percent Contrib | Dim / Tol Source & Calcs |
|---|---|---|---|---|---|---|---|---|---|
| Part with Groove | 123-002 | A | 1 | Profile: Edge Along Pt A | | | +/- 1.2500 | 83% | Profile 2.5, A, B, C (Upper Segment) |
| | | | 2 | Datum Feature Shift | | | +/- 0.0000 | 0% | N/A |
| | | | 3 | Dim: Groove Wall - Datum B | 20.0000 | | +/- 0.0000 | 0% | 20 Basic on Dwg |
| | | | 4 | Perpendicularity (Datum Feature B) | | 0.2500 | +/- 0.2500 | 17% | Perpendicularity 0.5, A on Dwg - See Notes |
| | | | 5 | Datum Feature Shift | | | +/- 0.0000 | 0% | N/A |
| | | | | Dimension Totals | 20.0000 | 0.2500 | | | |
| | | | | Nominal Distance: Pos Dims - Neg Dims = | 19.7500 | | | | |

**RESULTS:**

| | Nom | Tol | Min | Max |
|---|---|---|---|---|
| Arithmetic Stack (Worst Case) | 19.7500 | +/- 1.5000 | 18.2500 | 21.2500 |
| Statistical Stack (RSS) | 19.7500 | +/- 1.2748 | 18.4752 | 21.0248 |
| Adjusted Statistical: 1.5*RSS | 19.7500 | +/- 1.9121 | 17.8379 | 21.6621 |

Notes: - The Upper Segment Profile Tolerance is used in this Tolerance Stackup.
- It must be understood that the Perpendicularity tolerance applied to Datum Feature B allows portions of the Datum Feature to tilt and / or have form error relative to Datum B, which is perfectly perpendicular to Datum A. Therefore the Perpendicularity tolerance should be included in the Tolerance Stackup. The Perpendicularity tolerance only allows the distance between Datum Feature B and the Groove to decrease, so it must be accompanied by a negative Mean Shift. The Perpendicularity tolerance is added as an equal-bilateral tolerance of +/-0.25, with a Mean Shift of 0.25, which is half the Perpendicularity tolerance value. The Mean Shift is indicated by placing the 0.25 value in the "- Dims" column on the same line as the Perpendicularity tolerance. See Chapter 20 for more information.

Assumptions:

Suggested Action:

**FIGURE 9.9** Composite profile Tolerance Stackup groove distance.

datum reference frame. The Tolerance Stackup sketch for this example is shown in Fig. 9.8. The Tolerance Stackup report for this example is shown in Fig. 9.9. The upper segment profile tolerance would also be used if the goal of the Tolerance Stackup was to determine the distance between one of the groove walls and the ∅6 ± 0.1 hole, because both features are related to the same datum reference frame. In that example, the chain of Dimensions and Tolerances would start at the groove wall, pass through the datum reference frame and terminate at the hole.

### Example 9.2

Again refer back to the part shown in Fig. 9.7. In this example the goal of the Tolerance Stackup is to determine the minimum and maximum allowable groove width. Since both groove walls are toleranced with the same composite

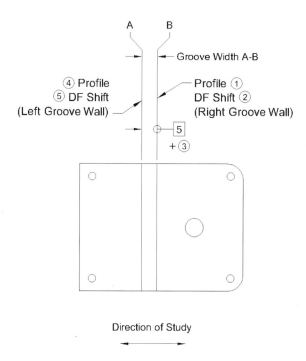

Tolerance Stackup Between Groove Walls to
Determine the Minimum and Maximum Groove Width
Lower Segment Tolerance is Used for Line Items 1 & 4

**FIGURE 9.10**  Composite profile Tolerance Stackup sketch.

Tolerance Stack                                                      Release 1.2a

Program: Tolerance Analysis and Stackup Manual

Product:   Part Number  -  | Rev  Description
                              A   Part with Groove

Stack Information:
Stack No.  Figure 9-11
Date       07/04/02
Revision   A

Problem: It is Important to Know the Minimum and Maximum Groove Width

Direction:  Horizontal
Author:     BR Fischer

Objective: Determine the Minimum and Maximum Groove Width

| Description of Component / Assy | Part Number | Rev | Item | Description | + Dims | - Dims | Tol | Percent Contrib | Dim / Tol Source & Calcs |
|---|---|---|---|---|---|---|---|---|---|
| Part with Groove | 123-002 | A | 1 | Profile Edge Along Pt A | | | +/- 0.5000 | 50% | Profile 1, A (Lower Segment) |
| | | | 2 | Datum Feature Shift | | | +/- 0.0000 | 0% | N/A |
| | | | 3 | Dim: Right Groove Wall - Left Groove Wall | 5.0000 | | +/- 0.0000 | 0% | 5 Basic on Dwg |
| | | | 4 | Profile Edge Along Pt A | | | +/- 0.5000 | 50% | Profile 1, A (Lower Segment) |
| | | | 5 | Datum Feature Shift | | | +/- 0.0000 | 0% | N/A |
| | | | | Dimension Totals | 5.0000 | 0.0000 | | | |
| | | | | Nominal Distance: Pos Dims - Neg Dims = | 5.0000 | | | | |

RESULTS:

| | Nom | Tol | Min | Max |
|---|---|---|---|---|
| Arithmetic Stack (Worst Case) | 5.0000 | +/- 1.0000 | 4.0000 | 6.0000 |
| Statistical Stack (RSS) | 5.0000 | +/- 0.7071 | 4.2929 | 5.7071 |
| Adjusted Statistical 1.5*RSS | 5.0000 | +/- 1.0607 | 3.9393 | 6.0607 |

Notes:

Assumptions:

Suggested Action:

**FIGURE 9.11**  Composite profile Tolerance Stackup groove width.

profile feature control frame, the walls are considered a pattern, and the profile tolerance specified in the lower segment is used in the Tolerance Stackup. Notice that the chain of Dimensions and Tolerances does not pass through the datum reference frame and only includes the 5-mm basic dimension between the groove walls and their respective lower segment profile tolerances and datum feature shift. The Tolerance Stackup sketch for this example is shown in Fig. 9.10. The Tolerance Stackup report for this example is shown in Fig. 9.11.

## POSITIONAL TOLERANCES

Positional tolerances can also be translated into ± tolerances. Positional tolerances specify a cylindrical or total width tolerance zone for features of size (FOS), such as holes or slots as in Fig. 9.12. Positional tolerances may be included in a Tolerance Stackup for a number of reasons. It may be because

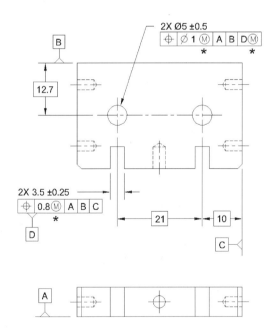

\* The Material Condition Modifiers associated
with the tolerance and the Datum Feature
References must be considered carefully
when tolerancing parts.

FIGURE 9.12 Part with positional Tolerancing.

| MATERIAL CONDITION MODIFIERS |
| --- |
| (M)   MAXIMUM MATERIAL CONDITION - MMC |
| (L)   LEAST MATERIAL CONDITION - LMC |
| (S)   REGARDLESS OF FEATURE SIZE - RFS |

Usually RFS is not specified explicitly. Absense of a modifier
infers RFS for geometric tolerances specified per ASME Y14.5M-1994.

**FIGURE 9.13**   Material condition modifiers.

there is a fit relationship between two or more parts, or it may be to determine the distance between the edge of a feature of size and another surface, such as the distance between a hole and the edge of a part or between a shaft and an adjacent part.

Care must be taken when converting positional tolerances into ± tolerances, as the material condition modifier associated with the positional tolerance must also be considered. Material condition modifiers are shown in Fig. 9.13. These include RFS, MMC, and LMC.

- Regardless of feature size (RFS). RFS is the default condition if no material condition modifier is present. The RFS tolerance specifies a fixed size tolerance zone. The tolerance zone size does not change with the size of the as-produced feature (see Fig. 9.14). A positional Tolerance specified RFS is the total width or diameter of a tolerance zone. So, a positional Tolerance specified RFS with a width value of 1, or a cylindrical value of $\varnothing 1$, can be quickly converted into ±0.5, regardless of the size of the as-produced feature.
- Maximum material condition (MMC). When an MMC modifier is applied to a positional Tolerance, the tolerance zone varies directly with size of the as-produced feature. When the feature is produced at its maximum material condition (smallest internal, largest external), it must be within the tolerance zone specified. As the size of the feature deviates from its maximum material condition, the tolerance zone increases proportionately (see Fig. 9.15).
- Least material condition (LMC). When an LMC modifier is applied to a positional Tolerance, the tolerance zone varies directly with size of the as-produced feature. When the feature is produced at its least material condition (largest internal, smallest external), it must be within the specified Tolerance. As the size of the feature deviates from its LMC, the tolerance zone increases proportionately (see Fig. 9.16).

## THIS ON THE DRAWING:

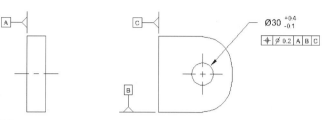

## MEANS THIS:

| Internal Feature of Size Specified RFS | | | | |
|---|---|---|---|---|
| Material Condition | Ø Hole | Positional Tolerance | Inner Boundary | Outer Boundary |
| LMC | 30.4 | 0.2 | 30.2 | 30.6 |
| | 30.3 | 0.2 | 30.1 | 30.5 |
| | 30.2 | 0.2 | 30 | 30.4 |
| | 30.1 | 0.2 | 29.9 | 30.3 |
| | 30 | 0.2 | 29.8 | 30.2 |
| MMC | 29.9 | 0.2 | 29.7 | 30.1 |

As shown in the chart, the Inner Boundary for an Internal Feature of Size is determined by subtracting the Positional Tolerance from the feature Ø. The Inner and Outer Boundaries vary with the as-produced size of the feature. The tolerance zone does not vary in size with the as-produced size of the feature.

## INTERNAL FEATURE OF SIZE, WITH POSITIONAL TOLERANCE RFS

## THIS ON THE DRAWING:

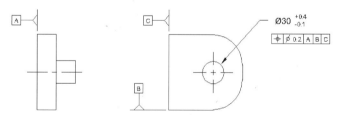

## MEANS THIS:

| External Feature of Size Specified RFS | | | | |
|---|---|---|---|---|
| Material Condition | Ø Pin | Positional Tolerance | Inner Boundary | Outer Boundary |
| MMC | 30.4 | 0.2 | 30.2 | 30.6 |
| | 30.3 | 0.2 | 30.1 | 30.5 |
| | 30.2 | 0.2 | 30 | 30.4 |
| | 30.1 | 0.2 | 29.9 | 30.3 |
| | 30 | 0.2 | 29.8 | 30.2 |
| LMC | 29.9 | 0.2 | 29.7 | 30.1 |

As shown in the chart, the Outer Boundary for an External Feature of Size is determined by adding the Positional Tolerance to the feature Ø. The Inner and Outer Boundaries vary with the as-produced size of the feature. The tolerance zone does not vary in size with the as-produced size of the feature.

## EXTERNAL FEATURE OF SIZE, WITH POSITIONAL TOLERANCE RFS

**FIGURE 9.14** RFS Tolerance chart.

## THIS ON THE DRAWING:

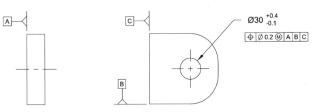

### MEANS THIS:

As shown in the chart, the Virtual Condition for an Internal Feature of Size at MMC is determined by subtracting the MMC Positional Tolerance from the MMC feature Ø (Smallest Ø). The Virtual Condition is a fixed value. In this case, it represents the minimum inner boundary of the feature.

The tolerance zone increases in size directly proportional to the size of the as-produced feature.

| Internal Feature of Size Specified @ MMC | | | | |
|---|---|---|---|---|
| Material Condition | Ø Hole | Positional Tolerance | Virtual Condition | Resultant Condition |
| LMC | 30.4 | 0.7 | | 31.1 |
| | 30.3 | 0.6 | | 30.9 |
| | 30.2 | 0.5 | 29.7 | 30.7 |
| | 30.1 | 0.4 | | 30.5 |
| | 30 | 0.3 | | 30.3 |
| MMC | 29.9 | 0.2 | | 30.1 |

**INTERNAL FEATURE OF SIZE, WITH POSITIONAL TOLERANCE @ MMC**

## THIS ON THE DRAWING:

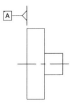
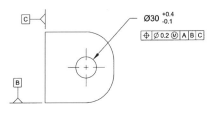

### MEANS THIS:

As shown in the chart, the Virtual Condition for an External Feature of Size at MMC is determined by adding the MMC Positional Tolerance to the MMC feature Ø (Largest Ø). The Virtual Condition is a fixed value. In this case, it represents the maximum outer boundary of the feature.

The tolerance zone increases in size inversely proportional to the size of the as-produced feature.

| External Feature of Size Specified @ MMC | | | | |
|---|---|---|---|---|
| Material Condition | Ø Pin | Positional Tolerance | Virtual Condition | Resultant Condition |
| MMC | 30.4 | 0.2 | | 30.2 |
| | 30.3 | 0.3 | | 30 |
| | 30.2 | 0.4 | 30.6 | 29.8 |
| | 30.1 | 0.5 | | 29.6 |
| | 30 | 0.6 | | 29.4 |
| LMC | 29.9 | 0.7 | | 29.2 |

**EXTERNAL FEATURE OF SIZE, WITH POSITIONAL TOLERANCE @ MMC**

**FIGURE 9.15**   MMC Tolerance chart.

## THIS ON THE DRAWING:

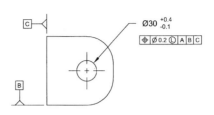

Ø30 +0.4/-0.1

⊕ Ø0.2 Ⓛ A B C

## MEANS THIS:

| Internal Feature of Size Specified @ LMC | | | | |
|---|---|---|---|---|
| Material Condition | Ø Hole | Positional Tolerance | Virtual Condition | Resultant Condition |
| LMC | 30.4 | 0.2 | | 30.2 |
| | 30.3 | 0.3 | | 30 |
| | 30.2 | 0.4 | 30.6 | 29.8 |
| | 30.1 | 0.5 | | 29.6 |
| | 30 | 0.6 | | 29.4 |
| MMC | 29.9 | 0.7 | | 29.2 |

As shown in the chart, the Virtual Condition for an Internal Feature of Size toleranced at LMC is calculated by adding the LMC Positional Tolerance to the LMC feature Ø (Largest Ø). The Virtual Condition is a fixed value. In this case, it represents the maximum outer boundary of the feature.

The tolerance zone increases in size inversely proportional to the size of the as-produced feature.

## INTERNAL FEATURE OF SIZE, WITH POSITIONAL TOLERANCE @ LMC

## THIS ON THE DRAWING:

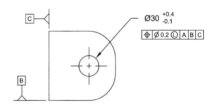

Ø30 +0.4/-0.1

⊕ Ø0.2 Ⓛ A B C

## MEANS THIS:

| Internal Feature of Size Specified @ LMC | | | | |
|---|---|---|---|---|
| Material Condition | Ø Pin | Positional Tolerance | Virtual Condition | Resultant Condition |
| MMC | 30.4 | 0.7 | | 31.1 |
| | 30.3 | 0.6 | | 30.9 |
| | 30.2 | 0.5 | 29.7 | 30.7 |
| | 30.1 | 0.4 | | 30.5 |
| | 30 | 0.3 | | 30.3 |
| LMC | 29.9 | 0.2 | | 30.1 |

As shown in the chart, the Virtual Condition for an External Feature of Size toleranced at LMC is calculated by subtracting the LMC Positional Tolerance from the LMC feature Ø (Smallest Ø). The Virtual Condition is a fixed value. In this case, it represents the minimum inner boundary of the feature.

The tolerance zone increases in size directly proportional to the size of the as-produced feature.

## EXTERNAL FEATURE OF SIZE, WITH POSITIONAL TOLERANCE @ LMC

**FIGURE 9.16** LMC Tolerance chart.

## POSITIONAL TOLERANCE, ASSEMBLY SHIFT, AND MISALIGNMENT

Positional Tolerances are often specified using the maximum material condition modifier. This allows the greatest latitude in manufacturing since the positional tolerance zone increases with the size of the feature, permitting more parts to pass inspection, potentially lowering part cost. This method of tolerancing is based on the premise that the only functional concern is for a fastener, shaft, etc., is to pass through the hole, and the additional positional tolerance as the hole increases in size from MMC to LMC is not objectionable. This is not always the case, however.

Specifying a positional tolerance at MMC for holes leads to the greatest possible mislocation when the hole is produced at LMC. This is due to the bonus tolerance (the additional allowable positional error as the holes deviate from MMC toward LMC). Assembly shift is also greatest when the holes are their largest size, which is LMC.

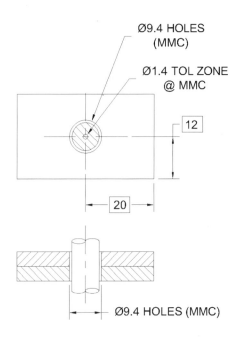

Misalignment with Holes @ MMC:
MMC Holes @ Nominal Position

**FIGURE 9.17**  Misalignment with holes at MMC #1, nominal.

In cases where alignment is also a functional concern, MMC tolerancing may not be the best choice, as the additional (bonus) positional tolerance may be functionally detrimental. RFS may be a better choice, even though it appears to be a tighter tolerance than MMC, due to the lack of bonus tolerance. All functional considerations should be weighed when determining which tolerancing method is most appropriate. The following examples describe how positional tolerance, bonus tolerance, and assembly shift contribute to part misalignment. Avoid the temptation to always use MMC. It is not always the right choice. Specifying geometric tolerances at MMC *may* reduce part costs; the part cost should be lower in cases where the bonus tolerance allows more parts to pass inspection and scrap is reduced. However, specifying geometric tolerances to apply at MMC also adds variation to the part at the assembly level, as seen in Figs. 9.17 through 9.24.

This additional variation may manifest itself as misaligned part features at assembly, causing assembled components to fail to meet their geometric

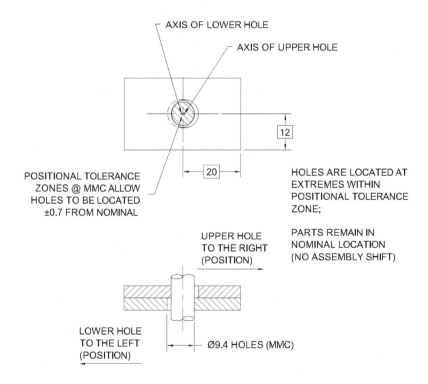

Misalignment with Holes @ MMC: Holes with Positional Error - No Assembly Shift

**FIGURE 9.18** Misalignment with holes at MMC #2, with positional error.

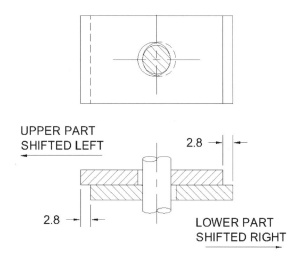

Maximum Misalignment with Holes @ MMC

**FIGURE 9.19**  Misalignment with holes at MMC #3, maximum misalignment.

requirements. Ultimately the cost of fixing the assembly will probably be greater than the part cost savings. See Chapter 12 for more information on specifying material condition modifiers.

## COMPOSITE POSITIONAL TOLERANCE

Composite positional tolerances are specified in composite (multiple segment) feature control frames as shown in Fig. 9.25 on p. 166. The positional tolerance specified in the uppermost segment of the feature control frame represents the total allowable variation in location of the feature(s) to a datum reference frame. Typically the tolerance defined in the uppermost segment is used in Tolerance Stackups. The positional tolerance defined in the uppermost segment of a composite feature control frame is exactly the same as a single segment positional tolerance feature control frame with the same contents.

The tolerance zone(s) defined in the uppermost segment of a composite position feature control frame are basically located to a datum reference frame, so the feature(s) are also located to the datum reference frame.

The tolerance zone(s) defined in the lower segments of a composite feature control frame are not basically located to a datum reference frame—they may only be basically oriented to a datum reference frame and are basically located to each other in the case of a pattern.

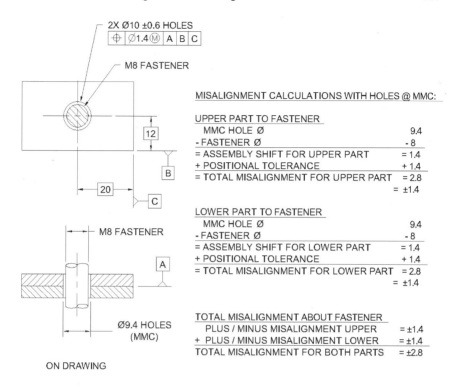

2X Ø10 ±0.6 HOLES

M8 FASTENER

MISALIGNMENT CALCULATIONS WITH HOLES @ MMC:

| UPPER PART TO FASTENER | |
|---|---|
| MMC HOLE Ø | 9.4 |
| - FASTENER Ø | - 8 |
| = ASSEMBLY SHIFT FOR UPPER PART | = 1.4 |
| + POSITIONAL TOLERANCE | + 1.4 |
| = TOTAL MISALIGNMENT FOR UPPER PART | = 2.8 |
| | = ±1.4 |

| LOWER PART TO FASTENER | |
|---|---|
| MMC HOLE Ø | 9.4 |
| - FASTENER Ø | - 8 |
| = ASSEMBLY SHIFT FOR LOWER PART | = 1.4 |
| + POSITIONAL TOLERANCE | + 1.4 |
| = TOTAL MISALIGNMENT FOR LOWER PART | = 2.8 |
| | = ±1.4 |

| TOTAL MISALIGNMENT ABOUT FASTENER | |
|---|---|
| PLUS / MINUS MISALIGNMENT UPPER | = ±1.4 |
| + PLUS / MINUS MISALIGNMENT LOWER | = ±1.4 |
| TOTAL MISALIGNMENT FOR BOTH PARTS | = ±2.8 |

M8 FASTENER

Ø9.4 HOLES
(MMC)

ON DRAWING

Misalignment with Holes @ MMC: Drawing and Calculations

**FIGURE 9.20** Misalignment with holes at MMC #4, drawing and calculations.

Even though there is a difference between the upper segment tolerance and the lower segment tolerance, both are included and formatted in the Tolerance Stackup report the same way as a single segment positional tolerance. Three lines are entered into the Tolerance Stackup report: the positional tolerance is entered on the first line, the bonus tolerance is entered on the second line, and datum feature shift is entered on the third line. The technique for including positional tolerance information in a Tolerance Stackup sketch and a Tolerance Stackup report is presented in Chapters 13 and 14.

In some cases, the positional tolerance specified in a lower segment may be of more importance, but it is less common. The positional tolerance specified in a lower segment may be used in a Tolerance Stackup if the Tolerance Stackup is between features in a pattern. The positional tolerance specified in a lower segment may also be used with more advanced tolerancing techniques, such as when "simultaneous requirements" is explicitly stated

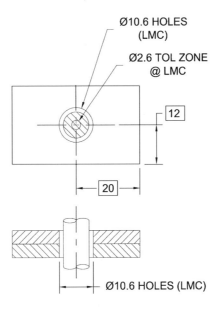

**FIGURE 9.21**    Misalignment with holes at LMC #1, nominal.

beneath two or more composite feature control frames related to the same datum reference frame. Using this technique combines the features toleranced by the composite feature control frames into a single pattern. The positional tolerance specified in a lower segment is also commonly used for the values of $T_1$ or $T_2$ in the fixed- and floating-fastener formulas discussed in Chapter 18. In fact, probably the most common application of composite positional tolerancing is where the tolerance in the upper segment locates a pattern of features to a datum reference frame with a relatively large tolerance, and the tolerance in the lower segment locates the features within the pattern with a smaller tolerance that is required for mating. Of course, this technique presumes that the feature-to-feature location within the pattern is more critical than the relationship of the pattern to the datum reference frame.

As a general rule, if the location of the pattern of features affects the distance being studied, the upper segment positional tolerance should be used in the Tolerance Stackup. The upper segment positional tolerance is also used in the Tolerance Stackup when the pattern of features toleranced with the composite feature control frame and a feature toleranced with a different feature control frame on the same part are included in the chain of Dimensions and Tolerances.

As a general rule, the positional tolerance defined in the lower segment of a composite feature control frame is used in the Tolerance Stackup when the chain of Dimensions and Tolerances only includes the features within the

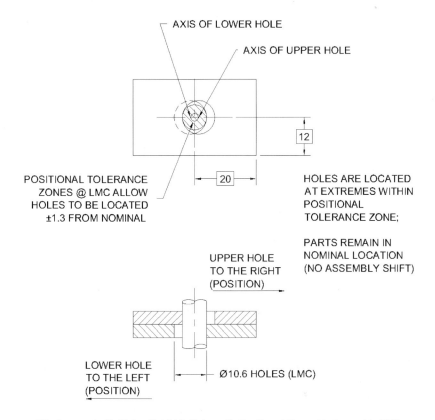

AXIS OF LOWER HOLE

AXIS OF UPPER HOLE

12

POSITIONAL TOLERANCE
ZONES @ LMC ALLOW
HOLES TO BE LOCATED
±1.3 FROM NOMINAL

20

HOLES ARE LOCATED
AT EXTREMES WITHIN
POSITIONAL
TOLERANCE ZONE;

PARTS REMAIN IN
NOMINAL LOCATION
(NO ASSEMBLY SHIFT)

UPPER HOLE
TO THE RIGHT
(POSITION)

LOWER HOLE
TO THE LEFT
(POSITION)

Ø10.6 HOLES (LMC)

Misalignment with Holes @ MMC: Holes with Positional Error - No Assembly Shift

**FIGURE 9.22**  Misalignment with holes at LMC #2, with positional error.

pattern, and does not include any other features on the same part. If the distance being studied is only affected by the feature-to-feature relationship within the pattern of features, the lower segment positional tolerance should be used in the Tolerance Stackup.

Several examples that show the uppermost segment and the lower segments of composite position feature control frames used in Tolerance Stackups follow.

*Example 9.1.*  Consider the back panel shown in Fig. 9.25. There are two patterns of four holes, each toleranced with a distinct composite feature control frame. The holes in one pattern are marked X and the holes in the other pattern are marked Y. A connector such as the one shown in Fig. 9.26 is mounted to each set of holes using M4 fasteners.

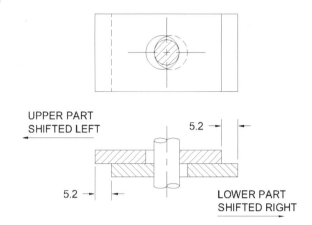

UPPER PART
SHIFTED LEFT

5.2

5.2

LOWER PART
SHIFTED RIGHT

Maximum Misalignment with Holes @ LMC

**FIGURE 9.23** Misalignment with holes at LMC #3 - max misalignment. The worst-case misalignment occurs with holes Toleranced at MMC and manufactured at LMC, which allows a larger positional Tolerance (due to bonus Tolerance) and larger assembly shift (due to larger holes).

Several Tolerance Stackups are required in this example.

The required size of the holes must be determined. The floating-fastener formula from Chapter 18 should be used to solve this problem. This formula states that the minimum size of the hole is based on the sum of the maximum fastener diameter and the positional tolerance on the clearance hole. The formula is $H = F + T$. This formula only determines what is required for mating, and without elaborating the point, the feature-to-feature tolerance is within the pattern. So, the lower segment tolerance will be used in the floating-fastener formula for $T$, which represents the positional tolerance applied to the minimum size hole. Remember, the lower segment tolerance defines the hole-to-hole relationship. Solving for $H$ by adding the values for $F$ and $T$, we see that $H = 4 + 1 = 5$. The minimum clearance hole diameter on both parts is 5; so the fasteners will pass through the holes.

It is also important that the two connectors remain separated after assembly—there must be a gap between the connectors. (See Fig. 9.27.) The gap is highlighted by the dimension with the question, "What is the minimum gap?" In this example, the left connector is located by the pattern of holes marked Y, and the right connector is located by the pattern of holes marked X. Each pattern has been toleranced with its own composite position feature control frame. To determine the minimum distance between the connectors,

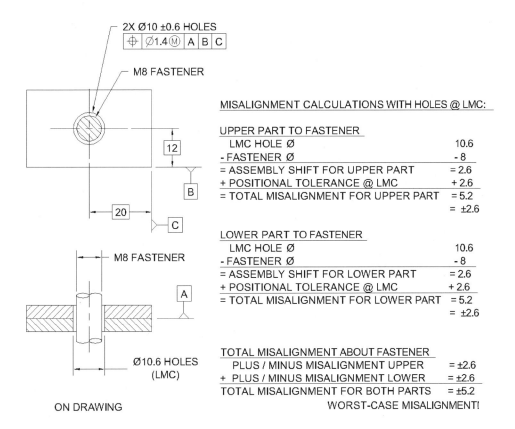

2X Ø10 ±0.6 HOLES

| ⊕ | Ø1.4Ⓜ | A | B | C |

M8 FASTENER

12

B

20

C

M8 FASTENER

A

Ø10.6 HOLES
(LMC)

ON DRAWING

MISALIGNMENT CALCULATIONS WITH HOLES @ LMC:

UPPER PART TO FASTENER
| | |
|---|---|
| LMC HOLE Ø | 10.6 |
| - FASTENER Ø | - 8 |
| = ASSEMBLY SHIFT FOR UPPER PART | = 2.6 |
| + POSITIONAL TOLERANCE @ LMC | + 2.6 |
| = TOTAL MISALIGNMENT FOR UPPER PART | = 5.2 |
| | = ±2.6 |

LOWER PART TO FASTENER
| | |
|---|---|
| LMC HOLE Ø | 10.6 |
| - FASTENER Ø | - 8 |
| = ASSEMBLY SHIFT FOR LOWER PART | = 2.6 |
| + POSITIONAL TOLERANCE @ LMC | + 2.6 |
| = TOTAL MISALIGNMENT FOR LOWER PART | = 5.2 |
| | = ±2.6 |

TOTAL MISALIGNMENT ABOUT FASTENER
| | |
|---|---|
| PLUS / MINUS MISALIGNMENT UPPER | = ±2.6 |
| + PLUS / MINUS MISALIGNMENT LOWER | = ±2.6 |
| TOTAL MISALIGNMENT FOR BOTH PARTS | = ±5.2 |
| | WORST-CASE MISALIGNMENT! |

Misalignment Error with Holes @ LMC: Drawing and Calculations

**FIGURE 9.24**   Misalignment with holes at LMC #4 - dwg and calculations.

the chain of Dimensions and Tolerances must start at the inside edge of the left connector, pass through its mounting holes, through the mating holes on the back panel, to the mating holes for the right connector on the back panel, through the mounting holes for the right connector, and terminate at the inside edge of the right connector. It must be stated that the chain of Dimensions and Tolerances actually should have included the basic dimensions from the left pattern of holes to the datum feature C and back to the right pattern of holes. Because the patterns are directly related by the 15-mm basic dimension and the datum features are not datum features of size, these extra dimensions have been omitted. Note that there is no change in the result.

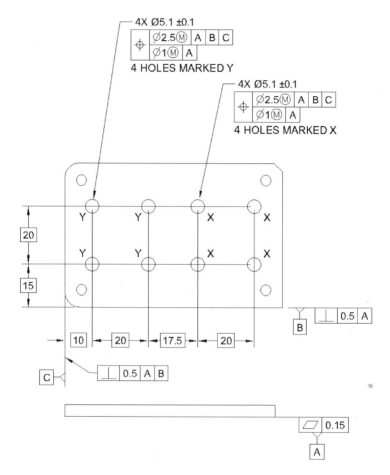

**FIGURE 9.25** Composite position back panel detail: Option 1. In this example, each four hole pattern has its own composite position feature control frame. Using this method makes the two patterns distinct.

It should also be noted that this Tolerance Stackup could have been solved by following the chain of Dimensions and Tolerances in the reverse order.

The Tolerance Stackup sketch for this problem is shown in Fig. 9.28, and the Tolerance Stackup report is shown in Fig. 9.29. In this example, the upper segment positional tolerances are used for line items 9 and 13, as the two patterns are both located to the datum reference frame by their respective upper segment positional tolerances. Notice that using this tolerancing scheme there is a worst-case interference of 0.8 between the connectors.

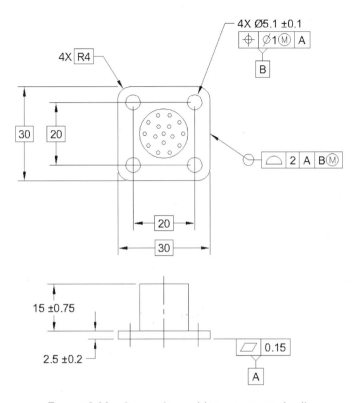

**FIGURE 9.26** Composite position, connector detail.

*Example 9.2.* Consider the back panel shown in Fig. 9.30. In this example the two patterns of four holes are grouped together into a pattern of eight holes and toleranced with a single composite feature control frame. As in Example 9.1, a connector is mounted to each set of holes using M4 fasteners.

In Example 9.1 the lower segment positional tolerance was used in the floating-fastener formula to determine the required size for the clearance holes. The lower segment tolerance is the same in Example 9.2 as in Example 9.1, so there is no change required in the size of the holes—the hole-to-hole relationship in each four hole pattern is still subject to the same diameter 1 positional tolerance.

The Tolerance Stackup to determine if the connectors remain separated after assembly is almost exactly the same as in Example 9.1. As in Example 9.1, there are 23 line items in the Tolerance Stackup. The only difference is that since both patterns of four holes on the back panel are toleranced together

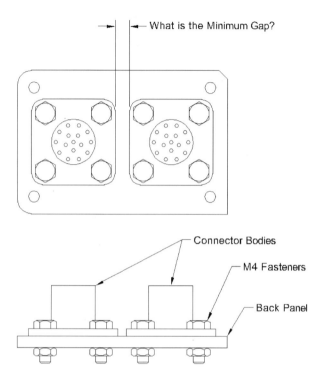

**FIGURE 9.27**   Composite position, connector module assembly: Option 1.

with the same composite feature control frame, the lower segment tolerance will be used in the chain of Dimensions and Tolerances. The Tolerance Stackup sketch for the option 2 parts is shown in Fig. 9.31. The lower segment positional tolerances appear on lines 9 and 13 in the option 2 Tolerance Stackup report shown in Fig. 9.32. Notice that using this tolerancing scheme there is no longer any interference; there is now a worst-case clearance of 0.7 between the connectors. It is important to recognize that in Example 9.2 the connector mounting holes in the back panel are toleranced as a single pattern. As stated earlier in this section, the lower segment positional tolerance is used when the Tolerance Stackup is within a pattern of features.

Consider Example 7.6, the complex bolted assembly. In this example, the $\varnothing 6.3 \pm 0.3$ mounting holes in items 1 and 2 locate item 2 to item 1. The location of the holes and assembly shift affect the distance (Gap A-B) being studied. In Fig. 9.33, the assembly from Problem 6 is repeated, but with positional tolerances applied to the mounting holes in items 1 and 2. Detail drawings of the base plate and the bracket are shown with GD&T applied in Fig. 9.34. Notice that the dimensions between the mounting holes in the base plate and between the flanges and the mounting holes in the brackets have

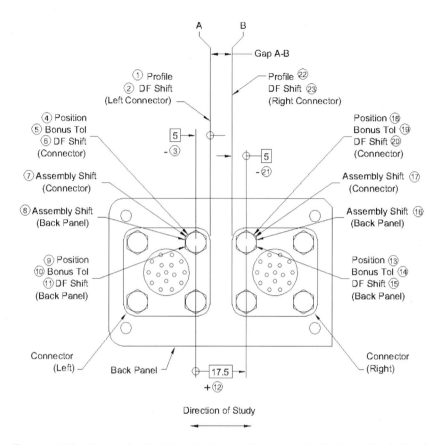

**FIGURE 9.28** Composite Position, Tolerance Stackup with Option 1 Back Panel: chain of Dimensions and Tolerances upper segment Tolerances used for line items 9 & 13.

been changed to basic dimensions. The holes are also 0.6 mm smaller, which is due to the floating-fastener formula, and using GD&T leads to a less ambiguous dimensioning and tolerancing scheme. The Tolerance Stackup sketch for this problem is shown in Fig. 9.35. Notice the addition of bonus tolerance and datum feature shift in the Tolerance Stackup report in Fig. 9.36. This Tolerance Stackup report is very similar to the Tolerance Stackup report form that will be discussed in Chapter 13. Every dimension, geometric tolerance, bonus tolerance, and datum feature shift that may contribute to the Tolerance Stackup is shown in the Tolerance Stackup sketch and the Tolerance Stackup report. Notice that some of values are set equal to zero, because they don't actually affect the Tolerance Stackup. The techniques and

Tolerance Stack     Release 1.2a

| Program: | Tolerance Analysis and Stackup Manual | | | | | | | Stack Information: | |
|---|---|---|---|---|---|---|---|---|---|
| Product: | Part Number: Opt-1   Rev A   Description: Connector Module Assembly, Option 1 | | | | | | | Stack No: Figure 9-29   Date: 07/04/02   Revision: A | |
| Problem: | The Connectors Must not Contact Each Other at Assembly | | | | | | | Direction: Horizontal | |
| Objective: | Determine If Connectors Make Contact at Assembly | | | | | | | Author: BR Fischer | |

| Description of Component / Assy | Part Number | Rev | Item | Description | + Dims | - Dims | Tol | Percent Contrib | Dim / Tol Source & Calcs |
|---|---|---|---|---|---|---|---|---|---|
| Connector (Left) | 123-002 | A | 1 | Profile Edge Along Pt A | | | +/- 1.0000 | 12.0% | = Profile 2, A, B/m |
| | | | 2 | Datum Feature Shift (DF_B @ LMC - DFS_B) / 2 | | | +/- 0.6000 | 7.2% | = (5.1 + 0.1 - (5.1 - 0.1 - 1)) / 2 |
| | | | 3 | Dim. Edge of Connector - Datum B | 5.0000 | | +/- 0.0000 | 0% | = (30 Basic - 20 Basic) / 2 on Dwg |
| | | | 4 | Position: DF_B Holes | | | +/- 0.0000 | 0% | N/A - (See Note 3) |
| | | | 5 | Bonus Tolerance | | | +/- 0.0000 | 0% | N/A - (See Note 3) |
| | | | 6 | Datum Feature Shift | | | +/- 0.0000 | 0% | N/A - DF_A not a Feature of Size |
| | | | 7 | Assembly Shift (Mounting Holes_LMC - F) / 2 | | | +/- 0.6000 | 7.2% | = ((5.1 + 0.1) - 4) / 2 (See Note 2) |
| Back Panel | 123-001 | A | 8 | Assembly Shift (Mounting Holes_LMC - F) / 2 | | | +/- 0.6000 | 7.2% | = ((5.1 + 0.1) - 4) / 2 (See Note 2) |
| | | | 9 | Position (Holes on Left) | | | +/- 1.2500 | 15.1% | Position dia 2.5 @ MMC A, B, C (Upper Segment) |
| | | | 10 | Bonus Tolerance | | | +/- 0.0100 | 1.2% | = (0.1 + 0.1) / 2 |
| | | | 11 | Datum Feature Shift | | | +/- 0.0000 | 0.0% | N/A - (See Note 1) |
| | | | 12 | Dim. CL Left Holes - CL Right Holes | 17.5000 | | +/- 0.0000 | 0% | 17.5 Basic on Dwg |
| | | | 13 | Position (Holes on Right) | | | +/- 1.2500 | 15.1% | Position dia 2.5 @ MMC A, B, C (Upper Segment) |
| | | | 14 | Bonus Tolerance | | | +/- 0.1000 | 1.2% | = (0.1 + 0.1) / 2 |
| | | | 15 | Datum Feature Shift | | | +/- 0.0000 | 0.0% | N/A - (See Note 1) |
| | | | 16 | Assembly Shift (Mounting Holes_LMC - F) / 2 | | | +/- 0.6000 | 7.2% | = ((5.1 + 0.1) - 4) / 2 (See Note 2) |
| Connector (Right) | 123-002 | A | 17 | Assembly Shift (Mounting Holes_LMC - F) / 2 | | | +/- 0.6000 | 7.2% | = ((5.1 + 0.1) - 4) / 2 (See Note 2) |
| | | | 18 | Position: DF_B Holes | | | +/- 0.0000 | 0% | N/A - (See Note 3) |
| | | | 19 | Bonus Tolerance | | | +/- 0.0000 | 0% | N/A - (See Note 3) |
| | | | 20 | Datum Feature Shift | | | +/- 0.0000 | 0% | N/A - DF_A not a Feature of Size |
| | | | 21 | Dim. Datum B - Edge of Connector | | 5.0000 | +/- 1.0000 | 0% | = (30 Basic - 20 Basic) / 2 on Dwg |
| | | | 22 | Profile Edge Along Pt B | | | +/- 1.0000 | 12.0% | Profile 2, A, B/m |
| | | | 23 | Datum Feature Shift (DF_B @ LMC - DFS_B) / 2 | | | +/- 0.6000 | 7.2% | = (5.1 + 0.1 - (5.1 - 0.1 - 1)) / 2 |
| | | | | Dimension Totals | 17.5000 | 10.0000 | | | |
| | | | | Nominal Distance: Pos Dims - Neg Dims = | | 7.5000 | | | |

**RESULTS:**

| | Nom | Tol | Min | Max |
|---|---|---|---|---|
| Arithmetic Stack (Worst Case) | 7.5000 | +/- 8.3000 | -0.8000 | 15.6000 |
| Statistical Stack (RSS) | 7.5000 | +/- 2.7028 | 4.7972 | 10.2028 |
| Adjusted Statistical 1.5*RSS | 7.5000 | +/- 4.0542 | 3.4458 | 11.5542 |

Notes:

1 - Datum Feature Shift is not included for the Back Panel in this Tolerance Stackup because Datum Features A, B & C are not Features of Size.

2 - M4 Screw Dimensions; Used 4mm as Major Diameter of Threads

3 - The Positional Tolerance on the Connector's Datum Feature B Holes does not contribute to the Stackup. Because the holes are the secondary Datum Feature, they are the basis from which all other features on the part are located in the direction of the Stackup.

Assumptions:

Suggested Action: - Using the tolerance in the Upper Segment on Lines 9 & 13, the worst-case Tolerance Stackup result is 0.8 interference.

**FIGURE 9.29**   Composite Position: Option 1 Tolerance Stackup report.

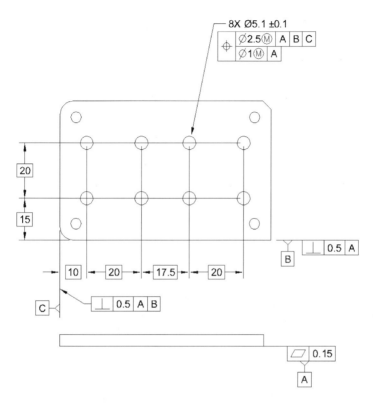

**FIGURE 9.30** Composite Position, back panel detail: Option 2. In this example, all eight holes are toleranced with a single Composite Position Feature Control Frame. Using this method treats the two patterns as a single pattern.

rules for completing the Tolerance Stackup report and creating the Tolerance Stackup sketch are presented in detail in Chapters 13 and 14.

The positional tolerances assigned to the 5.7 ± 0.3 holes in items 1 and 2 were converted to equal bilateral Tolerances. Since the positional tolerances are specified to apply when the features are at their maximum material condition (MMC, smallest hole), bonus tolerance must be added to the Tolerance Stackup. The bonus tolerance, which has no effect on mating and allows more parts to pass inspection, has a negative effect on overall alignment, as the holes are allowed to be farther out of position when they are produced at their largest LMC size.

It should be noted that the tolerancing scheme and tolerance values in this example are not exactly equivalent to Example 7.6. Bonus tolerance, datum feature shift, and the profile tolerances applied to the bracket flange

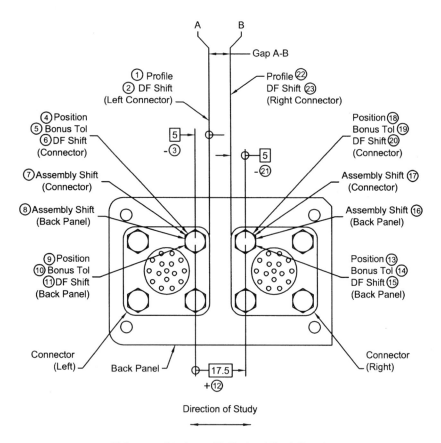

A    B

Gap A-B

① Profile                    Profile ㉒
② DF Shift                   DF Shift ㉓
(Left Connector)             (Right Connector)

④ Position                                    Position ⑱
⑤ Bonus Tol                                   Bonus Tol ⑲
⑥ DF Shift                                    DF Shift ⑳
(Connector)                                   (Connector)

⑦ Assembly Shift                              Assembly Shift ⑰
(Connector)                                   (Connector)

⑧ Assembly Shift                              Assembly Shift ⑯
(Back Panel)                                  (Back Panel)

⑨ Position                                    Position ⑬
⑩ Bonus Tol                                   Bonus Tol ⑭
⑪ DF Shift                                    DF Shift ⑮
(Back Panel)                                  (Back Panel)

Connector                                     Connector
(Left)         Back Panel                     (Right)

17.5

+⑫

Direction of Study

Tolerance Stackup with Option 1 Back Panel:
Chain of Dimensions and Tolerances
Lower Segment Tolerances Used for Line Items 9 & 13

**FIGURE 9.31**   Composite Position: Option 2 Tolerance Stackup sketch.

faces are not included in Example 7.6. These additional tolerances affect the
result of this Tolerance Stackup, but this result is still less than the result in
Example 7.6. This does not mean that by changing the Dimensioning and
Tolerancing scheme from ± to GD&T always leads to a smaller Tolerance
Stackup result, but functional Dimensioning and Tolerancing schemes usually result in less overall variation. Indeed that is the exact reason for using
functional Dimensioning and Tolerancing schemes. Different datum features
could have been chosen and referenced RFS to even further reduce the
variation, eliminating all the occurrences of bonus tolerance and datum
feature shift.

Tolerance Stack

Release 1.2a

| Program: | Tolerance Analysis and Stackup Manual |
|---|---|
| Product: | Part Number: Opt-2 Rev: A Description: Connector Module Assembly Option 2 |
| Problem: | The Connectors Must not Contact Each Other at Assembly |
| Objective: | Determine if Connectors Make Contact at Assembly |

Stack Information:
Stack No: Figure 9-32
Date: 07/04/02
Revision: A
Direction: Horizontal
Author: BR Fischer

| Description of Component / Assy | Part Number | Rev | Item | Description | + Dims | - Dims | Tol | Percent Contrib | Dim / Tol Source & Calcs |
|---|---|---|---|---|---|---|---|---|---|
| Connector (Left) | 123-002 | A | 1 | Profile Edge Along Pt A | | | +/- 1.0000 | 14.7% | Profile 2, A, Bm |
| | | | 2 | Datum Feature Shift (DF $_B$ @LMC - DFS$_B$)/2 | | | +/- 0.6000 | 8.8% | = (5.1 + 0.1 - (5.1 - 0.1 - 1))/2 |
| | | | 3 | Dim Edge of Connector - Datum B | | 5.0000 | +/- 0.0000 | 0% | = (30 Basic - 20 Basic)/2 on Dwg |
| | | | 4 | Position: DF $_B$ Holes | | | +/- 0.0000 | 0% | N/A - (See Note 3) |
| | | | 5 | Bonus Tolerance | | | +/- 0.0000 | 0% | N/A - (See Note 3) |
| | | | 6 | Datum Feature Shift | | | +/- 0.0000 | 0% | N/A - DF $_A$ not a Feature of Size |
| | | | 7 | Assembly Shift (Mounting Hole$_{LMC}$ - F)/2 | | | +/- 0.6000 | 8.8% | = ((5.1 + 0.1) - 4)/2 (See Note 2) |
| Back Panel | 123-001 | A | 8 | Assembly Shift (Mounting Hole$_{LMC}$ - F)/2 | | | +/- 0.6000 | 8.8% | = ((5.1 + 0.1) - 4)/2 (See Note 2) |
| | | | 9 | Position (Holes on Left) | | | +/- 0.5000 | 7.4% | Position dia 1 @ MMC A (Lower Segment) |
| | | | 10 | Bonus Tolerance | | | +/- 0.1000 | 1.5% | = (0.1 + 0.1)/2 |
| | | | 11 | Datum Feature Shift | | | +/- 0.0000 | 0.0% | N/A - (See Note 1) |
| | | | 12 | Dim CL Left Holes - CL Right Holes | 17.5000 | | +/- 0.0000 | 0% | 17.5 Basic on Dwg |
| | | | 13 | Position (Holes on Right) | | | +/- 0.5000 | 7.4% | Position dia 1 @ MMC A (Lower Segment) |
| | | | 14 | Bonus Tolerance | | | +/- 0.1000 | 1.5% | = (0.1 + 0.1)/2 |
| | | | 15 | Datum Feature Shift | | | +/- 0.0000 | 0.0% | N/A - (See Note 1) |
| | | | 16 | Assembly Shift (Mounting Hole$_{LMC}$ - F)/2 | | | +/- 0.6000 | 8.8% | = ((5.1 + 0.1) - 4)/2 (See Note 2) |
| Connector (Right) | 123-002 | A | 17 | Assembly Shift (Mounting Hole$_{LMC}$ - F)/2 | | | +/- 0.6000 | 8.8% | = ((5.1 + 0.1) - 4)/2 (See Note 2) |
| | | | 18 | Position: DF $_B$ Holes | | | +/- 0.0000 | 0% | N/A - (See Note 3) |
| | | | 19 | Bonus Tolerance | | | +/- 0.0000 | 0% | N/A - (See Note 3) |
| | | | 20 | Datum Feature Shift | | | +/- 0.0000 | 0% | N/A - DF $_A$ not a Feature of Size |
| | | | 21 | Dim Datum B - Edge of Connector | | 5.0000 | +/- 0.0000 | 0% | = (30 Basic - 20 Basic)/2 on Dwg |
| | | | 22 | Profile Edge Along Pt B | | | +/- 1.0000 | 14.7% | Profile 2, A, Bm |
| | | | 23 | Datum Feature Shift (DF $_B$ @LMC - DFS$_B$)/2 | | | +/- 0.6000 | 8.8% | = (5.1 + 0.1 - (5.1 - 0.1 - 1))/2 |

Dimension Totals: 17.5000 | 7.5000
Nominal Distance = Pos Dims - Neg Dims = 10.0000

RESULTS:

| | Nom | Tol | Min | Max |
|---|---|---|---|---|
| Arithmetic Stack (Worst Case): | 7.5000 | +/- 6.8000 | 0.7000 | 14.3000 |
| Statistical Stack (RSS): | 7.5000 | +/- 2.1633 | 5.3367 | 9.6633 |
| Adjusted Statistical 1.5*RSS: | 7.5000 | +/- 3.2450 | 4.2550 | 10.7450 |

Notes:
1 - Datum Feature Shift is not included for the Back Panel in this Tolerance Stackup because Datum Features A, B & C are not Features of Size.
2 - M4 Screw Dimensions. Used 4mm as Major Diameter of Threads
3 - The Positional Tolerance on the Connector's Datum Feature B Holes does not contribute to the Stackup. Because the holes are the secondary Datum Feature, they are the basis from which all other features on the part are located in the direction of the Stackup.

Assumptions:

Suggested Action:
- Using the tolerance in the Lower Segment on Lines 9 & 13, the worst-case Tolerance Stackup result is 0.7 Clearance

**FIGURE 9.32** Composite Position: Option 2 Tolerance Stackup report.

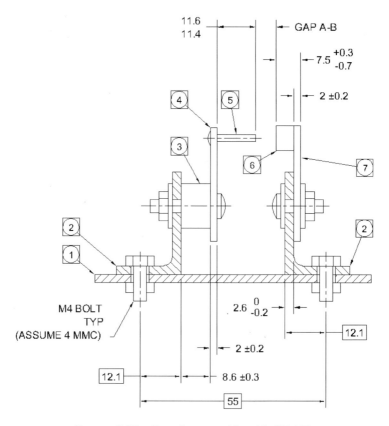

**Figure 9.33** Complex assembly with GD&T.

In Example 7.6 there was a 55 ± 1 dimension and tolerance between the holes in the base plate. The holes had to be sized to account for the possibility that all of the ±1 variation may apply to a single hole, one hole shifted and/or tilted and the other hole perfect. Using ± leads to many possible interpretations, such as where all the tolerance applies to one hole or where the tolerance is split evenly between the holes in the pattern. Both are equally legitimate interpretations, which should help the reader to see why ± should not be used to locate features. GD&T is far superior, as it provides unambiguous specifications. Positional tolerance zones are easy to understand and have one meaning.

In Fig. 9.34 the positional tolerance is diameter 1.4 for the holes in the base plate. This is pretty much equivalent to the ±1 between the holes in Example 7.6. The ±1 tolerance was split between the holes, each hole in the base plate in Example 7.6 could be considered to be ±0.5 from its exact location. When the left-hand (LH) and right-hand (RH) holes are considered

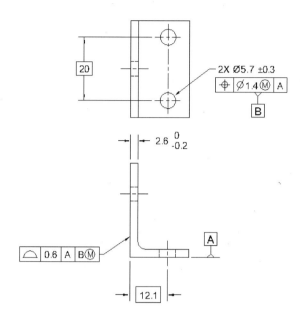

PART 2 - ANGLE BRKT

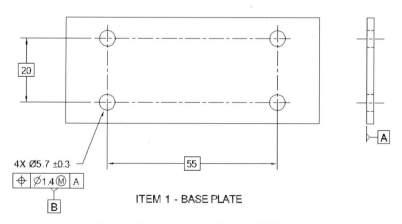

ITEM 1 - BASE PLATE

Parts for Complex Assembly with GD&T

**FIGURE 9.34** Part details for complex assembly with GD&T.

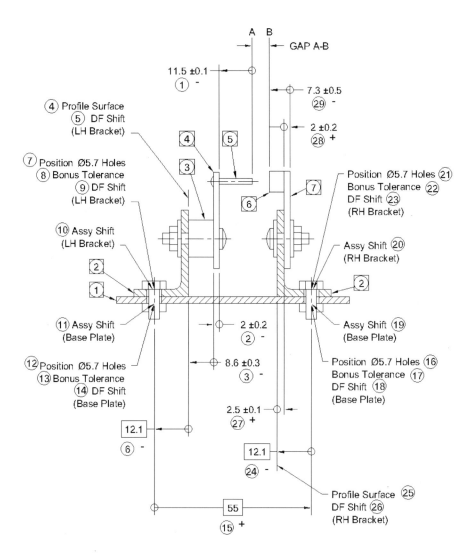

Tolerance Stackup Sketch with Chain of Dimensions
and Tolerances for the Assembly Shown in Figure 9-33

**FIGURE 9.35**  Tolerance Stackup sketch for complex assembly with GD&T.

Statistical Tolerance Stackup

| Dim No | Part No | + | - | +/- | Description |
|---|---|---|---|---|---|
| 1 | 5 | | 11.5 | +/- 0.1 | Dim: Pin Length |
| 2 | 4 | | 2 | +/- 0.2 | Dim: LH Plate Thickness |
| 3 | 3 | | 8.6 | +/- 0.3 | Standoff Thickness |
| 4 | 2 | | | +/- 0.3 | Profile of Flange Face on LH Angle Brkt |
| 5 | 2 | | | +/- 1 | Datum Feature Shift: ((5.7 + 0.3) - (5.7 - 0.3 - 1.4)) / 2 = +/-1 |
| 6 | 2 | | 12.1 | +/- 0 | Dim: Flange Face - CL DFB Holes on LH Angle Brkt  (Basic) |
| 7 | 2 | | | +/- 0 | Position of Dia 5.7 DF$_{B}$ Holes on LH Angle Brkt: N/A |
| 8 | 2 | | | +/- 0 | Bonus Tolerance: N/A |
| 9 | 2 | | | +/- 0 | Datum Feature Shift: N/A |
| 10 | 2 | | | +/- 1 | Assembly Shift: LH Angle Brkt Holes @ LMC: 6 (H) - 4 (F) = 2 / 2 = +/-1 |
| 11 | 1 | | | +/- 1 | Assembly Shift: Base Plate LH Holes @ LMC: 6 (H) - 4 (F) = 2 / 2 = +/-1 |
| 12 | 1 | | | +/- 0.7 | Position of LH Dia 5.7 DF$_{B}$ Holes on Base Plate |
| 13 | 1 | | | +/- 0.3 | Bonus Tolerance: (0.3 + 0.3) / 2 = +/-0.3 |
| 14 | 1 | | | +/- 0 | Datum Feature Shift: N/A - DF$_{A}$ not a Feature of Size |
| 15 | 1 | 55 | | +/- 0 | Dim: CL LH DF$_{B}$ Holes - CL RH DF$_{B}$ Holes on Base Plate  (Basic) |
| 16 | 1 | | | +/- 0.7 | Position of RH Dia 5.7 DF$_{B}$ Holes on Base Plate |
| 17 | 1 | | | +/- 0.3 | Bonus Tolerance: (0.3 + 0.3) / 2 = +/-0.3 |
| 18 | 1 | | | +/- 0 | Datum Feature Shift: N/A - DF$_{A}$ not a Feature of Size |
| 19 | 1 | | | +/- 1 | Assembly Shift: Base Plate LH Holes @ LMC: 6 (H) - 4 (F) = 2 / 2 = +/-1 |
| 20 | 2 | | | +/- 1 | Assembly Shift: RH Angle Brkt Holes @ LMC: 6 (H) - 4 (F) = 2 / 2 = +/-1 |
| 21 | 2 | | | +/- 0 | Position of Dia 5.7 DF$_{B}$ Holes on RH Angle Brkt: N/A |
| 22 | 2 | | | +/- 0 | Bonus Tolerance: N/A |
| 23 | 2 | | | +/- 0 | Datum Feature Shift: N/A |
| 24 | 2 | | 12.1 | +/- 0 | Dim: CL DFB Holes - Flange Face on RH Angle Brkt  (Basic) |
| 25 | 2 | | | +/- 0.3 | Profile of Flange Face on RH Angle Brkt |
| 26 | 2 | | | +/- 1 | Datum Feature Shift: ((5.7 + 0.3) - (5.7 - 0.3 - 1.4)) / 2 = +/-1 |
| 27 | 2 | 2.5 | | +/- 0.1 | RH Angle Brkt Flange Thickness |
| 28 | 7 | 2 | | +/- 0.2 | Thickness of RH Plate |
| 29 | 6 & 7 | | 7.3 | +/- 0.5 | Thickness of RH Plate & Boss |
| | | 59.5 | 53.6 | +/- 10 | Worst Case Tolerance |
| | | | | +/- 2.79 | RSS Tolerance |
| | | | | +/- 4.18 | Adjusted RSS Tolerance ( RSS * 1.5) |

| | | |
|---|---|---|
| Positive Total | 59.5 | |
| Negative Total | -53.6 | |
| Nominal Gap | 5.9  +/- 4.18 | Adjusted RSS Tolerance |

| | | |
|---|---|---|
| Max Gap | 10.08 | Clearance |
| Min Gap | 1.72 | Clearance |

**FIGURE 9.36** Tolerance Stackup report for complex assembly with GD&T.

together, each $\pm 0.5$ from their nominal position, they give $(\pm 0.5) + (\pm 0.5)$ = $\pm 1$. Assuming there was a $\pm 0.5$ tolerance relating the holes to the edge in the perpendicular direction, a square $\pm 0.5 = 1$ tolerance zone would exist. Using a circumscribed tolerance zone for conversion, a square $\pm 0.5$ tolerance zone converts to a cylindrical tolerance zone of diameter 1.4. Strictly speaking, since the base plate in Example 7.6 is incompletely dimensioned and toleranced and there is no $\pm 0.5$ tolerance in the perpendicular direction, the $\pm 1$ tolerance zone could have been converted to diameter 1. But that is a faulty approach because if the part was completely dimensioned and toleranced, there would be a tolerance in the perpendicular direction. The method of converting $\pm$ square and rectangular tolerance zones to cylindrical positional tolerance zones is covered in Chapter 10.

In Example 7.6 there was a $\pm 1$ dimension and tolerance between the holes in the bracket and the flange face. Because of the imprecision of the plus/minus system, it is unclear whether the $\pm 1$ tolerance applies to the holes or to the flange. There are several ways to interpret the specifications. For the sake of argument, this example splits the $\pm 1$ (2 mm total) tolerance between the holes and the flange. As with the base plate, a diameter 1.4 tolerance is specified for the holes. This leaves 0.6 for the profile tolerance applied to the flange face. In this case, it is exactly clear how the tolerances apply to the features, as each tolerance is clearly specified.

Using this approach the result is lower than the result using $\pm$. The adjusted RSS result shown in Fig. 9.36 is 4.18 vs. 4.79 in Example 7.6.

## CONVERTING POSITIONAL TOLERANCES TO EQUAL BILATERAL $\pm$ TOLERANCES

Positional tolerances are relatively easy to convert into equivalent $\pm$ location tolerances. The method used to convert a positional tolerance depends on the material condition modifier (RFS, MMC, or LMC) applied to the tolerance and whether the tolerance is applied to features that affect the location of other features in the Tolerance Stackup.

## POSITIONAL TOLERANCE CONVERSION

Converting positional tolerances specified at MMC to equal bilateral $\pm$ Tolerances, where the location of the features affects the Tolerance Stackup result:

1. Convert the specified positional tolerance at MMC.

   a. Divide the specified positional tolerance by 2.
   b. This is the equivalent equal bilateral $\pm$ positional tolerance at MMC.

2. Convert the ± size tolerance into the bonus tolerance.

   a. Add the + and − tolerance values to obtain the bonus tolerance.
   b. This is the total bonus tolerance.
   c. Divide the result by 2.
   d. This is the equivalent equal bilateral ± bonus tolerance.

3. Together these represent the positional tolerance at LMC. They are to be entered as separate line items in the Tolerance Stackup. See Chapter 14 for more information.

*Example.* Given the feature control frame in Fig. 9.37:

1. Calculate the equivalent ± equal bilateral positional tolerance at MMC.

   a. Positional tolerance at MMC = 2.
   b. Positional tolerance/2 = 2/2 = ±1.
      Equivalent ± equal bilateral positional tolerance at MMC = ±1.

2. Calculate the equivalent equal bilateral ± bonus tolerance.

   a. Size tolerance = ±0.3 = 0.3 + 0.3 = 0.6.
   b. Bonus tolerance = 0.6.
   c. Bonus tolerance/2 = 0.6/2 = ±0.3.
      Equivalent ± equal bilateral bonus tolerance = ±0.3

Converting positional tolerances specified at MMC to equal bilateral ± tolerances, where the location of the features does not affect the Tolerance Stackup result.

1. Convert the specified positional Tolerance at MMC.

   a. Divide the specified positional tolerance by 2.
   b. This is the equivalent equal bilateral ± positional tolerance at MMC.

$$\varnothing 6.3 \pm 0.3$$

| $\oplus$ | $\varnothing 2 \text{ⓜ}$ | A |

**FIGURE 9.37** Feature control frame for positional tolerance conversion #1.

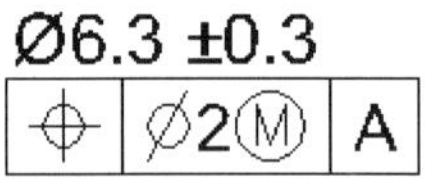

$$\varnothing 6.3 \pm 0.3$$

$$\boxed{\ \oplus\ }\boxed{\ \varnothing 2\text{\large{M}}\ }\boxed{\ \text{A}\ }$$

**FIGURE 9.38** Feature control frame for positional tolerance conversion #2.

*Example.* Given the feature control frame in Fig. 9.38.

1. Calculate the equivalent ± equal bilateral positional tolerance at MMC.
    a. Positional tolerance at MMC = 2.
    b. Positional tolerance/2 = 2/2 = ±1.
    c. Equivalent ± equal bilateral positional tolerance at MMC = ±1.

The bonus tolerance is not considered in these applications, as it has no effect on the variation between the features being studied. Care must be taken when determining which of the methods to employ, as these rules of thumb are not catchalls that work in every situation. Unfortunately for those of you desiring easy rules of thumb, the tolerance Analyst must think carefully about the problem being studied and recognize the variables that affect the total tolerance. Time and practice make it easier to recognize when and where to include the bonus tolerance.

Here is a suggestion for making it easier to determine whether the bonus tolerance should be included in the tolerance Stackup: Consider the feature of size with the positional tolerance and visualize if its location affects the distance being studied. If so, add the bonus tolerance value to the tolerance Stackup; if not, set the bonus tolerance value to zero.

Similar methods are used for features toleranced at their least material condition. Features toleranced regardless of feature size (RFS) are always converted using the second method, as there is no bonus tolerance associated with RFS.

## DATUM FEATURE SHIFT

Datum feature shift represents the variation that may be introduced when inspecting features related to datum features of size specified at LMC or MMC. When datum features of size are referenced at MMC or LMC, their datum feature simulators may be smaller or larger than the datum features of size, which allows the part to shift or move relative to the datum feature simulators. The worst-case difference in size between the datum features and their simulators is the amount of datum feature shift. The term *datum feature shift* means the datum features can shift during the inspection process—there is not a one-to-one relationship between the datum features and the datum feature simulators.

According to paragraph 2.11.3 in the ASME Y4.5M-1994 standard, datum features of size are to be simulated at their applicable virtual condition size, LMC size, or MMC size, whichever is applicable. Two considerations must be made to determine which datum feature simulator size is appropriate:

1. Determine whether LMC or MMC is specified.

2. Determine if there is a geometric tolerance specified that controls the datum feature of size's center geometry per the rules below:

    a. If a datum feature of size *is not* toleranced with a geometric tolerance that controls the datum feature's center geometry (such as orientation or position), then the datum feature of size is simulated at its appropriate LMC or MMC size.

    b. If a datum feature of size *is* toleranced with a geometric tolerance that controls the datum feature's center geometry (such as orientation or position), then the datum feature of size is simulated at its appropriate virtual condition size.

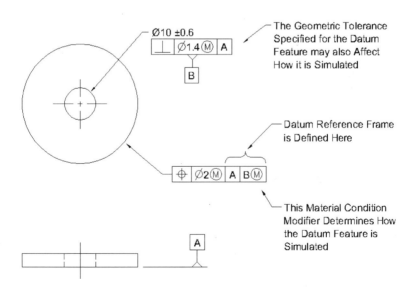

Datum Feature of Size Specified at MMC

**FIGURE 9.39** Datum feature shift: Meaning of material condition modifiers.

The most common application is where datum features of size are specified at MMC. Locating holes and slots, clearance holes, pins and studs used as datum features are all examples where MMC may be specified.

It is less common to see datum features of size specified at LMC. However, there are places it is necessary and functionally beneficial. Unfortunately, specifying datum features of size at LMC leads to problems at inspection, as the datum feature simulator may not fit within or around a datum feature of size produced at MMC. For example, the LMC virtual condition of an internal datum feature of size such as a hole is simulated by pin that is larger than the hole! Such simulation can only be done using virtual gaging techniques, such as a coordinate measuring machine (CMM.)

The material condition modifier following the datum feature reference in a feature control frame determines how the datum feature is simulated (see Fig. 9.39). (Remember, rule 2 in ASME Y4.5M-1994 states that RFS is implied when no material condition modifier is specified.)

### Datum Feature Shift: Datum Feature of Size Simulated at Its MMC Size

A datum feature of size is referenced at MMC in the positional tolerance feature control frame shown in Fig. 9.40. The datum feature is simulated at its MMC size because there are no applicable geometric tolerances controlling its center geometry.

### Datum Feature Shift: Calculations for the Part in Fig. 9.40:

- The specifications in Fig. 9.40 show that the datum feature of size should be simulated at its MMC size.

$$\text{MMC size} = \frac{\varnothing 10.0 \text{ nominal size}}{-0.2 \text{ size tolerance}}$$
$$= \varnothing 9.8 \text{ MMC size}$$

This is the size of the Datum feature simulator.

$$\text{LMC (largest) size of the hole} = \varnothing 10.0 \text{ nominal size}$$
$$\frac{+0.2 \text{ size tolerance}}{= \varnothing 10.2 \text{ LMC size}}$$

$$\text{Datum feature shift} = \frac{\varnothing 10.2 \text{ LMC Size}}{-\varnothing 9.8 \text{ MMC Size}}$$
$$= 0.4 \text{ Datum Feature Shift}$$

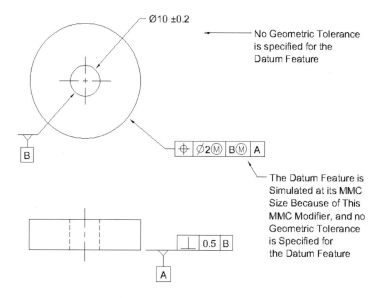

Datum Feature of Size Specified at MMC:
To Be Simulated at Its MMC Size

**FIGURE 9.40** Datum feature shift: Datum feature of size to be simulated at MMC.

- Divide the datum feature shift by 2 = 0.4/2 = ±0.2.
  This is the equal bilateral ± equivalent

Figure 9.41 shows the LMC datum feature of size in its nominal position on the MMC size datum feature simulator. The part can move or shift as much as the clearance between the datum feature and the datum feature simulator allows.

Figure 9.42 shows the part with the datum feature of size shifted about its datum feature simulator. The datum feature shift is relatively small because the datum feature of size has a small size tolerance and the datum feature simulator is sized at the datum feature's MMC size.

Datum feature shift is calculated similar to how assembly shift is calculated. The largest possible clearance between the datum feature simulator and the datum feature is calculated diametrally and divided by 2, which gives the ± equal bilateral equivalent. In the example above, the datum feature shift is merely the difference between the LMC and MMC sizes of the datum feature. Where certain geometric tolerances are applied to the datum feature, their specified value must also be subtracted from or added to the total as seen in Figs. 9.43 and 9.44 and the calculations that follow.

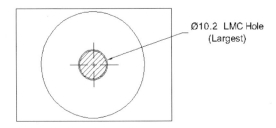

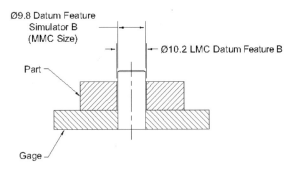

Datum Feature of Size Specified at MMC
With MMC Datum Feature Simulator - Not Shifted

**FIGURE 9.41**  Datum feature shift: Datum feature of size to be simulated at MMC with datum feature simulator, no shift.

## Datum Feature Shift: Datum Feature of Size Simulated at MMC Virtual Condition Size

Datum features of size are referenced at MMC in the profile feature control frame shown in Fig. 9.43. The datum features are simulated at their MMC virtual condition size because the datum features of size have a positional tolerance controlling their center geometry.

Figure 9.44 shows the LMC datum features of size in their nominal position on the MMC virtual condition size datum feature simulators. The part can move or shift as much as the clearance between the datum features and the datum feature simulators allows.

Figure 9.45 shows the part with the datum features of size shifted about their datum feature simulators. The datum feature shift is relatively large because the datum feature of size has a larger size tolerance and the datum feature simulator is sized at the datum feature's MMC virtual condition size.

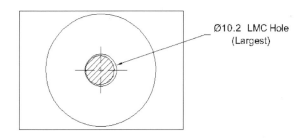

Ø10.2 LMC Hole
(Largest)

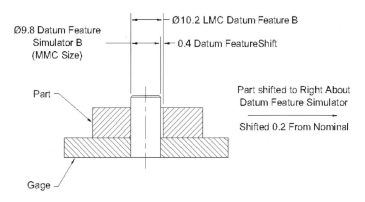

Ø9.8 Datum Feature
Simulator B
(MMC Size)

Ø10.2 LMC Datum Feature B

0.4 Datum FeatureShift

Part

Gage

Part shifted to Right About
Datum Feature Simulator

Shifted 0.2 From Nominal

Datum Feature of Size Specified at MMC
Shifted About MMC Size Datum Feature Simulator

**FIGURE 9.42**  Datum feature shift: Datum feature of size to be simulated at MMC with datum feature simulator, shifted.

### Datum Feature Calculations for the Part in Fig. 9.43:

- The specifications in Fig. 9.43 show that the datum features of size should be simulated at their MMC virtual condition size.

The MMC virtual condition size

$$
\begin{aligned}
&= \quad \varnothing 10.0 \text{ nominal size} \\
&\quad - 0.6 \text{ size tolerance} \\
&\quad \underline{- \varnothing 1.4 \text{ positional tolerance}} \\
&= \quad \varnothing 8.0 \text{ MMC virtual condition size}
\end{aligned}
$$

This is the size of the datum feature simulators.

$$
\begin{aligned}
\text{The LMC (largest) size of the hole} &= \varnothing 10.0 \text{ nominal size} \\
&\quad \underline{+ 0.6 \text{ size tolerance}} \\
&= \varnothing 10.6 \text{ LMC size}
\end{aligned}
$$

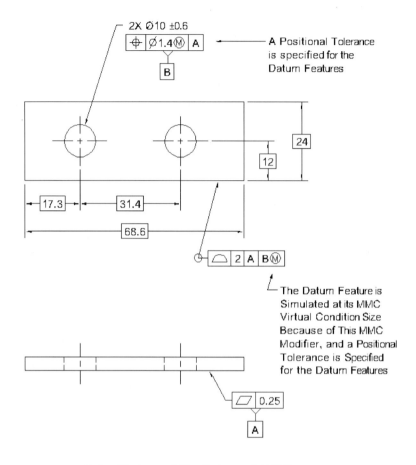

Datum Features of Size Referenced at MMC:
To Be Simulated at Their MMC Virtual Condition Size

**FIGURE 9.43** Datum feature shift: Datum feature of size to be simulated at MMC VC.

$$\text{Datum feature shift} = \quad \varnothing 10.6 \text{ LMC size}$$
$$\underline{- \varnothing 8.0 \text{ MMC virtual condition size}}$$
$$= \quad 2.6 \text{ Datum feature shift}$$

- Divide the datum feature shift by 2: $2.6/2 = \pm 1.3$.
  This is the equal bilateral $\pm$ equivalent.

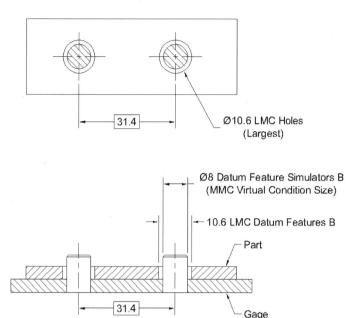

Datum Features of Size Referenced at MMC:
With MMC Virtual Condition Datum Feature Simulators - Not Shifted

**FIGURE 9.44** Datum feature shift: Datum feature of size to be simulated at MMC VC with datum feature simulator, no shift.

Datum feature shift is a tolerance Stackup contributor that is frequently overlooked. There are two ways it is added to a tolerance Stackup, each depending either on the tolerancing defaults employed or on the specific notations added to the drawing.

These defaults are simultaneous requirements and separate requirements. As discussed in the next section, simultaneous requirements is the default condition for drawings prepared to ASME Y14.5M-1994. It may be overridden by local notes as defined 5.3.6.2 in ASME Y14.5M-1994, or it may be overridden by a corporate or global standard.

It is important for the tolerance Analyst to be absolutely sure which default is in place when doing a tolerance Stackup. How datum feature shift is treated and how many times it is added to the tolerance Stackup depends on which default is in place as seen in the following section.

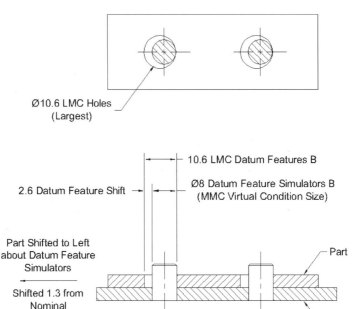

Datum Features of Size Referenced at MMC:
Shifted About MMC Virtual Condition Datum Feature Simulators

**FIGURE 9.45**   Datum feature shift: Datum feature of size to be simulated at MMC VC with datum feature simulator, shifted.

## SIMULTANEOUS AND SEPARATE REQUIREMENTS

### Simultaneous Requirements

*Simultaneous requirements* is the default condition for drawings prepared using the ASME Y14.5M-1994 standard. Unless specified otherwise, simultaneous requirements applies to all single segment feature control frames and the uppermost segment of all composite feature control frames related to the same datum reference frame.

The same datum reference frame means the same datum features are accompanied by the same modifiers referenced in exactly the same order of precedence in each feature control frame. This includes material condition modifiers and other modifiers. For example, all feature control frames related to datum reference frame A (primary), B at MMC (secondary), C at MMC (tertiary) are considered a simultaneous requirement. Datum reference frame

A (primary), B at MMC (secondary), C at MMC (tertiary) is not same datum reference frame as A (primary), B RFS (secondary), C RFS (tertiary); A (primary), B at MMC (secondary); or B at MMC (primary) , C at MMC (secondary), A (tertiary).

Simultaneous requirements means that all part features related to the same datum reference frame must be within tolerance at the same time without changing the part's relationship to the datum reference frame. Simultaneous requirements makes all the features related to the same datum reference frame by geometric tolerances into a pattern. To put it in inspection

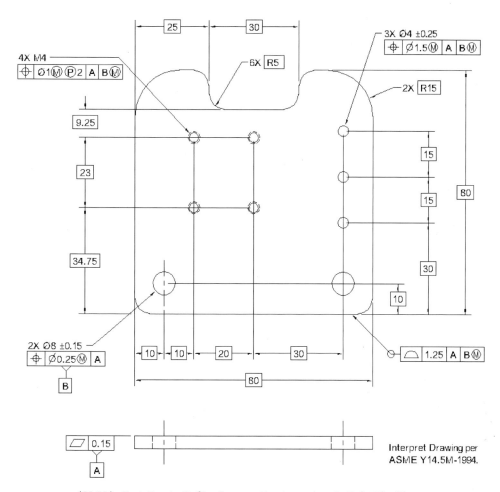

123-001 - Back Panel with Simultaneous Requirements as the Default Condition

**FIGURE 9.46** Simultaneous requirements as the default condition.

terms, all applicable geometric tolerances related to the same datum reference frame are to be inspected in a single setup.

For example, consider the part shown in Fig. 9.46. Three feature control frames specify geometric tolerances related to datum reference frame A (primary), B at MMC (secondary). These are the positional tolerance specified for the 4X M4 holes, the positional tolerance specified for the 3X $\varnothing 4 \pm 0.25$ holes and the profile tolerance specified all around the periphery of the part.

All surfaces, holes, etc., on the part toleranced relative to datum reference frame A, B at MMC must be in tolerance at the same time (simultaneously). All features related to datum reference frame A, B at MMC must be inspected in a single setup, without adjusting the part during inspection. The part may be adjusted to find the optimal relationship between the datum features and the datum feature simulators *before* the features are inspected, and the part may be adjusted *after* the features are inspected—the part just cannot be adjusted while the features are being inspected. If the part must be adjusted to bring a noncompliant feature into tolerance during inspection, then all related features must be inspected again with the part in its new location on the simulator.

As stated above, simultaneous requirements combines all the features related to the same datum reference frame by geometric tolerances into a single pattern. Datum feature shift does not affect the feature-to-feature relationship within a pattern; datum feature shift only affects the relationship between the pattern and referenced datum features of size.

Datum feature B is a pattern of two holes referenced at MMC in Fig. 9.46. As stated in the previous section, paragraph 2.11.3 in the ASME Y4.5M-1994 standard requires that the datum feature simulators for datum features B shall be sized at the MMC virtual condition size of the datum features ($\varnothing 7.6$) and will therefore be smaller than the holes. This of course means there will be datum feature shift: datum features B may shift about datum feature simulators B. When the datum features are produced at their LMC (largest) size, there is the possibility for $\pm 0.275$ datum feature shift.

The concept of *simultaneous requirements* is very important for tolerance Stackups as it relates directly to datum feature shift. When performing tolerance Stackups on parts where simultaneous requirements applies, datum feature shift is only added once or not at all for each datum reference frame per the following rules.

## Rules for Simultaneous Requirements and Datum Feature Shift

If the chain of Dimensions and Tolerances for a part in the tolerance Stackup only includes features related to a single datum reference frame with datum features of size at MMC or LMC and the chain of Dimensions and Tolerances does not include the referenced datum features of size, then datum feature shift is not added to the tolerance Stackup for the tolerances related to that datum reference frame. An example can be seen in Examples 16.2 and 16.3 in Chapter 16, where the inside surfaces of the enclosure are all toleranced to the same datum reference frame and datum feature shift is not added. This is because the chain of Dimensions and Tolerances only includes the toleranced features and does not pass through the datum features of size.

If the chain of Dimensions and Tolerances for a part in the tolerance Stackup only includes features related to a single datum reference frame with datum features of size at MMC or LMC, and the chain of Dimensions and Tolerances includes the referenced datum features of size, then datum feature shift is only added once to the tolerance Stackup. Datum feature shift is only included with the first tolerance in the chain of Dimensions and Tolerances related to the datum reference frame. For example, if a tolerance Stackup was done to determine the distance between the center of datum feature B and the upper surface of the part in Fig. 9.46, datum feature shift would be added to the tolerance Stackup once.

If the chain of Dimensions and Tolerances for a part in the tolerance Stackup includes features related to more than one datum reference frame, at least one of the datum reference frames includes datum features of size at MMC or LMC, and the chain of Dimensions and Tolerances passes through those datum features of size, then datum feature shift is added once to the tolerance Stackup for the first tolerance related to each datum reference frame with datum features of size at MMC or LMC on the part.

If the chain of Dimensions and Tolerances for a part in the tolerance Stackup includes features related to a single datum reference frame with datum features of size at MMC or LMC, and the chain of Dimensions and Tolerances passes through the referenced datum features of size to a mating part, then datum feature shift is only added once to the tolerance Stackup. Datum feature shift is only included with the first tolerance in the chain of Dimensions and Tolerances related to the datum reference frame.

If the chain of Dimensions and Tolerances for a part in the tolerance Stackup includes features related to more than one datum reference frame, one of the datum reference frames includes datum features of size at MMC or LMC, and the chain of Dimensions and Tolerances passes through the referenced datum features of size to a mating part, then datum feature shift is added once to the tolerance Stackup for the first tolerance related to each datum reference frame with datum features of size at MMC or LMC on the part. Any time a datum reference frame contains datum features of size specified at MMC or LMC, the part may be shifted about the datum feature simulators during inspection to find a location where the toleranced features are within specification.

Consider the following example from Fig. 9.46: a positional tolerance related to datum reference frame A, B at MMC is applied to the pattern of four M4 holes. Datum feature B is also a pattern of holes. Since Datum features B are referenced at MMC, there may be datum feature shift. Once the part is staged on the datum feature simulators, it may be adjusted to find a location where all four M4 holes are within their positional tolerance at the same time. The part may not be adjusted such that three of the holes are within tolerance but the fourth hole is out of tolerance and then shifted so the first three holes are out of tolerance but the fourth hole is within tolerance. By definition, this pattern of holes is not within its positional tolerance.

Datum feature shift is added to the tolerance Stackup because there is not a one-to-one relationship between the specified datum features and their datum feature simulators. The part can move relative to the datum reference frame (or vice versa), and it is possible that a toleranced feature may be inspected with a different relationship to the datum reference frame than encountered at assembly.

The concept of *simultaneous requirements* does not eliminate datum feature shift. Simultaneous requirements merely reduces the effect of datum feature shift by allowing it to occur at most once for each datum reference frame in the tolerance Stackup. If simultaneous requirements were not in effect, datum feature shift would be added after each tolerance in the tolerance Stackup related to the same datum reference frame with datum features of size at MMC or LMC, which is the condition of separate requirements.

## Separate Requirements

*Separate requirements* is the opposite condition of simultaneous requirements. During inspection, the relationship between the datum features of size and their simulators may be changed (the part may be shifted about the datum feature simulators) for each feature or group of features related to each feature control frame specifying the same datum reference frame. Where specified, separate requirements apply between distinct feature control frames. The relationship of the datum features to their simulators must be maintained while inspecting the features toleranced by any one feature control frame, but may be changed between feature control frames, even if they reference the same datum reference frame.

As with simultaneous requirements, all features related to any one feature control frame must be within tolerance at the same time. However, features toleranced by one feature control frame are not required to have the same relationship to the datum reference frame as the features toleranced by other feature control frames that reference the same datum reference frame.

Datum feature shift must be added to each geometric tolerance in the tolerance Stackup that references a datum reference frame with datum features of Size at MMC or LMC. If there are three geometric tolerances in the tolerance Stackup related to datum reference frame A, B at MMC, then datum feature shift about datum feature B must be added to the tolerance Stackup three times. If simultaneous requirements were in effect, datum feature shift about datum feature B would only be added at most once.

In Figure 9.47 the annotation "SEP REQT" has been specified beneath the three feature control frames related to datum reference frame A, B at MMC. This invokes separate requirements, overriding the simultaneous requirements default. Datum feature shift must be added each time any of

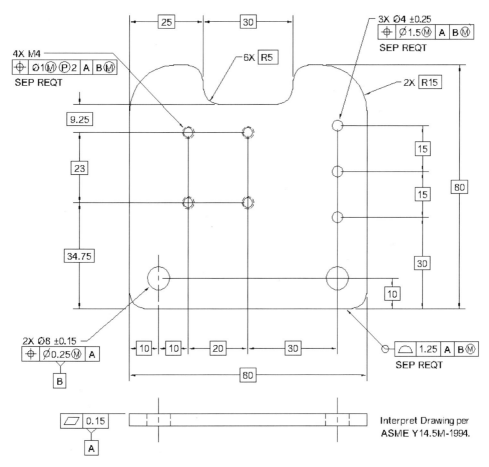

123-003 - Back Panel with Separate Requirements

**FIGURE 9.47** Separate requirements as the default condition.

the features related to datum reference frame A, B at MMC are included in a tolerance Stackup. Simultaneous requirements can also be overridden by a general note or in a referenced document.

Obviously datum feature shift plays a larger role in tolerance Stackups where separate requirements are in effect. The effect of separate requirements seems to imply that a part can be biased in more than one direction at the same time at assembly, that the part can be in more than one location at a time. Obviously this is not the case. After a bit of careful consideration, the reader may ask, "Why would anyone want to invoke separate requirements? It adds variation and doesn't reflect the physical reality of a part's as-assembled condition." Indeed, separate requirements rarely (if ever) reflect functional considerations and are usually specified for other reasons.

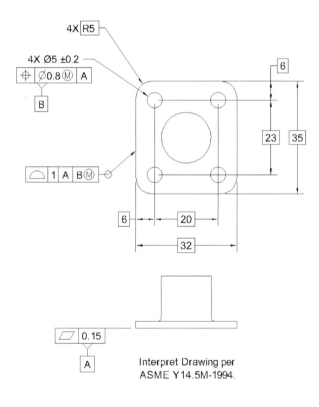

**FIGURE 9.48**   Switch carrier.

Typically the reason that separate requirements are specified is a good idea carried out in a bad way—it is based on the idea that the datum feature shift between features or patterns of features allows more parts to pass inspection. It assumes that there is no critical relationship between the tolerances specified for various patterns or features. There are many better ways to accomplish a similar goal. Composite tolerances, multiple datum reference frames, and larger tolerance values are all examples of more sensible ways feature relationships can be effectively toleranced.

Several tolerance Stackup examples follow. Both tolerance Stackups are the same except that simultaneous requirements is the default condition in the

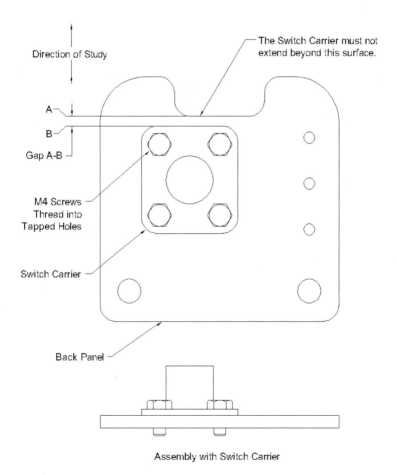

Assembly with Switch Carrier

**FIGURE 9.49** Simultaneous requirements assembly with switch carrier.

| | | | | | | Stack Information | | |
|---|---|---|---|---|---|---|---|---|

Program Tolerance Analysis and Stackup Manual

Product: Part Number Rev Description
Assembly 32(a) A Back Panel with Switch Carrier with SIMULTANEOUS REQUIREMENTS

Stack No: Figure 9-50
Date 07/04/02
Revision A

Problem: Switch Carrier Must Not Interfere with Mating Parts. It Must Not Hang Over Edge of Back Panel

Objective Determine if Switch Carrier Protrudes Beyond Upper Edge of Back Panel

Direction: Vertical
Author BR Fischer

| Description of Component / Assy | Part Number | Rev | Item | Description | + Dims | - Dims | Tol | Percent Contrib | Dim. / Tol Source & Calcs |
|---|---|---|---|---|---|---|---|---|---|
| Switch Carrier | 123-002 | A | 1 | Profile Upper Edge Along Pt B | | | +/- 0.5000 | 17.7% | Profile 1, A, Bm |
| | | | 2 | Datum Feature Shift (DF$_8$@mc - DFS$_8$) / 2 | | | +/- 0.6000 | 21.2% | = (5 + 0.2 - (5 - 0.2 - 0.8)) / 2 |
| | | | 3 | Dim. Upper Edge of Switch Carrier - Datum B | | 6.0000 | +/- 0.0000 | 0% | 6 Basic on Dwg |
| | | | 4 | Position: DF$_8$ Holes | | | +/- 0.0000 | 0% | N/A - (See Note 3) |
| | | | 5 | Bonus Tolerance | | | +/- 0.0000 | 0% | N/A - (See Note 3) |
| | | | 6 | Datum Feature Shift: | | | +/- 0.0000 | 0% | N/A - DF, not a Feature of Size |
| | | | 7 | Assembly Shift: (Mounting Hole$_8$mc - F) / 2 | | | +/- 0.6000 | 21.2% | = ((5 + 0.2) - 4) / 2  (See Note 2) |
| Back Panel | 123-001 | A | 8 | Position: M4 Holes | | | +/- 0.5000 | 17.7% | Position dia 1 @ MMC A, Bm |
| from Figure 30(a) | | | 9 | Bonus Tolerance | | | +/- 0.0000 | 0.0% | N/A - Assume Threads are self-centering |
| | | | 10 | Datum Feature Shift (DF$_8$@mc - DFS$_8$) / 2 | | | +/- 0.0000 | 0.0% | N/A - SIM REQTS - (See Note 1) |
| | | | 11 | Dim. CL Holes - Edge of Base Plate | 9.2500 | | +/- 0.0000 | 0% | 9.25 Basic on Dwg |
| | | | 12 | Profile Edge Along Pt A | | | +/- 0.6250 | 22.1% | Profile 1.25, A, Bm |
| | | | 13 | Datum Feature Shift (DF$_8$@mc - DFS$_8$) / 2 | | | +/- 0.0000 | 0% | N/A - SIM REQTS - (See Note 1) |

| | | |
|---|---|---|
| Dimension Totals | 9.2500 | 6.0000 |
| Nominal Distance: Pos Dims - Neg Dims = | 3.2500 | |

| RESULTS: | Nom | Tol | Min | Max |
|---|---|---|---|---|
| Arithmetic Stack (Worst Case) | 3.2500 | +/- 2.8250 | 0.4250 | 6.0750 |
| Statistical Stack (RSS) | 3.2500 | +/- 1.2691 | 1.9809 | 4.5191 |
| Adjusted Statistical 1.5*RSS | 3.2500 | +/- 1.9037 | 1.3463 | 5.1537 |

Notes: 1 - Datum Feature Shift is not included for the Back Panel in this Tolerance Stackup because Simultaneous Requirements is the Default and the Chain of Dimensions does not go through or include the Datum Features of Size on the Back Panel. The Upper Surface and the M4 holes on the Back Panel are considered a pattern, because Simultaneous Requirements is the Default on the Back Panel drawing.
2 - M4 Screw Dimensions. Used 4mm as Major Diameter of Threads
3 - In this example the Positional Tolerance on the Switch Carrier's Datum Feature B Holes does not contribute to the Stackup. Because the holes are the secondary Datum Feature, they are the basis from which all other features on the part are located in the direction of the Stackup.

Assumptions:

Suggested Action: - None. There is 0.425 clearance with Simultaneous Requirements in effect on the Back Panel

**FIGURE 9.50** Tolerance Stackup with Simultaneous requirements.

| Program: | Tolerance Analysis and Stackup Manual | | | | Stack Information: | |
|---|---|---|---|---|---|---|
| Product: | Part Number | Rev | Description | | Stack No: | Figure 9-51 |
| | Assembly 32(c) | A | Back Panel with With Switch Carrier with SEPARATE REQUIREMENTS | | Date | 07/04/02 |
| Problem: | Switch Carrier Must Not Interfere with Mating Parts: It Must Not Hang Over Edge of Back Panel | | | | Revision | A |
| Objective | Determine if Switch Carrier Protrudes Beyond Upper Edge of Back Panel | | | | Direction: | Vertical |
| | | | | | Author: | ER Fischer |

| Description of Component / Assy | Part Number | Rev | Item | Description | + Dims | - Dims | Tol | Percent Contrib | Dim / Tol Source & Calcs |
|---|---|---|---|---|---|---|---|---|---|
| Switch Carrier | 123-002 | A | 1 | Profile: Upper Edge Along Pt B | | | +/- 0.5000 | 14.8% | Profile 1, A, Bm |
| | | | 2 | Datum Feature Shift (DF$_B$ @ LMC – DF$_{SB}$) / 2 | | | +/- 0.6000 | 17.8% | = (5 + 0.2 – (5 – 0.2 – 0.8)) / 2 |
| | | | 3 | Dim: Upper Edge of Switch Carrier – Datum B | | 6.0000 | +/- 0.0000 | 0% | 6 Basic on Dwg |
| | | | 4 | Position: DF$_B$ Holes | | | +/- 0.0000 | 0% | N/A – (See Note 3) |
| | | | 5 | Bonus Tolerance | | | +/- 0.0000 | 0% | N/A |
| | | | 6 | Datum Feature Shift | | | +/- 0.0000 | 0% | N/A – DF$_A$ not a Feature of Size |
| | | | 7 | Assembly Shift (Mounting Holes$_{LMC}$ – F) / 2 | | | +/- 0.6000 | 17.8% | = ((5 + 0.2) – 4) / 2   (See Note 2) |
| Back Panel | 123-003 | A | 8 | Position: M4 Holes | | | +/- 0.5000 | 14.8% | Position dia 1 @ MMC A, Bm |
| from Figure 30(c) | | | 9 | Bonus Tolerance | | | +/- 0.0000 | 0.0% | N/A – Assume Threads are self-centering |
| | | | 10 | Datum Feature Shift (DF$_B$ @ LMC – DF$_{SB}$) / 2 | | | +/- 0.2750 | 8.1% | = ((8 + 0.15) – (8 – 0.15 – 0.25)) / 2 |
| | | | 11 | Dim: CL Holes – Edge of Base Plate | 9.2500 | | +/- 0.0000 | 0% | 9.25 Basic on Dwg |
| | | | 12 | Profile: Edge Along Pt A | | | +/- 0.6250 | 18.5% | Profile 1.25, A, Bm |
| | | | 13 | Datum Feature Shift (DF$_B$ @ LMC – DF$_{SB}$) / 2 | | | +/- 0.2750 | 8.1% | = ((8 + 0.15) – (8 – 0.15 – 0.25)) / 2  (SEP REQTS) |
| | | | | Dimension Totals | 9.2500 | 6.0000 | | | |
| | | | | Nominal Distance: Pos Dims – Neg Dims = | 3.2500 | | | | |

RESULTS:

| | Nom | Tol | Min | Max |
|---|---|---|---|---|
| Arithmetic Stack (Worst Case) | 3.2500 | +/- 3.3750 | -0.1250 | 6.6250 |
| Statistical Stack (RSS) | 3.2500 | +/- 1.3274 | 1.9226 | 4.5774 |
| Adjusted Statistical 1.5*RSS | 3.2500 | +/- 1.9910 | 1.2590 | 5.2410 |

Notes:  1 - Datum Feature Shift is included for the the Positional and Profile tolerances on the Back Panel in this Tolerance Stackup because Separate Requirements is the Default on the Back Panel drawing.
          The Upper Surface and the M4 holes on the Back Panel are not considered a pattern, because Separate Requirements is the Default.
       2 - M4 Screw Dimensions: Used 4mm as Major Diameter of Threads
       3 - In this example the Positional Tolerance on the Switch Carrier's Datum Feature B Holes does not contribute to the Stackup. Because the holes are the secondary Datum Feature, they are the
          basis from which all other features on the part are located in the direction of the Stackup.

Assumptions:

Suggested Action:   - With Separate Requirements in effect on the Back Panel the Switch Carrier Overlap is 0.125

**FIGURE 9.51**  Tolerance Stackup with Separate requirements.

first tolerance Stackup and separate requirements is the default condition in the second tolerance Stackup.

## Tolerance Stackup with Simultaneous Requirements

In this example, the switch carrier shown in Fig. 9.48 is mounted onto the back panel shown in Fig. 9.46, which has simultaneous requirements as the default condition. The assembly is shown in Fig. 9.49. The object of this tolerance Stackup is to determine if the switch carrier protrudes beyond the cutout in the edge of the back panel. The simultaneous requirements Tolerance Stackup Report is shown in Fig. 9.50. Simultaneous requirements only affects the back panel, as there are two tolerances in the tolerance Stackup that are toleranced relative to datum reference frame A, B at MMC on the back panel. Note: For simplicity, 4 mm was used as the size of the M4 threads in the assembly shift calculations in both tolerance Stackups.

Datum feature shift shows up in the Tolerance Stackup Report in Fig. 9.50 three times: once for the switch carrier and twice for the back panel.

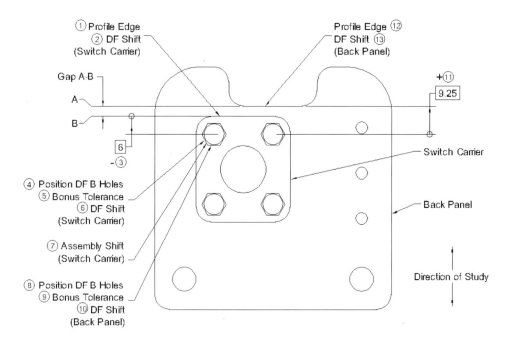

Chain of Dimensions and Tolerances

**FIGURE 9.52**  Simultaneous requirements-Tolerance Stackup sketch for Figs. 9.50 and 9.51.

Simultaneous requirements do not affect the switch carrier's contribution to the tolerance Stackup because only one of the switch carrier's tolerances in the tolerance Stackup is related to a datum reference frame that may have datum feature shift. Simultaneous requirements do affect the back panel's contribution to the tolerance Stackup because two of the back panel's tolerances in the tolerance Stackup are related to the same datum reference frame, which may have datum feature shift (A, B at MMC).

The separate requirements Tolerance Stackup Report is shown in Fig. 9.51. The difference between these Tolerance Stackups is in lines 10 and 13: with simultaneous requirements, datum feature shift is set to zero on both lines because the datum features of size on the back panel are not part of the chain of Dimensions and Tolerances with separate requirements, datum feature shift is added in lines 10 and 13.

The result of the tolerance Stackup shown in Fig. 9.50 shows there is a 0.425 worst-case smallest gap with simultaneous requirements as the default condition for the back panel. The tolerance Stackup Report shown in Fig. 9.51 shows there is 0.125 worst-case overlap with separate requirements as the default condition for the back panel. In these examples switching from simultaneous requirements to separate requirements causes a "no-build" condition; the result indicates that the switch carrier may protrude beyond the back panel and interfere with mating parts.

This is an extreme example; often the difference between a tolerance Stackup done with simultaneous requirements and separate requirements does not lead to a no-build condition. The chain of Dimensions and tolerances followed in these examples is shown in Fig. 9.52.

A final word of caution: Make sure you know whether simultaneous requirements or separate requirements as the default condition is known when performing a tolerance Stackup.

# 10

## Converting Plus/Minus Tolerancing to Positional Tolerancing and Projected Tolerance Zones

Plus/minus location tolerancing can be easily converted into positional tolerancing. The whole idea of *positional tolerancing* is based on the idea that a cylindrical feature of size, such as a hole, should be allowed to vary in location the same amount in any direction. This is true for most applications. If a cylindrical hole can be 1.4 mm diagonally from its nominal location and still function, then it should function when the hole is 1.4 mm from its nominal location in any direction. Hence a cylindrical tolerance zone allows a feature to vary equally in any direction from its nominal or basic location.

Plus/minus location tolerances state how much the location of a feature may vary in a specific direction. In the case of cylindrical features of size, such as holes and studs, the amount of variation (or tolerance) is linked to dimensions in two perpendicular directions, typically the horizontal and vertical directions as shown in Fig. 10.1.

In this example the tolerances are the same in both directions, which can be idealized as a square tolerance zone. The horizontal and the vertical tolerances for the hole are ± 0.5 mm. The hole may be displaced 0.5 mm left or right and 0.5 mm up or down. Using this method of tolerancing, the hole may be displaced a larger amount diagonally at the extremes of its

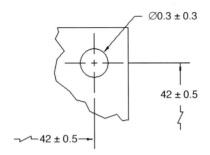

Standard ± Tolerancing
(Creates a rectangular tolerance zone)

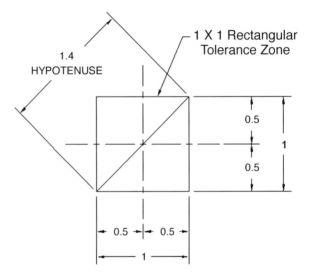

Rectangular Tolerance Zone with Hypotenuse

**FIGURE 10.1** Plus/minus tolerances in both directions. The Rectangular Tolerance Zone allows a total of 1 mm variation vertically and horizontally, but 1.4 mm diagonally, due to the square shape of the zone and the length of the hypotenuse. Plus/minus tolerancing creates a biased tolerance zone, allowing a greater tolerance diagonally than in any other direction.

tolerance zone. This amount may be obtained using the Pythogorean theorem or trigonometry.

Positional tolerancing with a cylindrical tolerance zone assumes the functional requirements of a hole are the same in any direction normal to its axis. The positional tolerance equivalent to an existing plus/minus tolerance is found by circumscribing a circle around the plus/minus tolerance zone (see Fig. 10.2). Note the 57% increase in the area of the tolerance zone. Assuming the part will still function with the larger tolerance zone, this increased tolerance zone may lead to more good parts, less scrap, and hopefully lower part costs.

When a feature has the same ± tolerance values in both perpendicular directions, as in the previous figures, the diameter of the positional tolerance zone can be found by multiplying either plus or minus tolerance by the square root of 2 ( ~1.414). See the following example.

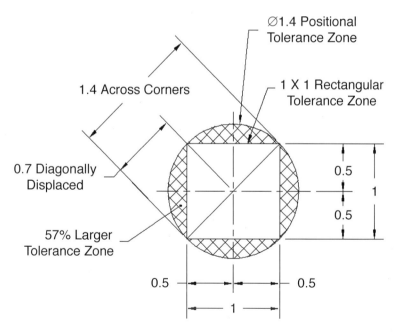

Overlaid Rectangular and Positional Tolerance Zones

**FIGURE 10.2** Positional tolerance zone circumscribed about ± zones.

### The Following Technique Works for Tolerance Values That are the *Same in Both Directions*:

Assume a hole that is toleranced $\pm$ 0.8 horizontally and vertically.

1. Given a $\pm$ tolerance (explicitly stated *tolerance* or *title block tolerance*), e.g., $\pm$ 0.8.
2. Multiply the tolerance by 2 to get the *total linear tolerance*, e.g., $2 \times 0.8 = 1.6$.
3. Multiply the total linear tolerance by $2^{1/2}$ ($\sim$1.414) to get the equivalent *diametral positional tolerance*, e.g., $2^{1/2} \times 1.6 = 2.29$.
4. $\varnothing$2.29 is the *equivalent positional tolerance*.

Two methods for calculating the equivalent cylindrical positional tolerance zone are shown in Fig. 10.3. Both methods show how to calculate the hypotenuse of the half-space triangle within a square $\pm$ tolerance zone.

### The Following Technique Works for *Any* Tolerance Values in Both Directions:

(Tolerance values may be different in the $X$ and $Y$ direction). Assume a hole that is toleranced $\pm$ 0.8 horizontally and $\pm$ 1.2 vertically.

1. Given a $\pm$ tolerance (explicitly stated tolerance or title block tolerance) in one direction, say $X$, e.g., $\pm$ 0.8.
2. Multiply the tolerance by 2 to get the total linear tolerance in the $X$ direction, e.g., $2 \times 0.8 = 1.6$.
3. Given a $\pm$ tolerance (explicitly stated tolerance or title block tolerance) in the other direction, say $Y$, e.g., $\pm$ 1.2.
4. Multiply the tolerance by 2 to get the total linear tolerance in the $Y$ direction, e.g., $2 \times 1.2 = 2.4$.
5. Use the Pythagorean theorem to determine the equivalent diametral positional tolerance, e.g., $(1.6^2 + 2.4^2)^{1/2} = (2.56 + 5.76)^{1/2} = 2.88$.
6. $\varnothing$2.88 is the equivalent positional tolerance.

The techniques presented above are for converting $\pm$ location tolerance zones to cylindrical positional tolerance zones. In cases where the $\pm$ location tolerances are the same in both directions, it makes the most sense to use the above approach, as a cylindrical feature should be allowed to vary the same amount in all directions and still function.

Plus/minus location tolerance zones can also be converted to cylindrical positional tolerance zones by inscribing a circle within the tolerance zone. Which method is used depends on the functional requirements of the design.

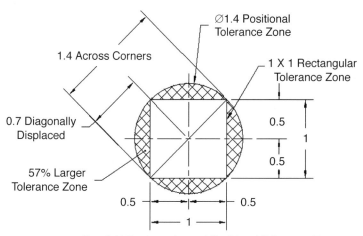

Overlaid Rectangular and Positional Tolerance Zones

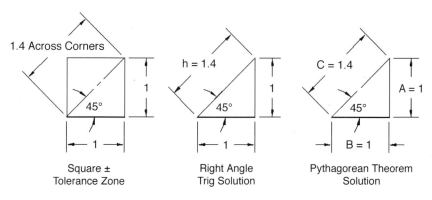

| Square ± Tolerance Zone | Right Angle Trig Solution | Pythagorean Theorem Solution |

Trigonometric Means of Converting ± Tolerance to Equivalent Positional Tolerance:

The hypotenuse of the 45° right triangle represents the worst case displacement within the ± tolerance zone. Using the above example:

Right Angle Trigonometric Solution:

$\sin 45° = $ opposite/hypotenuse $\Rightarrow \sin 45° = 1/h \Rightarrow h = 1/\sin 45° \Rightarrow h \approx 1/0.7071$
$\Rightarrow h \approx 1.414$

Pythagorean Theorem Solution:

$A^2 + B^2 = C^2 \Rightarrow 1^2 + 1^2 = C^2 \Rightarrow 1 + 1 = C^2 \Rightarrow C^2 = 2 \Rightarrow C = \sqrt{2} \Rightarrow C \approx 1.414$

1.414 is the maximum total displacement possible within the ± tolerance zone.
Positional Tolerance $= \varnothing 1.4$

**FIGURE 10.3** Plus/minus and positional tolerance zones with math.

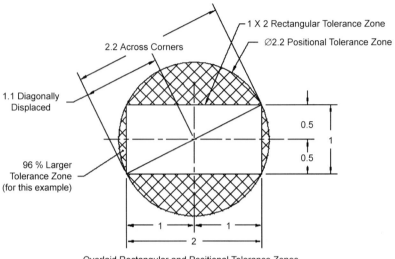

Overlaid Rectangular and Positional Tolerance Zones

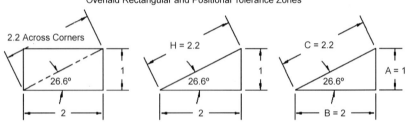

Rectangular ± Tolerance Zone     Right Angle Trig Solution     Pythagorean Theorem Solution

Trigonometric Means of Converting ± Tolerance to Equivalent Positional Tolerance:

The hypotenuse of the triangle shown represents the worst case displacement within the ± tolerance zone. Using the above example:

Trigonometric Solution:

Find the Angle:
$\tan \alpha = $ opposite / adjacent $\Rightarrow \tan^{-1}$ opp/adj $= \alpha \Rightarrow \alpha = 1/2 \Rightarrow \alpha = 0.5 \Rightarrow \alpha = 26.6°$

Find the length of the hypotenuse:

$\sin 26.6° = $ opposite/hypotenuse $\Rightarrow \sin 26.6° = 1/h \Rightarrow h = 1/\sin 26.6° \Rightarrow h = 1/0.4472$
$h \approx 2.236$

Pythagorean Theorem Solution:

$A^2 + B^2 = C^2 \Rightarrow 1^2 + 2^2 = C^2 \Rightarrow 1 + 4 = C^2 \Rightarrow C^2 = 5 \Rightarrow C = \sqrt{5} \Rightarrow C \approx 2.236$

Ø2.2 is the maximum total displacement possible within the ± tolerance zone. Ø2.2 is chosen as the diameter of the equivalent Positional Tolerance zone.

**FIGURE 10.4**   Rectangular ± and positional tolerance zones with math.

The circumscribed circle method is more commonly applied, but it must be confirmed that the resulting tolerance zone is functionally acceptable, as it allows larger deviation in the direction of the original dimensions and ± tolerances.

The circumscribed method is usually used because it is assumed that the ± tolerances merely represent the way things have been done for so many years, and the larger displacement possible diagonally within the ± tolerance zone is assumed to be functionally acceptable in any direction from nominal. The inscribed method is usually used because the ± tolerances are assumed to be carefully thought out and represent the maximum displacement possible within functional limits. The larger displacement possible diagonally within the ± tolerance zone is assumed to exceed the acceptable displacement allowed in the specified directions from nominal.

In either case, fixed-fastener calculations, floating-fastener calculations, or a more complex Tolerance Stackup must be performed to verify that the converted tolerances are acceptable within functional limits.

It would not be a good idea to convert the rectangular tolerance zone above to a cylindrical tolerance zone if the original ± tolerances were functionally necessary. An example is where the location of a feature is more important in one direction than in the perpendicular direction. If the original rectangular tolerance zone was required, then a rectangular or bidirectional positional tolerance zone should be specified. The techniques for specifying a bidirectional positional tolerance zone may be found in section 5.9 and seen in Fig. 5-41 of ASME Y14.5M–1994. The technique to convert from a rectangular to a cylindrical tolerance zone is shown in Fig. 10.4.

## PROJECTED TOLERANCE ZONES

Whenever a fastener mates with a threaded hole or a pin is pressed into a hole, or even when there is a very close fit, it is a good idea to specify a projected tolerance zone. Projected tolerance zones address the geometric effects of tilting within the tolerance zone. The projected tolerance zone extends from the mating surface through the maximum thickness of the mating part(s). In some cases, the tolerance zone must extend beyond the mating part to address assembly issues or where more than one part mates with the same fastener.

Projected tolerance zones are specified in a feature control frame and may be used with positional tolerances and orientation tolerances. The projection symbol follows the material condition modifier in the tolerance portion of the feature control frame. The distance of projection may be specified in the feature control frame right after the projection symbol, or the distance of projection may be represented on the drawing by a heavy chain line from the appropriate surface. The length of the chain line must be dimensioned.

When specifying a projected tolerance zone for a blind hole, it is obvious which surface is the interface. The tolerance zone projects from the surface the hole penetrates.

When specifying a projected tolerance zone for a through hole, it is not clear which surface is the interface; thus it is not clear from which end of the hole the tolerance zone must project. In most through hole applications, the direction of projection must be shown on the drawing. Everyone that reads the drawing must understand where the tolerance zone is. For example, if the direction of projection is not specified, the inspector may guess the tolerance zone projects in one direction, but the part may mate on the opposite surface. In this example, the geometric effect of tilting ends up being far worse than if a projected tolerance hadn't even been specified!

Many companies do not specify projected tolerance zones, primarily because of the reluctance of their manufacturing and inspection personnel. When using traditional inspection methods, a threaded plug gage must be used to "project" the tolerance zone of a threaded hole to the specified distance. The additional time and labor associated with this extra step leads many personnel to dislike the requirement. However, if a projected tolerance zone is not specified where it is needed, the results can be interference or unexpected radial variation. Using a CMM with a projected tolerance zone only requires the use of a different algorithm and should not affect the time or labor required to inspect features toleranced with projected tolerance zones. In cases where it is politically just too difficult get manufacturing or inspection to agree to the use of projected tolerance zones, there are alternatives where parts can still be properly toleranced.

Figures 10.5 and 10.6 show how a projected tolerance zone is specified and what it means. The Tolerance Stackup implications of not specifying projected tolerance zones where needed are many. As described in chapter 18, the fixed-fastener formula is based on having projected tolerance zones specified on the position of the tapped or pressed-fit holes in the Tolerance Stackup. If a projected tolerance zone is not specified where needed, the formula presented and solved in Fig. 10.7 must be used to determine the effects and necessary values. If an orientation tolerance such as perpendicularity is specified instead of a projected tolerance zone, the formula presented and solved in Figs. 10.8 and 10.9 must be used to determine the effects and necessary values.

The fixed- and floating-fastener worksheets available from Advanced Dimensional Management™ are excellent semiautomated tools for solving these problems and comparing the effects of specifying and not specifying projected tolerance zones.

THIS ON THE DRAWING:

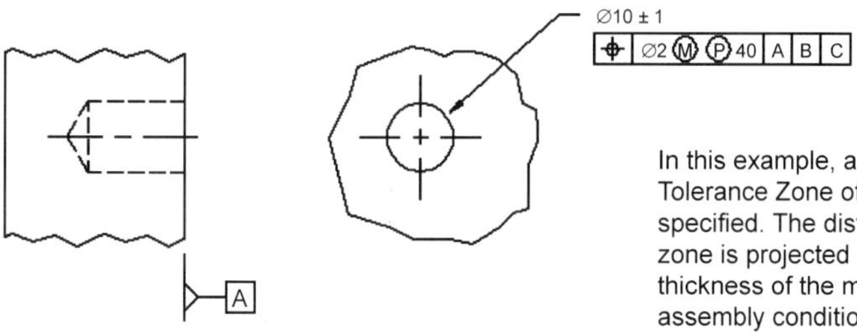

⌀10 ± 1

⊕ | ⌀2 Ⓜ Ⓟ 40 | A | B | C

In this example, a Projected Tolerance Zone of 40mm has been specified. The distance the tolerance zone is projected depends on the thickness of the mating parts and assembly conditions.

MEANS THIS:

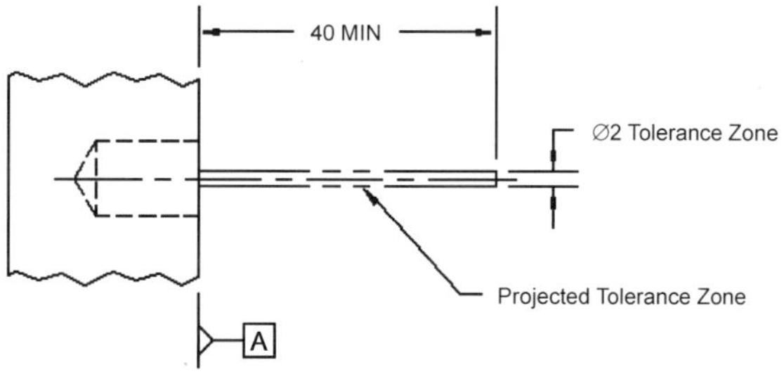

40 MIN

⌀2 Tolerance Zone

Projected Tolerance Zone

A

When a Projected Tolerance Zone is Specified, the tolerance zone lies entirely outside the part. The tolerance zone extends the specified distance from the surface the hole penetrates.

AND ALLOWS THIS:

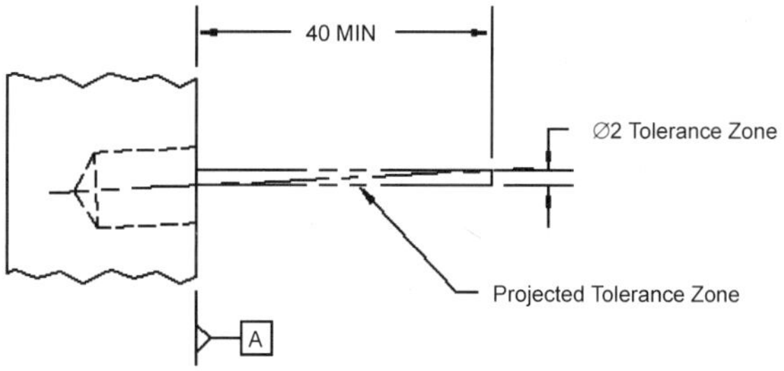

40 MIN

⌀2 Tolerance Zone

Projected Tolerance Zone

A

With a projected tolerance zone specified, the axis of the hole may shift and/or tilt within the projected zone. The axis of the hole is not constrained within the part itself. Notice that in this example the hole falls outside of the zone if the zone was inside the part—this is acceptable, as the axis is only required to be within the projected zone.

**FIGURE 10.5** Projected tolerance zones: Specification and meaning.

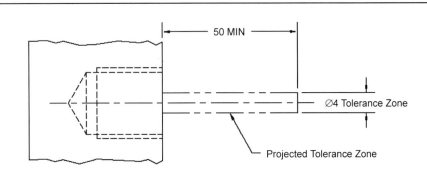

A Projected Tolerance Zone is Specified the tolerance zone lies entirely outside the part. The tolerance zone extends the specified distance from the surface the hole penetrates.

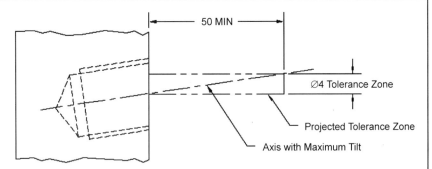

The axis of the hole may tilt and/or shift within the projected tolerance zone. This example shows maximum tilt within the projected zone. Notice where the axis of the tapped hole is within the part.

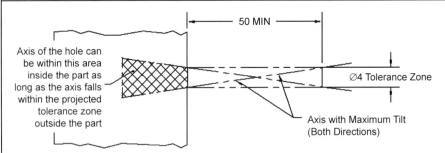

When a projected tolerance zone is specified, the location and orientation of the axis of the hole inside the part is limited only by the projected tolerance zone. The area the axis can occupy inside the part is shown above. However, the axis cannot tilt so much inside the part that it violates the projected tolerance zone outside the part.

**FIGURE 10.6**   Projected tolerance zones: Inside and outside the part.

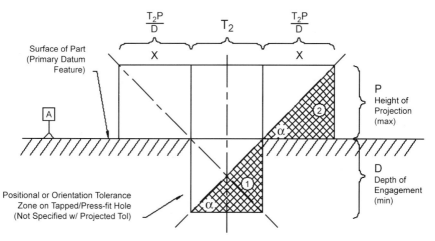

## Formula B5: When Projected Tolerance Zone is Not Specified: $H = F + T_1 + T_2(1 + \frac{2P}{D})$

Where:

| | |
|---|---|
| H = Smallest Clearance Hole Diameter (MMC) | $T_1$ = Tolerance Zone Diameter for Clearance Hole @ MMC |
| F = Largest Fastener Diameter (MMC) | $T_2$ = Tolerance Zone Diameter for Tapped/Press-fit Hole @ MMC |
| P = Maximum Projection of Fastener | D = Minimum Depth of Engagement in Threaded or Press-fil Hole |
| | $\alpha$ = Angle of Inclination (Tilt) Allowed by $T_2$ |

Proof:

Given the two triangles above representing the tolerance zone diameters:

Set up equality based on $\alpha = \alpha$ (same angle). Prove $X = \frac{T_2 P}{D}$:

① $\tan \alpha = \frac{D}{T_2}$  ② $\tan \alpha = \frac{P}{X}$

Substitute D/T2 for $\tan \alpha$ in Equation ②: $\frac{D}{T_2} = \frac{P}{X}$  Solve for X: $X = \frac{T_2 P}{D}$

As can be seen in the above figure, the Total Projected Tolerance Zone width = $X + T_2 + X = T_2 + 2X$

(X must be added twice because the axis of the tapped hole may tilt in any direction (left or right in the figure))

Substitute for X in the Total Projected Tolerance Zone Width: $T_2 + 2X = T_2 + 2\frac{T_2 P}{D} = T_2(1 + \frac{2P}{D})$

This tolerance zone value can be substituted into the Fixed Fastener Formula $H = F + T_1 + T_2$ to represent the Equivalent Tolerance Contributed by the Fixed Fastener.

Substitute the equivalent projected tolerance $T_2(1 + \frac{2P}{D})$ for $T_2$: $H = F + T_1 + T_2(1 + \frac{2P}{D})$

## Provision for Out-of-Squareness when Projected Tolerance Zone is Not Used

**FIGURE 10.7**  Projected tolerance zones: Formula B5.

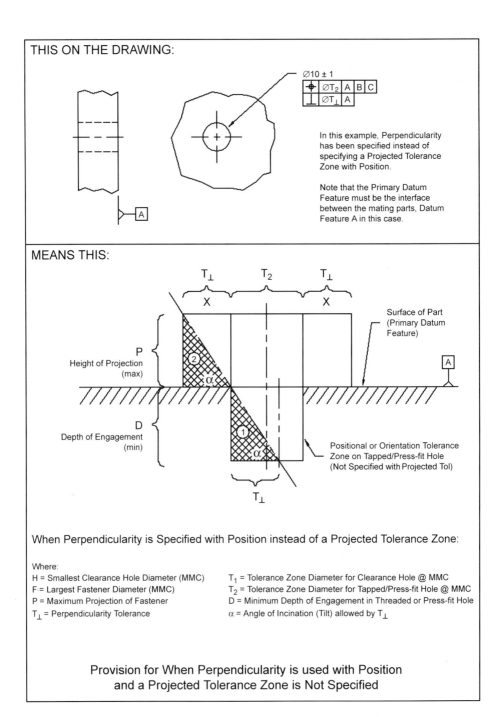

**FIGURE 10.8** Projected tolerance zones: Formula B5, modified, part 1.

## MEANS THIS (cont.):

Assumptions:

- Primary Datum Feature is the interface between the mating parts.

- $T_2$ and $T_\perp$ applied RFS.
  (Bonus tolerance associated with MMC and LMC present a problem. Not addressed here.)

Solution Requires a Modified Version of Formula B5: $H = F + T_1 + T_2 + \dfrac{2T_\perp P}{D}$

Per the previous figure, substitute X for the additional tolerance allowed by $T_\perp$:

$H = F + T_1 + T_2 + 2X$

Proof:

Given the two triangles above representing the tolerance zone diameters:

Set up equality based on $\alpha = \alpha$ (same angle). Prove $X = \dfrac{T_\perp P}{D}$ :

① $\tan \alpha = \dfrac{D}{T_\perp}$ ② $\tan \alpha = \dfrac{P}{X}$

Substitute $\dfrac{D}{T_\perp}$ for $\tan \alpha$ in Equation ② : $\dfrac{D}{T_\perp} = \dfrac{P}{X}$ Solve for X: $X = \dfrac{PT_\perp}{D}$

Substitute for X in modified formula B5 above: $H = F + T_1 + T_2 + 2X$

$H = F + T_1 + T_2 + \dfrac{2PT_\perp}{D}$

A comment about using Formula B5 and Formula B5 - Modified with MMC or LMC:

These formulas work when the tolerances are specified RFS. As can be seen by the figures on these pages, the reason these formulas are needed is to address the possible tilting of the axis within the tolerance zone, and the effect of that tilting within the mating part. If MMC or LMC were specified, the bonus tolerance would allow additional tilting of the axis due to the increase in size of the tolerance zone. The geometric relationship between the parts, the depth of engagement and the extent of projection creates a situation where this additional tilting is not accounted for in the formulas. The tilting allowed by bonus tolerance leads to potential interference as the toleranced feature deviates from its specified material condition.

### Provision for When Perpendicularity is used with Position and a Projected Tolerance Zone is Not Specified (cont'd)

**FIGURE 10.9** Projected tolerance zones: Formula B5, modified, part 2.

# 11

# Diametral and Radial Tolerance Stackups

Often it is necessary to calculate the maximum coaxial error or eccentricity between nominally coaxial features. When dealing with three-dimensional parts and assemblies, we are most often concerned about the coaxial error or coaxiality between related parts or features.

Rarely are we truly interested in the eccentricity (or conversely, concentricity) between parts and part features. Except in the case of spheres, concentricity and eccentricity are two-dimensional geometric conditions. Measurements are taken at cross sections of a part feature to obtain center points, and the variation of those center points from a Datum axis is measured. Usually these two-dimensional measurements and geometric conditions are of little interest to the design engineer when dealing with parts in the three-dimensional world. However, due to its colloquial use and people's comfort with the term *eccentricity*, it is incorrectly used in many three-dimensional applications. This text shall discuss coaxial error or variation, as that is typically of greater concern to the engineer due to its functional implications.

Examples of where the coaxial error between features may be important are the relationship between the head of a bolt and the screw thread, between diameters on a turned shaft, such as the ends of a camshaft or a crankshaft, between a hole and its counterbore, between an o-ring groove and a shaft, between a ring groove and the OD of a piston, or between the stepped-down diameters of a flow nozzle, to name a few.

It is absolutely necessary to use GD&T to relate coaxial features. In the past, features were drawn coaxially on the drawing, and only their sizes were toleranced (see Fig. 11.1). In this example, the head and the body of the pin are shown coaxially, sharing common centerlines. The sizes are toleranced, but the features are not located to one another. How far apart may the axes of these features be? There is no answer to this rhetorical question. A common misconception is since they are turned on a lathe or screw machine, they will be coaxial, because of how they are made! Although this is a nice thought, it is not legally defendable, the allowable coaxial error can not be quantified, and the parts must be accepted even if they are received with a large coaxial error between the features. As discussed in the beginning of this text, drawings toleranced in this manner force the inspector and manufacturer to guess how closely the features must be related. The allowable misalignment between the features must be specified using a tolerance such as position or runout.

Given that features are properly related using GD&T, calculating their maximum coaxial error is a straightforward process. One or more coaxial features are selected as datum features, and other nominally coaxial features are related to the associated datum reference frame. Diametral dimensions and tolerances are used in the calculations and converted into radial ± values.

Nominally coaxial features are often related with positional tolerances. Typically, these are specified to apply at the maximum material condition of the features. The amount the axis of a controlled feature may vary from the axis of a datum feature is unambiguously specified in a feature control frame.

Although a positional tolerance may be specified at the maximum material condition, the variation possible at both the maximum and least material conditions must be considered in many applications. The designer must verify that no detrimental effects result when the features are produced

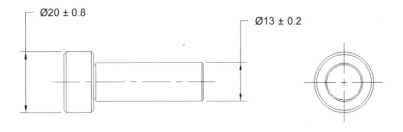

COAXIALITY BETWEEN THESE DIAMETERS IS NOT SPECIFIED.

**FIGURE 11.1**   Coaxial pin without GD&T.

at their least material conditions, which leads to the worst-case possible coaxial error. Remember, MMC is usually specified for reasons of fit, not reasons of alignment.

As stated above, the allowable coaxial error between these diameters may be properly defined using positional tolerancing, as shown in Fig. 11.2. GD&T offers several methods to relate the features, such as position, profile, runout, symmetry, and concentricity. Only position will be discussed here. It should be noted that, strictly speaking, the methods presented in this chapter calculate the coaxial error between the diameters, not their allowable coaxial variation. Coaxial error indicates that the centerlines of the features may be misaligned due to variables relating to the toleranced feature and the datum feature, whereas the coaxial variation is a function of only the size and location of the toleranced feature.

## COAXIAL ERROR AND POSITIONAL TOLERANCING

Figures 11.3 through 11.10 depict nominally coaxial features related using positional tolerancing and their resulting maximum possible coaxial error. One feature is specified as a datum feature, and the other feature is positionally toleranced to the datum axis derived from the datum feature. This text presents four common positional tolerancing applications for this type of part, in which the datum feature, the toleranced feature, or both are specified at MMC or RFS. Figures 11.3, 11.5, 11.7, and 11.9 represent the part as toleranced on the drawing. Figures 11.4, 11.6, 11.8, and 11.10 depict the maximum coaxial error possible for Figures 11.3, 11.5, 11.7, and 11.9, respectively. All these figures depict the same part with the same tolerance

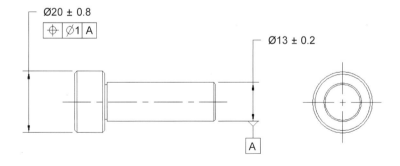

THE COAXIALITY BETWEEN THESE DIAMETERS IS SPECIFIED.

**FIGURE 11.2** Coaxial pin with GD&T.

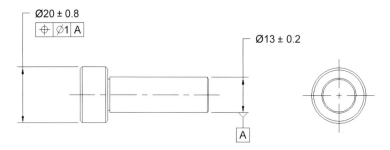

**FIGURE 11.3**  Coaxial pin with GD&T.
Feature Reference: RFS, Datum Feature Reference: RFS.

values—the only differences between the figures are the material condition modifiers specified in each example.

The variables and formula for calculating the maximum possible coaxial error follow:

Variables:

$DFS_A$ = $\varnothing$ of datum feature simulator A.

$DF_A$  = worst-case $\varnothing$ of datum feature A (LMC).

$PT_L$  = toleranced feature's positional tolerance zone $\varnothing$ when it's produced @ LMC.

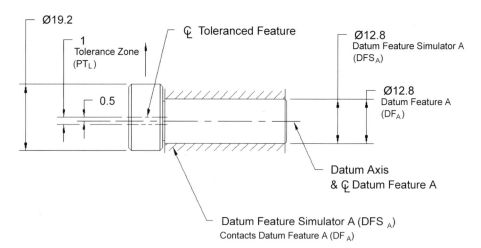

**FIGURE 11.4**  Maximum coaxial error for Pin in Figure 12.3.

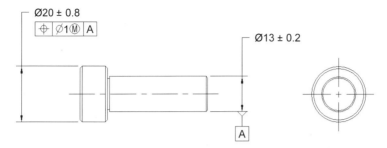

**FIGURE 11.5** Coaxial pin with GD&T.
Feature Reference: MMC, Datum Feature Reference: RFS.

Formula:

$$\text{Max coaxial error} = \frac{\text{DFS}_A - \text{DF}_A + \text{PT}_L}{2}$$

The above formula works where datum features are referenced RFS or MMC, and/or positional tolerances are specified to apply RFS or MMC.

The expression $\text{DFS}_A - \text{DF}_A$ in the numerator of the formula represents the possible datum feature shift. It represents the worst-case difference between the sizes of the datum feature simulator and the as-produced datum feature. Datum feature shift is discussed in Chapter 9. Where datum features are referenced at their LMC and/or positional tolerances are specified to apply at the feature's LMC, the same approach may be used. In this case, the worst-case coaxial error is possible when the features are produced at their MMC. Therefore, the MMC sizes would be used in the formula instead of the LMC sizes.

The feature control frame in Fig. 11.3 references the datum feature regardless of feature size (RFS), and specifies that the positional tolerance applies to the toleranced feature regardless of feature size (RFS).

The datum feature is simulated by its actual mating envelope at its actual mating size, meaning the simulator is a perfect cylinder that contacts the datum feature and is considered to be the same size as the datum feature (no datum feature shift).

As shown in Fig. 11.4, the datum feature simulator and the datum feature are in contact, and their axes are considered to be coaxial. Their relationship does not contribute to the coaxial error possible between the features (no datum feature shift).

The tolerance zone of the $\varnothing 20 \pm 0.8$ feature is related to the axis of the datum feature simulator, which is the datum axis. The tolerance zone is specified as $\varnothing 1$ regardless of feature size. It does not increase in size as the feature size approaches LMC, and it remains $\varnothing 1$ regardless of the as-produced size of the feature. Therefore, the size of the tolerance zone when the feature is produced at its LMC is still $\varnothing 1$.

$$\text{Max coaxial error} = \frac{\text{DFS}_A - \text{DF}_A + \text{PT}_L}{2} = \frac{12.8 - 12.8 + 1}{2}$$

$$= \frac{1}{2} = 0.5$$

Note that the LMC size of 12.8 was used for the as-produced size of datum feature A in the previous example. Because datum feature A is referenced RFS, it must be simulated by a simulator that is the same size as the datum feature. Since the simulator and the datum feature are the same size, their net contribution to the total coaxial error is zero (no datum feature shift). The formula yields the same result of 0.5 maximum coaxial error as if the datum feature and the datum feature simulator were not included in the formula, and the formula could be reduced to PT/2.

The feature control frame in Fig. 11.5 references the datum feature regardless of feature size (RFS) and specifies that the positional tolerance applies at the toleranced feature's maximum material condition (MMC) size.

The datum feature is simulated by its actual mating envelope at its actual mating size, meaning the simulator is a perfect cylinder that contacts the datum feature and is considered to be the same size as the datum feature (no datum feature shift).

As shown in Fig. 11.6, the datum feature simulator and the datum feature are in contact, and their axes are considered to be coaxial. Their relationship does not contribute to the coaxial error possible between the features (no datum feature shift).

The tolerance zone of the $\varnothing 20 \pm 0.8$ feature is related to the axis of the datum feature simulator, which is the datum axis. The tolerance zone is $\varnothing 1$ when the feature is produced at its MMC size and increases to a maximum of $\varnothing 2.6$ when the feature is produced at its LMC size. This is the size of the tolerance zone to use in the formula.

$$\text{Max coaxial error} = \frac{\text{DFS}_A - \text{DF}_A + \text{PT}_L}{2} = \frac{12.8 - 12.8 + 2.6}{2}$$

$$= \frac{2.6}{2} = 1.3$$

Note that the LMC size of 12.8 was used for the as-produced size of datum feature A in the previous example. Because datum feature A is referenced RFS, it must be simulated by a simulator that is the same size as the datum feature. Since the simulator and the datum feature are the same

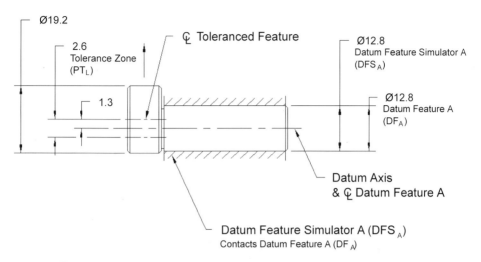

**FIGURE 11.6** Maximum coaxial error for Pin in Figure 12.5: MMC-RFS.

size, their net contribution to the total coaxial error is zero (no datum feature shift). The formula yields the same result of 1.3 maximum coaxial error as if the datum feature and the datum feature simulator were not included in the formula, and the formula could be reduced to $PT_L/2$.

The feature control frame in Fig. 11.7 references the datum feature at its maximum material condition (MMC) size, and specifies that the positional tolerance applies to the toleranced feature regardless of feature size (RFS).

As specified in Fig. 11.7, the MMC size of the datum feature is $\varnothing13.2$. Even if the feature is produced at its LMC size of $\varnothing12.8$, it is still simulated by a $\varnothing13.2$ datum feature simulator. This contributes the first portion of the maximum possible coaxial error. As shown in Fig. 11.8, datum feature A may

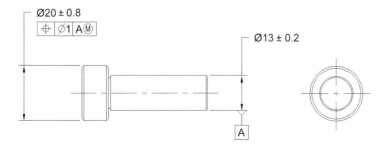

**FIGURE 11.7** Coaxial pin with GD&T.
Feature Reference: RFS, Datum Feature Reference: MMC.

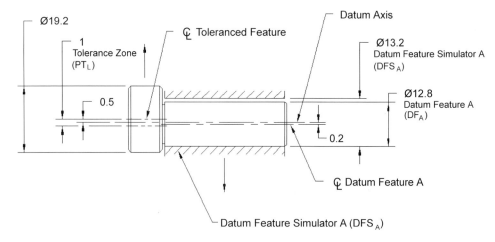

**FIGURE 11.8** Maximum coaxial error for Pin in Figure 10.7: RFS-MMC.

rest at the bottom of datum feature simulator A, their axes a maximum of 0.2 apart when the datum feature is produced at its LMC size. This is the worst-case datum feature shift.

The tolerance zone of the ∅20 ± 0.8 feature is related to the axis of the datum feature simulator, which is the datum axis. The tolerance zone is specified as ∅1 regardless of feature size. It does not increase in size as the feature size approaches LMC, and remains ∅1 regardless of the as-produced size of the feature. Therefore, the size of the tolerance zone when the feature is produced at its LMC is still ∅1.

$$\text{Max coaxial error} = \frac{\text{DFS}_A - \text{DF}_A + \text{PT}_L}{2} = \frac{13.2 - 12.8 + 1}{2}$$

$$= \frac{1.4}{2} = 0.7$$

The feature control frame in Fig. 11.9 references the datum feature at its maximum material condition (MMC) size and specifies that the positional tolerance applies at the feature's maximum material condition (MMC) size.

As specified in Fig. 11.9, the MMC size of the datum feature is ∅13.2. Even if the feature is produced at its LMC size of ∅12.8, it is still simulated by a ∅13.2 datum feature simulator. This contributes the first portion of the maximum possible coaxial error. As shown in Fig. 11.10, datum feature A may rest at the bottom of datum feature simulator A, their axes a maximum of 0.2 apart when the datum feature is produced at its LMC size. This is the worst-case datum feature shift.

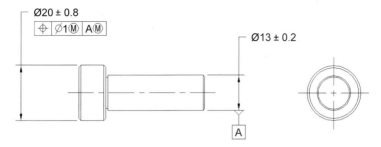

**FIGURE 11.9**  Coaxial pin with GD&T.
Feature Reference: MMC, Datum Feature Reference: MMC.

The tolerance zone of the ∅20 ± 0.8 feature is related to the axis of the datum feature simulator, which is the datum axis—the toleranced feature is not directly related to the datum feature's axis. The tolerance zone is ∅1 when the feature is produced at its MMC size and increases to a maximum of ∅2.6 when the feature is produced at its LMC size. This is the second portion of maximum possible coaxial error.

This example illustrates that the worst-case coaxial error for this part is possible when the datum feature reference and the positional tolerance are specified at MMC and the datum feature and the toleranced feature are produced at their LMC sizes (see Fig. 11.10).

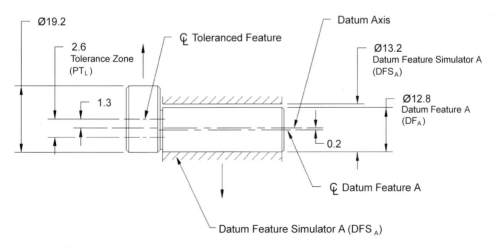

**FIGURE 11.10**  Maximum coaxial error for Pin in Figure 12.9: MMC-MMC.

$$\text{Max coaxial error} = \frac{\text{DFS}_A - \text{DF}_A + \text{PT}_L}{2} = \frac{13.2 - 12.8 + 2.6}{2}$$

$$= \frac{3}{2} = 1.5 \ (\textit{Worst--case!})$$

This extreme coaxial error (or misalignment) is not justification to throw MMC out the window and never use it again for these applications. MMC is typically specified for other reasons, such as fit, guaranteeing passage of a fastener through a hole or guaranteeing the head of the pin in the figures will fit into a counterbored hole. Care must be taken when assigning material condition modifiers to features and datum features. Both the fit and the alignment aspects must be considered as discussed in Chapter 12.

Unfortunately, MMC is often specified as somewhat of a default, without understanding the implications of what happens when the features are produced at the opposite material condition, LMC.

In these examples, a simple part was used to facilitate easier presentation and understanding of the method. Chances are that if the mating part was toleranced to work with this part, the MMC modifiers would make no difference, if the only consideration was fit.

# 12

## Specifying Material Condition Modifiers and Their Effect on Tolerance Stackups

When specifying certain geometric tolerances, the designer must determine which of the three material condition modifiers (RFS, MMC or LMC) is the correct choice for the given application. See Fig. 12.1. This is true when considering which material condition modifier should be associated with the tolerance zone and which material condition modifier(s) should be associated with the datum feature references in the feature control frame.

There seems to be a prevailing point of view that MMC is the best choice in all but a few applications. The purpose of this chapter is to clarify the criteria for selecting the correct material condition modifier for the application.

Selecting a material condition modifier is not an arbitrary decision—it is a functional decision. Other factors may influence the decision, such as fixturing, inspection practices, or company preference, but first and foremost it is a functional decision.

The effect of the material condition modifier on the function of the feature becomes crystal clear when performing a Tolerance Stackup. When performing a Tolerance Stackup, the Tolerance Analyst breaks out the amount of variation contributed by each tolerance and each material condition modifier onto separate lines in the Tolerance Stackup report. In this format, the effect of the material condition modifier is very clear. Seeing that an MMC

| (M) | MAXIMUM MATERIAL CONDITION |
| (L) | LEAST MATERIAL CONDITION |
| (S) | REGARDLESS OF FEATURE SIZE |

**FIGURE 12.1** Material condition modifiers. The symbol for RFS is obsolete, but may be reinvoked from ANSI Y14.5M-1982-see ASME Y14.5M-1994 para. 2.8(b.)

modifier adds 0.2 mm to the total variation may lead the designer to change the MMC specification to an RFS specification. This is especially true when the 0.2 mm contributes to a potential interference between important features. The same logic holds true for an LMC specification, where its additional tolerance adds unncessarily to the total variation.

So what is the difference between RFS, MMC and LMC?

RFS means *regardless of feature size*. It is the default condition on drawings prepared using ASME Y14.5M-1994, and it is not necessary to use a symbol to specify RFS. If desired, the symbol from ANSI Y14.5M-1982 may be used in certain contexts to make it clear that RFS applies. In terms of the tolerance zone, RFS means that the tolerance zone remains constant in size. The zone is the same size regardless of the size at which the toleranced feature is produced.

MMC means *maximum material condition*. To specify MMC on drawings prepared using ASME Y14.5M-1994, the MMC symbol must be specified. In terms of the tolerance zone, MMC means the tolerance zone increases in size proportionally with the feature. It increases directly proportional with the size of an internal feature, such as a hole, and indirectly proportional with the size of an external feature, such as a pin.

The tolerance zone inreases in size equal to the amount the size of an internal feature has increased from its maximum material condition. The maximum material condition of an internal feature is its smallest size. The bigger the hole, the bigger the tolerance zone.

The tolerance zone increases in size equal to the amount the size of an external feature has decreased from its maximum material condition. The maximum material condition of an external feature is its largest size. The smaller the pin, the bigger the tolerance zone.

LMC means Least Material Condition. To specify LMC on drawings prepared using ASME Y14.5M-1994, the LMC symbol must be specified. In terms of the tolerance zone, LMC means the tolerance zone increases in size proportionally with the feature. It increases indirectly proportional with the size of an internal feature, such as a hole, and directly proportional with the size of an external feature, such as a pin.

The tolerance zone increases in size equal to the amount the size of an internal feature has decreased from its least material condition. The least material condition of an internal feature is its largest size. The smaller the hole, the bigger the tolerance zone.

The tolerance zone increases in size equal to the amount the size of an external feature has increased from its least material condition. The least material condition of an external feature is its smallest size. The bigger the pin, the bigger the tolerance zone.

This increase in size of the tolerance zone is commonly referred to as *bonus tolerance*, and is viewed by many to be "extra tolerance for free." The point of this chapter is that depending on the situation, it may actually be extra tolerance for free, or as is shown in many Tolerance Stackups, it may just be a source of additional unwanted variation and detrimental to function. The designer must determine which of these categories the bonus tolerance falls into and select the correct material condition modifier accordingly.

RFS can be viewed as a subset of LMC or MMC. If a positional tolerance with a 2-mm-diameter tolerance zone RFS is applied to a hole, the tolerance zone remains 2 mm regardless of the size of the hole. If a positional tolerance with a 2-mm-diameter tolerance zone at MMC is applied to the same hole, the tolerance zone starts at 2 mm and increases as the size of the hole increases from its smallest size. If a positional tolerance with a 2-mm diameter tolerance zone at LMC is applied to the same hole, the tolerance zone starts at 2 mm and increases as the size of the hole decreases from its largest size.

Both the MMC and LMC tolerance zone start out at 2 mm, just like the RFS tolerance zone. The difference is that the size of the RFS zone remains constant, and the size of the MMC and LMC zones may increase. All three specifications, RFS, MMC, and LMC share the 2-mm zone.

## MATERIAL CONDITION MODIFIER SELECTION CRITERIA

Three factors must be considered when selecting a material condition modifier: fit, edge distance or wall thickness, and alignment. All three factors are functional concerns, as the material condition modifier selected may affect the functional requirements of the feature or related features. In some cases, the functional requirement leads to using a material condition modifier to address one functional concern, such as fit. In other cases, in fact, most cases, several functional concerns must be addressed and balanced, each with competing requirements, such as fit and wall thickness.

## Fit or Clearance

In most mating part applications, fasteners pass through holes in one or both parts. Fit is always a functional concern in these applications; the requirement is that the fastener fits through the hole at worst-case conditions. The worst-case conditions are when the fastener and hole are at their maximum material conditions (largest bolt, smallest hole) and the holes are at their worst case location and/or orientation. Assuming that fit is the only concern, and no other tolerances influence the location of the mating features being considered, the fixed-fastener and floating-fastener formulas may be used to ensure the fasteners will fit in the worst-case condition. Using the formulas, the designer can determine the size tolerance and location tolerance of the features. As will be discussed in Chapter 18, the size of the holes, the size of the fasteners, and the positional tolerance of the clearance holes and tapped holes are functionally interrelated and must be calculated together.

If fit is the *only* functional concern, then MMC is the correct material condition modifier to use. When tolerancing an internal feature, the only requirement is that a pin, fastener, shaft, etc., passes through the toleranced feature. When tolerancing an external feature, the only requirement is that a hole, sleeve, bushing, etc., passes over the toleranced feature. In such cases the bonus tolerance does not impact the function—bonus tolerance never *helps* the design, that is, it is never beneficial. There are merely some cases where it doesn't hurt. As an internal feature gets larger and its tolerance zone increases in size, or as an external feature gets smaller and its tolerance zone increases in size, the features will still fit together.

*Special Case.* In the case where a pin is press-fit into a hole, it must be assumed that the pin follows the hole. In such cases, it is the pin that interfaces with the hole in the mating part. The material condition of the press-fit hole is not of primary functional concern. From a fit-with-the-mating-parts point of view, it is the material condition of the pin that matters, not the hole. In such cases, RFS is the best material condition modifier for the hole, as any bonus tolerance associated with the hole in either direction is nonfunctional and detrimental.

## Maintaining Minimum Wall Thickness or Edge Distance (When at Least One of the Features Is an Internal Feature)

In some cases, the main functional concern is that the minimum edge distance between two features is preserved. This could be the distance between the edges of two adjacent holes, the distance between the edge of a hole and the edge of a part, the distance between the edge of a boss and the edge of a hole in the boss, or maintaining the minimum wall between the inside diameter and outside diameter of a tube. Take the case of a hole with the sole function of reducing the weight of a part. Nothing passes through the hole—there is no fit to consider. However, the hole is fairly close to an edge, and when the hole is largest, a minimum distance from the edge of the part must be maintained to ensure the strength of the part is not compromised.

If the *only* functional concern is that minimum wall thickness or edge distance is maintained at the worst-case condition, then LMC is the correct material condition modifier to use. When tolerancing an internal feature such as a hole or an external feature such as the boss described above, the only requirement is that a minimum edge distance is maintained between the largest hole and the smallest boss. In such cases the bonus tolerance does not impact the function—bonus tolerance never *helps* the design, that is, it is never beneficial. There are merely some cases where it doesn't hurt. As an internal feature gets smaller and its tolerance zone increases in size, or as an external feature gets larger and its tolerance zone increases in size, the minimum edge distance is not compromised.

*Special Case.*   In some cases, where two external features are adjacent to one another, such as two buttons on a keyboard, and the minimum edge distance between them must be maintained, MMC would be the correct material condition modifier to use. This assumes there are no other functional considerations, such as fit. The reason is that when the two adjacent external features are at their largest size, their position must be controlled with the smallest tolerance to ensure the minimum edge distance is not violated. As the adjacent external features decrease in size, the bonus tolerance allows them to be mislocated by the same amount, leaving the same minimum edge distance between them.

## Alignment

In many situations, the axes of features must be as close to their nominal location as possible. This could be the case of an electrical connector with multiple coaxial diameters, where each must make contact with a mating socket feature. Bonus tolerance in either direction, whether the features get larger or smaller, is nonfunctional, as the features must make contact all around regardless of their material condition.

Where alignment is the *only* functional concern, such as where the axes of two or more diameters must be aligned, RFS is the material condition modifier to use.

## Combination of Factors

A common example is where a part is located by one pattern of holes, the part is fastened through those holes first, and another pattern of features on the part must align with the mating part. This is especially true on large or heavy parts, where all fasteners cannot be tightened at the same time. The potential mislocation allowed by the part shifting about the first set of fasteners adds to the positional error of the second set of features. Traditionally, the first set of holes would be toleranced with an MMC modifier, as fit is apparently the primary concern. However, the bonus tolerance associated with the position of the first pattern of holes adds to the potential misalignment of the second pattern of holes and thus is detrimental to function. In situations such as this, RFS is best choice of material condition modifier for the first pattern of holes. Yes, fit is a functional concern for the first pattern of holes, as fasteners must pass through the holes. However, in this case fit is not the only concern. Assembly shift is also a concern. When the holes are largest, the part can shift the maximum amount about the fasteners. If the holes were toleranced using MMC, their positional tolerance zones would be largest when they were produced at their largest size. This would compound the problem of misalignment on the second pattern of holes.

Situations such as this are restricted on both sides by competing requirements. There is a fit requirement, which tells us that if MMC was specified, the bonus tolerance increase associated with larger holes would not affect the fit, but if LMC was specified, the bonus tolerance increase associated with smaller holes would negatively affect the fit; there is an alignment requirement, which tells us that if LMC was specified on the holes, the bonus tolerance increase associated with smaller holes would not affect the alignment of other features, but if MMC was specified, the bonus tolerance increase associated with larger holes would affect the alignment of other features. Truly these are contradictory requirements. Bonus tolerance is detrimental to function in both directions, whether the holes are produced at their MMC or LMC sizes. Consequently, in these cases RFS is the best modifier to use, as there is no bonus tolerance associated with RFS.

Another very common situation is where fit and wall thickness are both functionally important. Fit leads us to MMC as the correct choice; wall thickness leads us to LMC as the correct choice. Again, RFS is the best choice for such situations, as the MMC and LMC bonus tolerances are functionally detrimental: These are problems with boundary conditions on both sides. An increase in the size of the tolerance zone is functionally detrimental at both extremes.

These situations are very common and very often overlooked. The knee-jerk reaction that MMC will save the world with its "extra tolerance for free" leads many to make the mistake of using MMC where it is not the best choice. You must think about the application carefully, consider all the functional ramifications that are affected by the feature being toleranced, and decide which material condition modifier is best for your application.

As stated earlier, these issues become crystal clear when they are represented in a Tolerance Stackup. The effect of bonus tolerance is broken out as a separate line item in the Tolerance Stackup report and its contribution to the considered dimension is easily seen and quantified.

That said, MMC or LMC can still be used in situations where the apparent choice is another material condition modifier. A Tolerance Stackup must be done to quantify the effect of the bonus tolerance, and it must be determined whether the result is functionally acceptable. Other reasons, such as inspection methods, gaging, ease of assembly, or assembly methods can also affect the choice of the correct material condition modifier. However, the primary concern is always function.

# 13

## The Tolerance Stackup Sketch

The importance of creating a sketch of the parts and the chain of Dimensions and Tolerances that make up the Tolerance Stackup cannot be overstated. Creating the sketch is perhaps *the most important* event that must occur when performing a Tolerance Stackup. Experience has shown that the sketch helps the Tolerance Analyst visualize the chain of Dimensions and Tolerances, and it helps others understand the Tolerance Stackup after it is complete. Creating the Tolerance Stackup sketch should be the first step when starting a Tolerance Stackup. The sketch should always be done prior to filling out the Tolerance Stackup report form.

Visualizing the chain of Dimensions and Tolerances that makes up the Tolerance Stackup can be very difficult. I have found that a sketch was essential for catching all the contributing dimensions and tolerances for possibly 95% of the Tolerance Stackups I have attempted to solve. Many times I was in a hurry and started a Tolerance Stackup without a sketch, only to find later that several dimensions and tolerances were missed. I stopped, took my time, created a Tolerance Stackup sketch, and the missing dimensions and tolerances were obvious.

Three things are needed to create the Tolerance Stackup sketch:

- Detail drawings of all the manufactured components in the Tolerance Stackup
- Detail sheets and related dimension and tolerance information for catalog items
- An assembly drawing, and possibly a model of the assembly or the actual assembly.

The Tolerance Analyst must also have a clear understanding of the assembly process, preferably obtained from formally documented assembly procedures. As stated repeatedly in the text, the assembly sequence can have a profound effect on the total variation encountered at assembly. In cases where the assembly process is unknown, the author suggests taking a conservative approach and including any additional possible contributors that may arise from a faulty assembly procedure. In such cases it may be a good idea to do several Tolerance Stackups, each one representing a different possible assembly process. It is likely that each Tolerance Stackup will have different results due to different assembly variables. Using these results the Tolerance Analyst may be able to sway the assembly department to adopt a preferred assembly method that reduces the total potential variation.

The reason detail drawings of the manufactured components are required for the Tolerance Stackup should be obvious: it is from these detail drawings that the dimensions and tolerances are obtained for the Tolerance Stackup. Dimensioning and tolerancing schemes can vary, and some lead to easier Tolerance Stackups than others. In the simplest scenario, the distance in question is directly dimensioned (i.e., the width of the groove in Fig. 13.1).

In this example there is no need for a Tolerance Stackup: the dimension and tolerance are specified. Determining the minimum and maximum distance is straightforward and easy. The specified tolerance is subtracted from and added to the specified dimension to find the respective minimum and maximum limits.

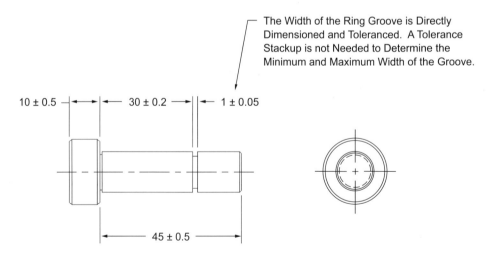

FIGURE 13.1 Pin with required dimension and tolerance. No tolerance stackup is needed.

It is far more likely that the distance under scrutiny is not directly dimensioned and toleranced and is a function of other dimensions, tolerances, and possibly assembly procedures. In fact, that is the reason for the Tolerance Stackup, to determine the limits between two features that are not directly specified. See Fig. 13.2.

In Figure 13.2 the width of the groove is not directly dimensioned and toleranced; so a Tolerance Stackup is required to find its minimum and maximum limits.

Detailed dimension and tolerance information is also required for catalog and purchased items in an assembly. Remember, this includes orientation and positional or location tolerances as well as size tolerances—many catalog detail sheets provide the size and the size tolerance but fall short in terms of relating the features. A good example is shown in Fig. 13.2, where the size and size tolerance of the shank and head diameters of the pin are given, but their coaxiality tolerance is undefined. Such information is often difficult to come by, but it is still necessary when performing a Tolerance Stackup. In such cases the Tolerance Analyst should contact the vendor for the required information. If the required information cannot be obtained or cannot be obtained soon enough, the Tolerance Analyst should make an educated guess and assume the coaxiality tolerance and explain that the tolerance value is an estimate. Refer to the material on assumptions in Chapter 7 for more on this important topic.

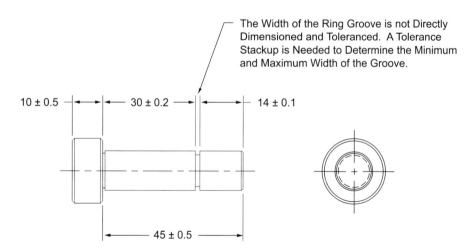

**FIGURE 13.2** Pin without required dimension and tolerance. A tolerance stackup is needed to determine the limits of the groove width.

The dimensioning and tolerancing schemes used on drawings determine which dimensions and tolerances must be included in the Tolerance Stackup and the Tolerance Stackup sketch. Performing a Tolerance Stackup with parts and assemblies that employ functional dimensioning and tolerancing schemes is much easier than with parts and assemblies that are dimensioned and toleranced poorly. Functionally dimensioned and toleranced parts have dimensions and tolerances arranged and related to the important features, such as mating surfaces and features that locate other parts. The dimensioning and tolerancing on such drawings is much more direct—in the end there are fewer dimensions and tolerances that contribute to the Tolerance Stackup. Often the result of a Tolerance Stackup that predicts excessive variation is to revise the drawings using a more robust and more direct dimensioning and tolerancing scheme, such as GD&T applied in a functional manner.

Assembly drawings are important because they show the parts in their as-assembled condition. This helps the Tolerance Analyst understand which parts contribute to the Tolerance Stackup and therefore must be included in the Tolerance Stackup sketch. Usually the assembly drawing is the source of the name, part number, and revision status of all the parts in the Tolerance Stackup. The assembly drawing also shows how parts are related to one another, for example, whether parts are located by mating faces or by tight fitting pins inserted into holes or whether parts are aligned horizontally or vertically. Lastly, the assembly drawing may have some dimensions and tolerances specified, controlling or limiting the potential variation allowed by the accumulated part feature dimensions and tolerances.

Dimensions and tolerances specified on an assembly drawing must be carefully considered, as sometimes such dimensions and tolerances cannot be achieved. In some cases there is no adjustment possible between the assembled parts, or the part feature tolerances contributing to an assembly gap or distance are greater than the stated assembly tolerance. These conditions render the stated assembly tolerance meaningless, as the sum of the part feature tolerances cannot be reduced at assembly. The Tolerance Analyst must make sense of such contradictions, get the drawings corrected, or explain how the problem was addressed in the Tolerance Stackup.

## TOLERANCE STACKUP SKETCH CONTENT

The Tolerance Stackup sketch provides a step-by-step pictorial explanation or road map of the Tolerance Stackup. It shows all the parts in the correct relationship with the chain of Dimensions and Tolerances that contribute to the Tolerance Stackup. The parts are identified, the contributing dimensions and tolerances are identified, their directions shown, and they are numbered to correspond with the line item numbers in the Tolerance Stackup report.

Clearly relating the Tolerance Stackup sketch to the Tolerance Stackup report is very important. Using this technique, the Tolerance Analyst can be sure that his or her report provides the greatest value to those that need to understand the report.

## Part and Assembly Geometry in the Tolerance Stackup Sketch

The Tolerance Stackup sketch does not have to be an exact reproduction of the part and assembly geometry, although using the exact geometry does seem to help many people visualize the Tolerance Stackup. The Tolerance Stackup sketch may be schematic if desired, a simplification of the actual geometry. Although simplified Tolerance Stackup sketches are easy to create and typically less cluttered than a fully detailed Tolerance Stackup sketch, sometimes the simplification may lead to omission of important geometric information. Good advice is to use the most accurate geometry possible, preferably right from the CAD models or CAD drawing files. The Tolerance Analyst must balance accuracy and completeness of detail with clarity and being overly complex when creating the Tolerance Stackup sketch.

The scale of the part geometry in the Tolerance Stackup sketch should be large enough to clearly show the pertinent part geometry, the distance or gap being studied should be clear, and the origin and terminus of each dimension and tolerance should be visible. The scale of the geometry shown in Fig. 13.3 is too small to adequately communicate the required information. The scale of the geometry shown in Fig. 13.4 is large enough to adequately

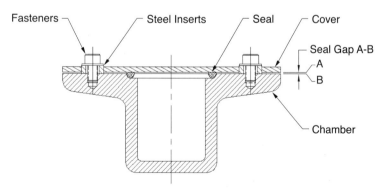

The Seal Gap must be Between 0.000 and 0.010.
(Dimensions and Tolerances Not Shown.)

**FIGURE 13.3**  Scale of Tolerance Stackup sketch is too small.

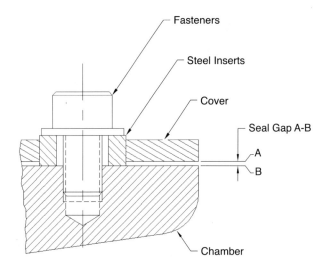

The Seal Gap must be Between 0.000 and 0.010.
(Dimensions and Tolerances Not Shown.)

**FIGURE 13.4**   Scale of Tolerance Stackup sketch is adequate.

communicate the required information, which makes the Tolerance Stackup sketch easier to understand.

Sometimes the size and geometry of the parts or assembly is such that details like the origin and termini of dimensions and tolerances and the distance being studied are not clear in the Tolerance Stackup sketch. In these cases, it may be advisable to include enlarged detail views with the Tolerance Stackup sketch, as shown in Fig. 13.5.

## Tolerance Stackup Sketch Annotation

Tolerance Stackup sketch annotation may include part identification, identification of the distance or gap being studied, identification of the Tolerance Stackup direction, ± Dimensions and Tolerances, converted angular Dimensions and Tolerances, geometric Dimensions and Tolerances, bonus tolerances, datum feature shift, assembly shift, item numbers, dimension direction signs (positive or negative direction), title and reference information. Other information may be included as well, such as any additional information that may help explain the Tolerance Stackup. Figure 13.6 shows some of the items listed above for an assembly of parts dimensioned and toleranced using the

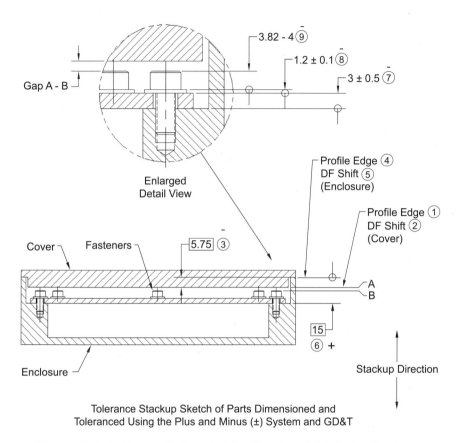

Gap A - B

3.82 - 4 ⑨

1.2 ± 0.1 ⑧

3 ± 0.5 ⑦

Enlarged
Detail View

Profile Edge ④
DF Shift ⑤
(Enclosure)

Profile Edge ①
DF Shift ②
(Cover)

Cover   Fasteners   5.75 ③

A
B

Enclosure

15
⑥ +

Stackup Direction

Tolerance Stackup Sketch of Parts Dimensioned and
Toleranced Using the Plus and Minus (±) System and GD&T

**FIGURE 13.5** Tolerance Stackup sketch with enlarged detail view for clarity.

plus/minus System. The Tolerance Stackup sketch in Fig. 13.6 accompanies the Tolerance Stackup report shown in Fig. 13.7.

The dimensions and tolerances in Fig. 13.6 are labeled using sequential item numbers inside circles, and the numbers correspond to the line item numbers in the Tolerance Stackup report in Fig. 13.7. This makes it easy to associate the Tolerance Stackup sketch with the Tolerance Stackup report. Equal bilateral tolerances associated with a dimension share the same item number as the dimension.

The direction of the Tolerance Stackup is identified in the Tolerance Stackup sketch. As Tolerance Stackups are linear, the direction is identified by an arrow or two arrows in opposite directions. Text such as "Stackup Direction" or "Direction of Study" may be added to clarify the meaning of the arrows. This can be seen in Figures 13.5 and 13.6.

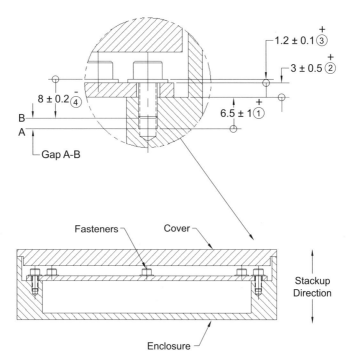

**FIGURE 13.6** Tolerance Stackup sketch with annotation: Parts Dimensioned and Toleranced using ±.

Each dimension is labeled as being in the positive or negative direction in two ways. The direction of each dimensions is shown by placing a dimension origin symbol at one end of the dimension and an arrowhead at the other end. The direction of the dimension is from the dimension origin symbol toward the arrowhead. The direction of each dimension is also shown by placing a positive ("+") or negative ("−") sign next to the dimension's item number. All the dimensions labeled as positive originate and terminate in the same direction. All the dimensions labeled as negative originate and terminate in the opposite direction. The dimension values are placed in the corresponding positive or negative dimension column in the associated Tolerance Stackup report.

Item numbers are placed in a circle adjacent to each dimension. Item numbers are also placed adjacent to each occurrence of assembly shift, geometric tolerances, bonus tolerances, and datum feature shift as applicable. Item numbers are assigned in the order each contributor is encountered as the chain is followed from point A to point B. It is important that the Item

Tolerance Stackup

Release 1.2a

| Program: | Electronics Packaging Program AV-11 | | | | | | | Stackup Information: | |
|---|---|---|---|---|---|---|---|---|---|
| Product: | Part Number 12345678-001 | Rev A | Description Ground Plate Enclosure Assembly | | | | | Stack No: Figure 13-7 Date: 07/04/02 Revision A | |
| Problem: | Screws Must Not Bottom Out in Tapped Holes | | | | | | | Direction: Z Axis | |
| Objective: | Determine if the M4 Holes in the Enclosure are Deep Enough | | | | | | | Author: BR Fischer | |

| Description of Component / Assy | Part Number | Rev | Item | Description | + Dims | - Dims | Tol | Percent Contrib | Dim / Tol Source & Calcs |
|---|---|---|---|---|---|---|---|---|---|
| Enclos_ure | 12345678-002 | A | 1 | Dim: Bottom M4 Tapped Hole - DF_A | 6.5000 | | +/- 1.0000 | 56% | 6.5 +/-.1 on Dwg |
| Ground Plate | 12345678-004 | A | 2 | Dim: Bottom Surface - Top Surface | 3.0000 | | +/- 0.5000 | 28% | 3 +/- 0.5 on Dwg |
| M4 Washer | | | 3 | Dim: Bottom Surface - Top Surface | 1.2000 | | +/- 0.1000 | 6% | 1.2 +/- 0.1 fm Machinery's Hdbk 23rd Ed. |
| M4 X 8 SHCS | | | 4 | Dim: Underside of Head - End of Screw | | 8.0000 | +/- 0.2000 | 11% | 8 +/-0.2 fm Vendor Dwg |
| | | | | Dimension Totals | 10.7000 | 8.0000 | | | |
| | | | | Nominal Distance: Pos Dims - Neg Dims = | 2.7000 | | | | |

RESULTS:

| | Nom | Tol | Min | Max |
|---|---|---|---|---|
| Arithmetic Stack (Worst Case) | 2.7000 | +/- 1.8000 | 0.9000 | 4.5000 |
| Statistical Stack (RSS) | 2.7000 | +/- 1.1402 | 1.5598 | 3.8402 |
| Adjusted Statistical: 1.5*RSS | 2.7000 | +/- 1.7103 | 0.9897 | 4.4103 |

Notes:

Assumptions: - Used Enclosure and Ground Plate Option 1 for this study.

Suggested Action:

**FIGURE 13.7** Tolerance Stackup report for Fig. 13.6.

numbers in the Tolerance Stackup sketch and the Tolerance Stackup report are in agreement.

Figure 13.8 shows a Tolerance Stackup sketch for an assembly of parts that were dimensioned and toleranced using GD&T. Notice that the basic dimensions and the geometric tolerances have distinct item numbers. This is common on Tolerance Stackups of parts dimensioned and toleranced using GD&T. This is also true in a Tolerance Stackup report based on GD&T.

In the Tolerance Stackup report shown in Fig. 13.7 each dimension and its ± tolerance are on the same line of the Tolerance Stackup report, and consequently, are given the same line item number. In the following example, parts are dimensioned and toleranced using GD&T. The basic dimensions and the geometric tolerances specified in the associated feature control frames are on separate lines in the Tolerance Stackup report and hence have distinct item numbers.

It is important to state that there will likely be some ± dimensions and tolerances on parts with GD&T. This is acceptable per the standard. That said, the author recommends that dimensions with ± tolerances only be used

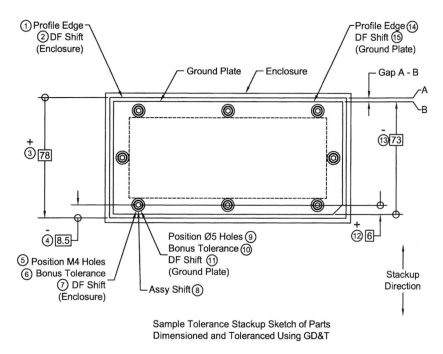

Sample Tolerance Stackup Sketch of Parts
Dimensioned and Toleranced Using GD&T

**FIGURE 13.8**   Tolerance Stackup Sketch with annotation: Parts Dimensioned and Toleranced using GD&T.

for the size of features of size, like holes and studs—from a functional point of view and from a Tolerance Stackup point of view, it is a far better approach to locate every feature of a part using GD&T.

A Tolerance Stackup sketch for parts using GD&T is structured similarly to the Tolerance Stackup report, and each geometric tolerance may be followed by bonus tolerance and/or datum feature shift. As applicable, the bonus tolerance and datum feature shift are located directly below the geometric tolerance on the Tolerance Stackup sketch. The name of the applicable part is shown in parentheses beneath each set of geometric tolerance information in the Tolerance Stackup sketch. This can be seen in Figures 13.8 and 13. 9.

### Steps for Creating a Tolerance Stackup Sketch on Parts and Assemblies Dimensioned and Toleranced Using the Plus/Minus (±) System:

1. Part identification:

   a. Identify each part in the Tolerance Stackup. This may be the part name, the part number, or other adequately descriptive information as desired.

2. Distance or gap being studied:

   a. Show a dimension across the distance or gap being studied.

   b. Label the distance or gap "A-B," "Gap A-B," or "Distance A-B," or descriptively, such as "Snap Ring Groove Width" or "Distance Between Flange Faces."

   c. Label one end point A and the other end point B.

3. Direction of the Tolerance Stackup:

   a. Draw an arrow or two arrows in opposite directions with text to show the direction of the Tolerance Stackup.

4. The chain of Dimensions and Tolerances:

   a. Add a Dimension starting at point A as described in Chapter 7.

   b. Place a dimension origin symbol at the start of the dimension and an arrowhead at the other end of the dimension.

   c. Complete the chain of Dimensions and Tolerances by adding dimensions with equal bilateral tolerances head to tail from point A to point B.

   d. Determine the positive direction for the Tolerance Stackup as described in Chapter 7.

Tolerance Stackup

| Program: | Electronics Packaging Program AV-11 | | | | | | | | Stackup Information |
|---|---|---|---|---|---|---|---|---|---|

| Product: | Part Number | Rev | Description | Stackup Information | |
|---|---|---|---|---|---|
| | 12345678-001 | A | Ground Plate Enclosure Assembly: Option 1 w Surfaces as Datum Features B & C | Stack No: | Figure 13.9 |
| Problem: | Edges of Ground Plate must not Touch Walls of Enclosure | | | Date: | 07/04/02 |
| Objective: | Option 1: Determine if Ground Plate Contacts Enclosure Walls | | | Revision: | A |
| | | | | Direction: | Y Axis |
| | | | | Author: | BR Fischer |

| Description of Component / Assy | Part Number | Rev | Item | Description | + Dims | - Dims | Tol | Percent Contrib | Dim / Tol Source & Calcs |
|---|---|---|---|---|---|---|---|---|---|
| Enclosure | 12345678-002 | A | 1 | Profile: Edge Along Pt A | | | ± 0.5000 | 18% | Profile 1, A, B, C |
| | | | 2 | Datum Feature Shift: | | | ± 0.0000 | 0% | N/A - DFs not a Features of Size |
| | | | 3 | Dim: Edge of Enclosure - Datum B | 78.0000 | | ± 0.0000 | 0% | 78 Basic on Dwg |
| | | | 4 | Dim: Datum B - CL M4 Holes | | 8.5000 | ± 0.0000 | 0% | 8.5 Basic on Dwg |
| | | | 5 | Position: M4 Holes | | | ± 0.3000 | 11% | Position dia 0.6 @ MMC A, B, C |
| | | | 6 | Bonus Tolerance | | | ± 0.0900 | 3% | = (3.422 - 3.242) / 2 |
| | | | 7 | Datum Feature Shift: | | | ± 0.0000 | 0% | N/A - DFs not a Features of Size |
| Ground Plate | 12345678-004 | A | 8 | Assembly Shift: (Mounting Holes$_{mc}$ - F$_{mc}$) / 2 | | | ± 0.8650 | 32% | = ((5.4 + 0.15) - 3.82) / 2 |
| | | | 9 | Position: Dia 5.4 ± 0.15 Holes | | | ± 0.3250 | 12% | Position dia 0.65 @ MMC A |
| | | | 10 | Bonus Tolerance | | | ± 0.1500 | 5% | = (0.15 + 0.15) / 2 |
| | | | 11 | Datum Feature Shift: | | | ± 0.0000 | 0% | N/A - DFs not a Features of Size |
| | | | 12 | Dim: CL Dia 5.4 Holes - Datum B | 6.0000 | | ± 0.0000 | 0% | 6 Basic on Dwg |
| | | | 13 | Dim: Datum B - Edge of Ground Plate | | 73.0000 | ± 0.0000 | 0% | 73 Basic on Dwg |
| | | | 14 | Profile: Edge Along Pt B | | | ± 0.5000 | 18% | Profile 1, A, B, C |
| | | | 15 | Datum Feature Shift: | | | ± 0.0000 | 0% | N/A - DFs not a Features of Size |

Dimension Totals: 84.0000    81.5000

Nominal Distance: Pos Dims - Neg Dims = 2.5000

**RESULTS:**

| | Nom | Tol | Min | Max |
|---|---|---|---|---|
| Arithmetic Stack (Worst Case) | 2.5000 | ± 2.7300 | -0.2300 | 5.2300 |
| Statistical Stack (RSS) | 2.5000 | ± 1.2143 | 1.2857 | 3.7143 |
| Adjusted Statistical: 1.5*RSS | 2.5000 | ± 1.8214 | 0.6786 | 4.3214 |

Notes:   - M4 Screw Dimensions: Major Dia: 4 / 3.82   - M4 Tapped Hole Dimensions: Minor Dia: 3.422 / 3.242
- Used min and max screw thread minor dia in Datum Feature Shift Calculations on line 2.
- Used smallest screw major dia in Shift Calculations on line 8.
- The positional tolerances on the clearance holes and the M4 holes are larger because they are toleranced relative to the edge surfaces, and manufacturing said that was the best they could do.
- The larger positional tolerances required the clearance holes to be larger, due to the Fixed Fastener Formula. This increased the Shift calculated in line 8

Assumptions:
- Although threads are typically assumed to be self centering, the Positional Tol applies to the Minor Diameter of the M4 holes. Use the min / max Minor Dia to calculate the Bonus Tolerance on line 6.

Suggested Action:
- May want to holes as locators instead of edges. See Stacks Opt - 2 & Opt - 3.

**FIGURE 13.9**   Tolerance Stackup report for Fig. 13.8.

e. Label the dimensions as positive ( + ) or negative ( − ) as described in Chapter 7.

f. Include each occurence of assembly shift as it appears in the chain.

    i. Indicate each occurence of assembly shift using a leader directed note.

    ii. Add the part name in parentheses beneath the assembly shift to show which part is shifting.

g. Indicate any angular Dimensions and Tolerances or other Tolerances that contribute to the Tolerance Stackup.

h. Assign an item number to each dimension with its equal bilateral tolerance, each angular or other tolerance, and each occurrence of assembly shift as it is encountered in the chain. Start with "1" using positive sequential whole numbers until every dimension, angular or other tolerance, and every occurrence of assembly shift is numbered.

i. Make sure the item numbers match the line item numbers in the associated Tolerance Stackup report.

j. Assign a positive or negative dimension direction sign to each dimension in the chain.

k. Make sure there are no interruptions in the chain between point A and point B.

5. Add a descriptive title to the Tolerance Stackup sketch.

6. Add any reference information that may be helpful to the reader.

Refer to Figs. 13.6 and 13.7 for examples of a Tolerance Stackup sketch and Tolerance Stackup report for parts dimensioned and toleranced using the plus/minus (±) system.

### Steps for Creating a Tolerance Stackup Sketch on Parts and Assemblies Dimensioned and Toleranced Using GD&T:

1. Part identification:

a. Identify each part in the Tolerance Stackup. This may be the part name, the part number, or other adequately descriptive information as desired.

2. Distance or gap being studied:

a. Show a dimension across the distance or gap being studied.

b.  Label the distance or gap "A-B," "Gap A-B," or "Distance
    A-B," or descriptively, such as "Snap Ring Groove Width" or
    "Distance Between Flange Faces."

c.  Label one end point A and the other end point B.

3.  Direction of the Tolerance Stackup:

    a.  Draw an arrow or two arrows in opposite directions with text
        to show the direction of the Tolerance Stackup.

4.  The chain of Dimensions and Tolerances:

    a.  If there is an applicable geometric tolerance with bonus tol-
        erance and/or datum feature shift specified at point A, in-
        dicate so as described in 4.g below.

    b.  Add a basic dimension starting at point A as described in
        Chapter 7.

    c.  Place a dimension origin symbol at the start of the dimension
        and an arrowhead at the other end of the dimension.

    d.  Complete the chain of Dimensions and Tolerances by adding
        basic dimensions head to tail from point A to point B.

    e.  Determine the positive direction for the Tolerance Stackup as
        described in Chapter 7.

    f.  Label the dimensions as positive ($+$) or negative ($-$) as de-
        scribed in Chapter 7.

    g.  Include each geometric tolerance as it appears in the chain.

        i.   Indicate the geometric tolerance (position, profile, etc.)
             using a leader-directed note.

        ii.  Include the bonus tolerance and/or datum feature shift
             beneath the geometric tolerance as applicable. Include
             these even if their value is zero per the rules in Chapter 9.

        iii. Add the part name in parentheses beneath the geo-
             metric tolerance, bonus tolerance, and/or datum
             feature shift to explain where the GD&T originated.

             (Note that the geometric tolerance, bonus tolerance,
             and/or datum feature shift are not directly related to a
             dimension.)

    h.  Include each occurrence of assembly shift as it appears in the
        chain.

        i.   Indicate the Assembly Shift using a leader-directed
             note.

ii. Add the part name in parentheses beneath the assembly shift to show which part is shifting.

i. Assign an item number to each basic dimension, geometric tolerance, bonus tolerance, datum feature shift, and each occurrence of assembly shift as it is encountered in the chain.

i. Start with "1" using positive sequential whole numbers until every basic dimension, geometric tolerance, bonus tolerance, datum feature shift, and every occurrence of assembly shift is numbered.

j. Make sure the item numbers match the line item numbers in the associated Tolerance Stackup report.

k. Assign a positive or negative dimension direction sign to each basic dimension in the chain.

l. Make sure there are no interruptions in the chain between point A and point B.

5. Add a descriptive title to the Tolerance Stackup sketch.

6. Add any reference information that may be helpful to the reader.

Refer to Figs. 13.8 and 13.9 for examples of a Tolerance Stackup sketch and Tolerance Stackup report for parts dimensioned and toleranced using GD&T.

The Tolerance Stackup sketching steps described above for plus/minus (±) and GD&T may be combined as needed where both dimensioning and tolerancing methods are used in the same Tolerance Stackup. Refer to Fig. 13.5 for an example of a Tolerance Stackup sketch where parts were dimensioned and toleranced using both methods.

## TOLERANCE STACKUP SKETCH RECAP

The Tolerance Stackup sketch is a critical part of the Tolerance Analysis and Tolerance Stackup process. In fact, it may be the most important step. The Tolerance Stackup sketch helps the Tolerance Analyst visualize the problem more clearly and helps to ensure no dimensions or tolerances are overlooked. The Tolerance Stackup sketch helps others that must interpret the Tolerance Stackup report understand which problem was solved, how the problem was solved, and whether the results are correct. Every dimension, ± tolerance, geometric tolerance, bonus tolerance, datum feature shift, and assembly shift that contributes to the Tolerance Stackup is shown and given a unique item number that coincides with the Tolerance Stackup report.

# 14

# The Tolerance Stackup
# Report Form

Briefly introduced in Chapter 8, this chapter completes the coverage of the Tolerance Stackup report form. Tolerance Stackup reporting as a whole will be discussed, and the importance of content, format, and the purpose of each field in the Tolerance Stackup report form will be explained.

The Tolerance Stackup report form is essential to all Tolerance Analysis activities. There are many things to consider when reporting the results of a Tolerance Stackup. It is important to recognize which information must be included, be it project related, product related, the source of the dimensions and tolerances, procedural (how the problem was solved), a sketch of the parts or assemblies being studied, the results, and any explanatory material or recommendations. It is also important to determine the best way to format and present the information to make it as easy as possible for someone else that must use the Tolerance Stackup report to make a decision—that is the point of a Tolerance Stackup, to assist in decision making.

Tolerance Stackups are typically done for any of these reasons:

- Determine if a new design will yield the desired results at final use or assembly.
- Determine if a change to existing design geometry will yield acceptable results at final use or assembly.
- Determine if a change in one or more dimensions or tolerances of an existing design will yield acceptable results at final use or assembly.

- Determine the amount of tolerance that may be allocated or distributed to parts in a new assembly.
- Determine the total tolerance possible for an existing assembly.
- Determine the root cause of a tolerance-related problem (such as an unwanted interference or excessively large gap) in an existing assembly.

Although this list includes examples from before the design is released and after production, all of the bulleted items share one thing in common: most likely the result of the Tolerance Stackup will need to be discussed with another individual, group, or even a client or customer. The purpose of a computer-based Tolerance Stackup report form is twofold: it is a semi-automated tool for solving Tolerance Stackups, and it is a tool for communicating the results of Tolerance Stackups. The Tolerance Stackup may also need to be kept for historical or legal reasons, as proof of how a particular problem was solved or how the tolerancing risk was assessed and dealt with. When someone less familiar with the study has to understand the Tolerance Stackup, the standardized Tolerance Stackup report form will make their job much easier. When time is the culprit, say when the Tolerance Stackup may sit dormant for some period of time, the same engineer that initially did the Tolerance Stackup may need to revisit the Stackup several years later as part of a redesign or as part of a warranty claim. Clear, complete, and standardized Tolerance Analysis reporting is essential for understanding the method and results in these situations.

It is critical that all Tolerance Analysis and Stackup activities are standardized. This has exactly the same basis and reasons as drawing standards—to avoid errors, avoid misunderstandings, avoid wasting time, and avoid wasting money. The necessity of standardized drawing practices has been understood for years—the concept is equally valid in a Tolerance Stackup reporting context. All Tolerance Stackup reporting within a given firm should be carried out in the same manner and should be presented using the same reporting format where possible. It is also important to recognize where an inadequate Tolerance Stackup reporting system is in use; in such cases the system should either be modified to include the information presented in this chapter or an updated and complete Tolerance Stackup report form should be adopted.

Using a consistent approach and a standard report format makes learning to perform complex Tolerance Stackups much easier. The problem is approached in the same way every time; its nuances are more easily recognized and addressed, information is gathered, the chain of Dimensions and Tolerances is documented, information is entered, calculations are made, and results are reported. Using the same approach every time will help the

Tolerance Analyst ensure that all the information has been captured, the procedure is followed and the results are correct. That is not to say that using a standardized Tolerance Stackup report form will eliminate errors. Using a standardized approach will eliminate the procedural issues, and allow the Tolerance Analyst to focus on solving the problem. A consistent approach also makes it easier for others to interpret your results and understand the work you have done.

As an example, Advanced Dimensional Management's™ Tolerance Stackup Reporting Tool is shown in Fig. 14.1. It is a versatile, semiautomated spreadsheet tool that works in Microsoft Excel. Standardized data entry, semiautomated problem solution, and automatic reporting of worst-case and statistical results on the same format make this is a very easy tool to use.

Like all good Tolerance Stackup reporting tools, this form contains many important pieces of information grouped and presented logically in an easy-to-read manner. This form is broken into six sections or blocks as described in Fig. 14.1:

Block 1. Tracking and title data block (manually entered)

Block 2. Data entry block (manually entered and automated)

Block 3. Results block (automated)

Block 4. Notes block (manually entered)

Block 5. Assumptions block (manually entered)

Block 6. Suggested action block (manually entered)

There are several standardized Tolerance Stackup reporting tools available from various tolerancing and dimensional management firms around the country. It is important that the form you choose contains places for all the information presented in this chapter. The more information that can be captured at the time a Tolerance Stackup is done the better. Anyone who needs to understand the Tolerance Stackup will find all of this information essential.

## FILLING OUT THE TOLERANCE STACKUP REPORT FORM

This section explains how to fill out the Tolerance Stackup report form field by field. The data entry procedures and requirements are explained for each field. Some of this may be redundant with the previous chapters, but it is valuable to present it in a more concise easy-to-reference format.

The fields in the Tolerance Stackup report form in Fig. 14.2 are labeled to coincide with the instructions that follow. The numbered references in each field match the numbered bullets. For example, item 1(a) in Fig. 14.2 is for item 1.a in the following list.

Tolerance Stack                                                                Release 1.2a

Program:

Product:    Part Number    Rev.  Description                              Stack Information:

Problem:                                                                  Stack No:
                                                                          Date
Objective:                                                                Revision
                                                                          Direction:
                                                                          Author:

## 1 - Tracking and Title Data Block

Description of
Component / Assy   Part Number   Rev   Item   Description

## 2 - Data Entry Block

+ Dims    – Dims    Tol         Percent
                                Contrib    Dim / Tol Source & Calcs

Dimension Totals
Nominal Distance: Pos Dims - Neg Dims =

RESULTS:                          Nom    Tol    Min    Max

Arithmetic Stack (Worst Case)
Statistical Stack (RSS)
Adjusted Statistical 1.5*RSS          ## 3 - Results Block

## 4 - Notes Block

## 5 - Assumptions Block

## 6 - Suggested Action Block

Notes:

Assumptions:

Suggested Action:

**FIGURE 14.1**   Sample Tolerance Stackup report format with blocks identified.

Tolerance Stack

Release 1.2a

| Program: | 1(a) | | | | |
|---|---|---|---|---|---|
| Product: | Part Number | Rev | Description | | |
| | 1(b) | 1(b) | 1(b) | | |
| Problem: | 1(c) | | | | |
| Objective: | 1(d) | | | | |

**Stack Information**

Stack No: 1(e)
Date: 1(f)
Revision 1(g)

Direction: 1(h)
Author: 1(i)

| Description of Component / Assy | Part Number | Rev | Item | Description | + Dims | – Dims | Tol | Percent Contrib | Dim / Tol Source & Calcs |
|---|---|---|---|---|---|---|---|---|---|
| 2(a) | 2(b) | 2(c) | 2(d) | 2(e) | 2(f) | 2(g) | ± 2(h) | 2(i) | 2(j) |
| | | | | | +| | | | |
| | | | | | +| | | | |
| | | | | | +| | | | |

| | | + Dims | – Dims | |
|---|---|---|---|---|
| Dimension Totals | | 2(k) | 2(k) | |
| Nominal Distance Pos Dims – Neg Dims = | | | 2(l) | |

**RESULTS:**

| | Nom | Tol | Min | Max |
|---|---|---|---|---|
| Arithmetic Stack (Worst Case) | 3(a) | ± 3(b) | 3(e) | 3(e) |
| Statistical Stack (RSS) | 3(a) | ± 3(c) | 3(f) | 3(f) |
| Adjusted Statistical 1.5*RSS | 3(a) | ± 3(d) | 3(g) | 3(g) |

Notes: 4(a)

Assumptions: 5(a)

Suggested Action: 6(a)

**FIGURE 14.2** Sample Tolerance Stackup report format with fields labeled.

| Program: | 1(a) | | |
|---|---|---|---|
| Product: | Part Number | Rev | Description |
| | 1(b) | 1(b) | 1(b) |
| Problem: | 1(c) | | |
| Objective: | 1(d) | | |

**FIGURE 14.3** Sample Tolerance Stackup report format: Tracking and title block data, left side enlarged.

Enlarged views of each Tolerance Stackup report block are included with the instructions that follow for easier reference.

1. Tracking and title data block: This portion of the report contains all of the information needed to describe which product or products are being studied: describe the problem being studied, explain the intent of the Tolerance Stackup, and capture all the tracking information about the Tolerance Stackup. See Fig. 14.3 for items 1.a–1.d.

   a. Program: Enter the program or project name and/or number in this field.

   b. Product: Enter the product data in these fields. Enter the part/assembly number, part/assembly revision, and part/assembly description in these three fields.

   c. Problem: Enter a problem statement (describe the problem) in this field.

   d. Objective: Enter the objective of the Tolerance Stackup in this field.

   Tolerance Stackup information (See Fig. 14.4 for items 1.e–1.i.):

   e. Stack No: Enter the Tolerance Stackup tracking number. This should be a formally assigned and tracked number

| Stack Information: | |
|---|---|
| Stack No: | 1(e) |
| Date: | 1(f) |
| Revision | 1(g) |
| | |
| Direction: | 1(h) |
| Author: | 1(i) |

**FIGURE 14.4** Sample Tolerance Stackup report format: Tracking and title block data, right side enlarged.

similar to the drawing number assigned to a drawing. This is the method used to track the Tolerance Stackup within a company's data management system.

f. Date: Enter the date the Tolerance Stackup was started or completed as determined by company policy. This should not be a field that automatically updates each time the spreadsheet is opened, as historical tracking information would be lost.

g. Revision: Enter the Tolerance Stackup revision (the revision of the Tolerance Stackup). This is not the revision of the product or assembly being studied. The Tolerance Stackup revision is important because Tolerance Stackups may be changed many times, as the first attempt may have been flawed or products may have changed. It is a good idea to keep historical copies of Tolerance Stackups until it is certain that they will no longer be of value.

h. Direction: Enter the Tolerance Stackup direction. This can be a local or global coordinate system direction such as "along $x$ axis," "positive 37.5 degrees from $z$ axis," descriptive "perpendicular to rear panel," or "between head of bolt #6 and feed cover," or it can be a combination of these. Vector notation could also be used if desired.

i. Author: Enter the name of the person performing the Tolerance Stackup. This is important for addressing questions and historical reasons.

2. Data entry block: Line-by-line Tolerance Stackup data is entered into this portion of the report. Information is entered in the order it is encountered in the chain of Dimensions and Tolerances. The report contains: the description, part number, and revision of each component (part) or subassembly in the chain of Dimensions and Tolerances; line item numbers for each line in the Tolerance Stackup; the description of each Dimension and Tolerance entered into the Tolerance Stackup; columns for the + (positive) and − (negative) direction dimensions; columns for the equivalent bilateral ± Tolerance values; the percent contribution of each Tolerance; and a column for calculations and to describe how the Dimension or Tolerance was obtained.

See Fig. 14.5 for items 2.a–2.e.

a. Description of component or Assembly: Enter the name of the part or assembly that is the source of the Dimension or Tolerance on that Line. This should be the name that appears

| Description of Component / Assy | Part Number | Rev | Item | Description |
|---|---|---|---|---|
| 2(a) | 2(b) | 2(c) | 2(d) | 2(e) |
| | | | | |
| | | | | |
| | | | | |
| | | | | |

**FIGURE 14.5**   Sample Tolerance Stackup report format: Data entry block, left side enlarged.

in the part or assembly drawing's title block or the number assigned by other formal means within the data management system. This may also be the name of a purchased item obtained from a catalog or a similar source.

b.   Part number: Enter the part or drawing number of the part or assembly that is the source of the Dimension or Tolerance on that Line. This should be the part number that appears in the part or assembly drawing's title block, or the number assigned by other formal means within the data management system. This may also be the catalog part number for a purchased item.

c.   Rev (revision): Enter the revision of the part or assembly that is the source of the Dimension or Tolerance on that Line. This should be the revision that appears in the part or assembly drawing's revision block or the number assigned by other formal means within the data management system.

d.   Item: Enter the line item number. Number the items in the Tolerance Stackup with consecutive positive integers starting with 1. Line item numbers are an important communication tool and useful for discussing results.

e.   Description: Enter the description of the Dimension and/or Tolerance. See the following section for a more detailed explanation of data format in this field.

See Fig. 14.6 for items 2.f–2.j.

f.   + Dims: Enter any positive direction dimension values in this field. These values may be taken directly from a drawing or a model where equal bilateral Dimensions and Tolerances were specified. If unequal bilateral or unilateral Dimensions and Tolerances were specified, these values are obtained by using the equal bilateral conversion techniques described earlier in the text.

| + Dims | − Dims | Tol | Percent Contrib | Dim / Tol Source & Calcs |
|--------|--------|--------|--------|--------|
| 2(f) | 2(g) | ± 2(h) | 2(i) | 2(j) |
|  |  | ± |  |  |
|  |  | ± |  |  |
|  |  | ± |  |  |
|  |  | ± |  |  |
|  |  | ± |  |  |

**FIGURE 14.6**  Sample Tolerance Stackup report format: Data entry block, right side enlarged.

g.  − Dims: Enter any negative direction Dimension values in this field. These values may be taken directly from a drawing or a model where equal bilateral Dimensions and Tolerances were specified. If unequal bilateral or unilateral Dimensions and Tolerances were specified, these values are obtained by using the equal bilateral conversion techniques described earlier in the text.

h.  Tol (tolerance): Enter the equal bilateral Tolerance value in this field. These values may be taken directly from a drawing where equal bilateral Tolerances were specified. If unequal bilateral or unilateral Tolerances were specified, these values are obtained by using the equal bilateral conversion techniques described earlier in the text.

i.  Percent contribution: This field is automatically calculated. It represents the percentage of the total worst-case tolerance that each tolerance contributes. This is very useful for determining which tolerances to change when the result of a Tolerance Stackup shows an undesirable condition. Obviously the tolerances that contribute the most to the total contribute the highest percentages to the total. These are often the best places to make a change to obtain the desired Tolerance Stackup result.

j.  Dim/Tol source and Calculations: Enter the source of the Dimension and/or Tolerance in this field. Also enter any calculations that were used to derive the Tolerance value. If a dimension was taken from the drawing, say so in this field. If a dimension was measured from a model, say so in this field. If a geometric tolerance is converted to an equal bilateral ± format, state the geometric tolerance in this field. If a standard line item is included in the Tolerance Stackup but does not contribute to the total tolerance, label it as N/A. Guidelines and many examples on how to enter data into this field are discussed in the following section.

| | | | ± |
|---|---|---|---|
| | | | ± |
| | | | ± |
| | | | ± |
| Dimension Totals | Σ(k) | Σ(k) | |
| Nominal Distance: Pos Dims - Neg Dims = | | Σ(l) | |

**FIGURE 14.7** Sample Tolerance Stackup report format: Data entry block, bottom side enlarged.

See Fig. 14.7 for items 2.k–2.l.

k.  Dimension totals: These fields are calculated automatically. The field on the left is the sum of the positive direction dimensions, and the field on the right is the sum of the negative direction dimensions.

l.  Nominal distance: This field is calculated automatically. It is the difference between the negative direction dimension total and the positive direction dimension total. If this value is negative, one or more dimension values were entered into the wrong column; that is, one or more positive dimensions were entered into the negative column, or the signs for the dimensions in the chain of Dimensions and Tolerances were chosen incorrectly.

3.  Results block (See Fig. 14.8 for items 3.a–3.g.)

a.  Nominal distance: These fields are calculated automatically, and all three fields are the same value. They are the nominal distance calculated in field 2(l) of Fig. 14.7.

b.  Arithmetic (worst-case) ± Tolerance value: This field is calculated automatically. It is the total equal bilateral Tolerance value that is subtracted from and added to the nominal distance to obtain the worst-case minimum and maximum distance values, respectively.

c.  Statistical (RSS) ± Tolerance value: This field is calculated automatically. It is the root-sum-square equal bilateral Tolerance value that is subtracted from and added to the

| | | Nom | Tol | Min | Max |
|---|---|---|---|---|---|
| **RESULTS:** | Arithmetic Stack (Worst Case) | 3(a) | ± 3(b) | 3(e) | 3(e) |
| | Statistical Stack (RSS) | 3(a) | ± 3(c) | 3(f) | 3(f) |
| | Adjusted Statistical: 1.5*RSS | 3(a) | ± 3(d) | 3(g) | 3(g) |

**FIGURE 14.8** Sample Tolerance Stackup report format: Results block.

nominal distance to obtain the statistical minimum and maximum distance values, respectively.

    d.   Adjusted statistical ± Tolerance value: This field is calculated automatically. By default, it is 1.5 times the root-sum-square equal bilateral Tolerance value in field 3(c). This value is subtracted from and added to the nominal distance to obtain the adjusted statistical minimum and maximum distance values, respectively. The 1.5 multiplier can be changed at the Tolerance Analyst's or the company's discretion.

    e.   Arithmetic (worst-case) minimum and maximum distance values: These fields are calculated automatically. The total equal bilateral worst-case Tolerance is subtracted from and added to the nominal distance to obtain the minimum and maximum worst-case distance values, respectively. These are the worst-case results of the Tolerance Stackup.

    f.   Statistical (RSS) minimum and maximum distance values: These fields are calculated automatically. The statistical root-sum-square Tolerance is subtracted from and added to the nominal distance to obtain the minimum and maximum statistical distance values, respectively. These are the statistical root-sum-square results of the Tolerance Stackup.

    g.   Adjusted statistical (RSS) minimum and maximum distance values: These fields are calculated automatically. The adjusted statistical root-sum-square Tolerance is subtracted from and added to the nominal distance to obtain the minimum and maximum adjusted statistical distance values, respectively. These are the adjusted statistical root-sum-square results of the Tolerance Stackup.

4.   Notes block

    a.   Notes area: Enter any pertinent notes in this area. Notes may be bulleted, numbered, or entered as a paragraph at the discretion of the Tolerance Analyst. Including notes is a good way to capture special information, sources of information or directions from clients or other groups. This

**FIGURE 14.9**   Sample Tolerance Stackup report format: Notes block, left side enlarged.

**FIGURE 14.10** Sample Tolerance Stackup report assumptions block, left side enlarged.

        is also a good place to explain special procedural information about how the Tolerance Stackup was approached and solved if needed. See Fig. 14.9.

5. Assumptions block

    a. Assumptions area: Enter any assumptions needed to solve the Tolerance Stackup in this area. Assumptions may be bulleted, numbered, or entered as a paragraph at the discretion of the Tolerance Analyst. This is a critical area of the Tolerance Stackup and must be filled out carefully for others to understand how the problem was approached and solved. It may also act as a flag to highlight the need for additional information. See Fig. 14.10.

6. Suggested action block

    a. Suggested action area: Enter any suggestions for correcting problems highlighted by the Tolerance Stackup in this area. Suggested action items may be bulleted, numbered, or entered as a paragraph at the discretion of the Tolerance Analyst. If the Tolerance Stackup does not highlight a problem, there is probably no need to suggest any action.

## GENERAL GUIDELINES FOR ENTERING DESCRIPTION, PART NUMBER, AND REVISION INFORMATION

The description, part number, and revision indicate from which part the dimension and tolerance data are taken. Each time a new part or assembly is encountered in the chain of Dimensions and Tolerances, new description, part number and revision data must be added to indicate the change of parts or assemblies.

    Description, part number, and revision information may not need to be included with each line item. Take the example in Fig. 14.11. There are two parts in this Tolerance Stackup; each occurs once in the chain of Dimensions

Tolerance Stack

Release 1.2a

| Program: | Electronics Packaging Program AV-11 | | | | | | | | |
|---|---|---|---|---|---|---|---|---|---|
| Product: | Part Number 12345678-001 | Rev A | Description Ground Plate Enclosure Assembly. Option 1 w/ 8 Holes as Datum Feature B | | | | | | |
| Problem: | Edges of Ground Plate must not Touch Walls of Enclosure | | | | | | | | |
| Objective: | Option 1: Determine If Ground Plate Contacts Enclosure Walls | | | | | | | | |

Stack Information:
Stack No: AV-11-010a
Date: 07/04/02
Revision: A
Direction: Along Plane of Ground Plate (Y Axis)
Author: BR Fischer

| Description of Component / Assy | Part Number | Rev | Item | Description | + Dims | − Dims | Tol | Percent Contrib | Dim / Tol Source & Calcs |
|---|---|---|---|---|---|---|---|---|---|
| Enclosure | 12345678-002 | A | 1 | Profile Edge Along Pt A | | | ±0.5000 | 19% | Profile 1, A, Bm |
| | | | 2 | Datum Feature Shift: $(DF_{B@MMC} - DFS_B)/2$ | | | ±0.2900 | 11% | = [3.422 − (3.242 − 0.4)]/2  (Shift within Minor Dia) |
| | | | 3 | Dim. Edge of Enclosure − Datum B | 8.5000 | | ±0.0000 | 0% | 8.5 Basic on Dwg |
| | | | 4 | Position $DF_B$ M4 Holes | | | ±0.2000 | 8% | Position dia 0.4 @ MMC A |
| | | | 5 | Bonus Tolerance | | | ±0.0000 | 0% | N/A − Threads |
| | | | 6 | Datum Feature Shift: $(DF_{B@MMC} - DFS_B)/2$ | | | ±0.0000 | 0% | N/A − $DF_A$ not a Feature of Size |
| Ground Plate | 12345678-004 | A | 7 | Assembly Shift: $(Mounting\ Hole_{S_{MC}} - F_{LMC})/2$ | | | ±0.6650 | 25% | = [(5 + 0.15) − 3.82]/2 |
| | | | 8 | Position, $DF_B$ Dia 5 ± 0.1 Holes | | | ±0.2250 | 9% | Position dia 0.45 @ MMC A |
| | | | 9 | Bonus Tolerance | | | ±0.1000 | 4% | = (0.1 + 0.1)/2 |
| | | | 10 | Datum Feature Shift: $(DF_{B@MMC} - DFS_B)/2$ | | | ±0.0000 | 0% | N/A − $DF_A$ not a Feature of Size |
| | | | 11 | Dim Datum B − Edge of Ground Plate | | 6.0000 | ±0.0000 | 0% | 6 Basic on Dwg |
| | | | 12 | Profile Edge Along Pt B | | | ±0.5000 | 19% | Profile 1, A, Bm |
| | | | 13 | Datum Feature Shift: $(DF_{B@MMC} - DFS_B)/2$ | | | ±0.1500 | 6% | = [(5 + 0.15) − (5 − 0.15)]/2 |

Dimension Totals: 8.5000 | 6.0000
Nominal Distance: Pos Dims − Neg Dims = 2.5000

**RESULTS:**

| | Nom | Tol | Min | Max |
|---|---|---|---|---|
| Arithmetic Stack (Worst Case) | 2.5000 | ±2.6300 | -0.1300 | 5.1300 |
| Statistical Stack (RSS) | 2.5000 | ±1.0721 | 1.4279 | 3.5721 |
| Adjusted Statistical 1.5*RSS | 2.5000 | ±1.6082 | 0.8918 | 4.1082 |

Notes:
- M4 Screw Dimensions: Major Dia 4/3.82    - M4 Tapped Hole Dimensions: Minor Dia 3.422/3.242
- Used min and max screw thread minor dia in Datum Feature Shift Calculations on line 2.
- Used smallest screw major dia in Assembly Shift Calculations on line 7

Assumptions:
- Assume threads are self centering. Do not include bonus tolerance on line 5.

Suggested Action:
- May want to use two holes as locators instead of all eight. See Stack Opt - 2.

**FIGURE 14.11** Sample Tolerance Stackup spreadsheet with sample solution and examples.

and Tolerances. Line items 1–6 are taken from the first part, the enclosure. Line item 7 is the assembly shift of the second part, the ground plate, about the fasteners. Line items 8–13 are taken from the second part, the ground plate. If the enclosure was encountered for a second time later in the Tolerance Stackup, its information would be added again to indicate that the dimensions and tolerances were from the enclosure.

As a general rule, the description of a component or assembly only needs to be stated on the first line that includes dimensional and tolerance data from that part or assembly. Looking at Fig. 14.11, we see that description, part number, and revision for each part in the assembly is only included once for each part. It is understood that all subsequent lines are for the same part until a new description, part number, and revision are encountered. This is not a necessity, however. This information could be included on each line of the Tolerance Stackup report form if desired. The author has just found that this approach makes for an easier-to-read report.

## DIMENSION AND TOLERANCE ENTRY

This section explains how to enter the data for plus/minus Dimensions and Tolerances, assembly shift, and for geometric tolerances, including material condition modifiers, bonus tolerance, datum feature shift, and datum reference frames into the Tolerance Stackup report form.

## GUIDELINES FOR ENTERING PLUS/MINUS DIMENSIONS AND TOLERANCES

This section explains how to enter plus/minus Dimension and Tolerance information into the Tolerance Stackup report form. Entering plus/minus Dimension and Tolerance values into the Tolerance Stackup report form is very easy. Unlike geometric dimensions and tolerances, plus/minus Dimension and Tolerance values are entered on the same line of the Tolerance Stackup report form.

The equal bilateral equivalent Dimension and Tolerance values are entered into the Tolerance Stackup report as follows:

First, the Tolerance Stackup sketch is created and the chain of Dimensions and Tolerances is identified.

Each dimension with its equal bilateral tolerance is numbered as it appears in the chain as described in Chapter 13. Other tolerances such as applicable angular tolerances are also numbered.

Positive or negative direction signs are assigned to each dimension in the Tolerance Stackup.

Each positive dimension value is entered into the + Dims column, and each negative dimension value is entered into the − Dims column; in accordance with the directions assigned in the Tolerance Stackup sketch. Each corresponding tolerance value is entered into the Tol column on the same line as the dimension value. The method for converting plus/minus Dimensional and Tolerance data into equal bilateral format is presented in Chapter 4.

Any other tolerances such as angular tolerances that contribute to the Tolerance Stackup are entered as they are encountered in the chain of Dimensions and Tolerances.

Assembly Shift is numbered and entered in the Tolerance Stackup report as it is encountered in the chain of Dimensions and Tolerances.

The source and original format of each plus/minus Dimension and Tolerance is entered in the Dim/Tol Source & Calcs column. See the following examples:

Two plus/minus Dimensions and Tolerances are part of a Tolerance Stackup.

The first dimension and tolerance is 2.5 ± 0.25. This dimension was determined to be in the positive direction in chain of Dimensions and Tolerances. It is already in equal bilateral format so no conversion is required. The quantity 2.5 is entered into the + Dims column and 0.25 is entered into the Tol column. "2.5 ± 0.25 on dwg" is entered into the Dim/Tol Source & Calcs column. This explains that the dimension and tolerance were taken right from the drawing and shows their original format. See Fig. 14.12.

The second dimension and tolerance is 5.65 + 0.35/−0.65. The equal bilateral equivalent dimension and tolerance value are 5.5 ± 0.5. This dimension was determined to be in the negative direction in chain of Dimensions and Tolerances. The quantity 5.5 is entered into the − Dims column and 0.5 is entered into the Tol column. "5.65 + 0.35/−0.65 on dwg" is entered into the Dim/Tol Source & Calcs column. This explains that

| + Dims | − Dims | Tol | Percent Contrib | Dim/Tol Source & Calcs |
|--------|--------|-----|-----------------|------------------------|
| 2.5000 |        | ± 0.2500 | 33% | 2.5 ± 0.25 on dwg |
|        | 5.5000 | ± 0.5000 | 67% | 5.65 +0.35/−0.65 on dwg |
|        |        | ±   |     |     |
|        |        | ±   |     |     |
|        |        | ±   |     |     |

**FIGURE 14.12** Sample Tolerance Stackup report format: Plus/minus Dimension and Tolerance data entry, right side enlarged.

the dimension and tolerance taken from the drawing were converted to equal bilateral format, and shows their original format. See Fig. 14.12.

The percent contribution column is automatically filled in as described in the previous section. Since both dimensions and tolerances are taken from the same part, the description, part number, and revision data are only entered on the first line. See Fig. 14.13.

Description of Plus/Minus Dimensions:

Each line in the Tolerance Stackup Report Form that includes a dimension value should be clearly labeled. In the Description field, the word *Dim:* is entered followed by the origin and terminus of the dimension. In Fig. 14.13, line item 1 is a dimension from the bottom surface to the mounting face on the sample part. The description in line 1 reads "Dim: Bottom Surface–Mounting Face." This tells the reader that the line item contains a dimension value, and shows that the dimension originates at the bottom surface and terminates at the mounting face. The description in line item 2 reads "Dim: Mounting Face–Top Surface."


## GUIDELINES FOR ENTERING GEOMETRIC DIMENSIONS AND TOLERANCES

This section explains how to enter geometric dimension and tolerance information into the Tolerance Stackup report form. Unlike plus/minus Dimensions and Tolerances, geometric dimensions and tolerances are entered onto separate lines on the Tolerance Stackup report form. Refer to Chapter 9 for methods of converting various geometric dimensions and tolerances into equal bilateral format.

Geometric dimension and tolerance values are entered into the Tolerance Stackup report as follows:

First, the Tolerance Stackup sketch is created and the chain of Dimensions and Tolerances is identified.

Each basic dimension is numbered as it appears in the chain as described in Chapter 13.

Positive or negative direction signs are assigned to each basic dimension in the Tolerance Stackup.

Each geometric tolerance is followed by lines for bonus tolerance and/or datum feature shift as applicable. Refer to the following section for more detailed instructions.

Each geometric tolerance, bonus tolerance, and datum feature shift is numbered as it is encountered in the chain of Dimensions and Tolerances.

| Description of Component / Assy | Part Number | Rev | Item | Description | + Dims | – Dims | Tol | Percent Contrib | Dim / Tol Source & Calcs |
|---|---|---|---|---|---|---|---|---|---|
| Sample Part | 123-ABC | A | 1 | Dim: Bottom Surface - Mounting Face | 2.5000 | | ± 0.2500 | 33% | 2.5 ± 0.25 on dwg |
| | | | 2 | Dim: Mounting Face - Top Surface | | 5.5000 | ± 0.5000 | 67% | 5.65 +0.35/–0.65 on dwg |
| | | | | | | | ±I | | |
| | | | | | | | ±I | | |
| | | | | | | | ±I | | |

**FIGURE 14.13** Sample Tolerance Stackup report format: Plus/minus Dimension and Tolerance data entry, full view.

Other tolerances such as applicable angular tolerances are also numbered as they are encountered in the chain of Dimensions and Tolerances.

Each positive basic dimension value is entered into the + Dims column, and each negative basic dimension value is entered into the − Dims column in accordance with the directions assigned in the Tolerance Stackup sketch.

Geometric tolerance values are entered into the Tol column on a different line than the basic dimension values.

Any other tolerances such as angular tolerances that contribute to the Tolerance Stackup are entered as they are encountered in the chain of Dimensions and Tolerances.

Assembly shift is numbered and entered in the Tolerance Stackup report as it is encountered in the chain of Dimensions and Tolerances.

## BASIC DIMENSIONS IN THE TOLERANCE STACKUP REPORT FORM

Basic dimension values are entered into the Tolerance Stackup report form on a separate line from the geometric tolerances. Like plus/minus Dimensions and Tolerances, where geometric dimensions and tolerances are specified as unequal bilateral or unilateral, they must be converted to equal bilateral format before they can be entered into the Tolerance Stackup report form. The methods of conversion are described in Chapter 9. These primarily occur with profile tolerancing.

If the applicable geometric tolerance specified for the feature located by the basic dimension is stated in equal bilateral format on the drawing, the basic dimension value from the drawing is entered into either the + Dims or − Dims column in accordance with the directions assigned in the chain of Dimensions and Tolerances. If the applicable geometric tolerance specified for the feature located by the basic dimension is not stated in equal bilateral format on the drawing, the basic dimension value must converted before it can be entered in the Tolerance Stackup report form.

Enter the source and original format of the basic dimension in the Dim/Tol Source & Calcs column. See the following examples:

Two surfaces located by basic dimensions are part of a Tolerance Stackup.

The first surface is located by a 50-mm basic dimension from datum feature A. It is toleranced with an equal bilateral profile tolerance of 1.2 mm to datum reference frame A. This dimension was determined to be in the positive direction in chain of Dimensions and Tolerances. It is already in equal bilateral format so no conversion is required. The value 50 is entered into the + Dims column. "50 Basic on dwg" is entered into the Dim/Tol Source & Calcs column. This explains that the dimension was taken right from the drawing and shows its original format. The associated profile tolerance information is included on the lines that precede the basic dimension information. See Fig. 14.14.

The second surface is located by a 22.5-mm basic dimension from datum feature A. It is toleranced with a unilaterally positive profile tolerance of 1 mm to datum reference frame A. Using the conversion techniques in Chapter 9, the equal bilateral equivalent dimension and tolerance value are 23 ± 0.5. This dimension was determined to be in the negative direction in chain of Dimensions and Tolerances. The value 23 is entered into the Dims column. "22.5 Basic on dwg (mean shift)" is entered into the Dim/Tol Source & Calcs column. This explains that the dimension taken from the drawing was converted to equal bilateral format, and shows its original format. The associated profile tolerance information is included on the lines that follow the basic dimension information. See Fig. 14.14.

Sometimes general notes or rules are applied to drawings that state "Unless Specified Otherwise, All Math Data is Basic" or something similar. The intent is to allow dimensions to be measured on the model rather than stated on the drawing. Sometimes the majority of a part's geometry is left undimensioned. The user is directed to the CAD file and the model geometry to obtain basic dimensional data. The user must open the CAD file and measure the point-to-point distance needed for the Tolerance Stackup. In some industries this is very common, and makes a lot of sense. Many part geometries are so complex that they cannot be completely dimensioned anyway, so it makes sense to go to the model for dimensional data. This technique is most commonly used with parts comprised of free-form or warped surfaces, such as automobile body panels, personal electronic device packaging, pump impellers, complex castings, and other unusual geometries.

| + Dims | − Dims | Tol | Percent Contrib | Dim / Tol Source & Calcs |
|---|---|---|---|---|
| | | ± 0.6000 | 55% | Profile 1.2, A |
| | | ± 0.0000 | 0% | N/A - DF$_A$ not a Feature of Size |
| 50.0000 | | ± 0.0000 | 0% | 50 Basic on dwg |
| | 23.0000 | ± 0.0000 | 0% | 22.5 Basic on dwg (Mean Shift) |
| | | ± 0.5000 | 45% | Profile 1, A (Unilateral Positive) |
| | | ± 0.0000 | 0% | N/A - DF$_A$ not a Feature of Size |

**FIGURE 14.14** Sample Tolerance Stackup report format: Basic dimension entry, right side enlarged.

Care must be taken when obtaining basic dimensional data for Tolerance Stackups directly from the model.

The rule when measuring basic dimensions from a CAD model is the same as when the basic dimension is stated explicitly on the drawing. If the applicable geometric tolerance specified for the feature located by the measured basic dimension is not stated in equal bilateral format on the drawing, the basic dimension value must converted before it can be entered in the Tolerance Stackup report form.

## DESCRIPTION OF BASIC DIMENSIONS

Each line in the Tolerance Stackup report form that includes a basic dimension value should be clearly labeled. In the Description field, the word *Dim:* is entered followed by the origin and terminus of the dimension. In Fig. 14.15, line item 3 is a dimension from the bottom surface to datum feature A on the sample part. The description in line 3 reads "Dim: Bottom Surface–Datum Feature A." This tells the reader that the line item contains a dimension value, and shows that the dimension originates at the bottom surface and terminates at datum feature A. The description in line item 4 reads "Dim: Datum Feature A–Top Surface."

## GENERAL GUIDELINES FOR ENTERING GD&T INFORMATION

GD&T is different from plus/minus Dimensions and Tolerances in a number of ways. From a Tolerance Stackup point of view, a big difference is that the dimension value and geometric tolerance information are entered on separate lines in the Tolerance Stackup report form. Another difference is that material condition modifiers can be applied to some geometric tolerances, allowing the tolerance zone to increase in size, leading to bonus tolerance. Another difference is that many geometric tolerances are related to a datum reference frame, which is essential in Tolerance Stackups. Datum feature references may be modified by material condition modifiers, creating the possibility of datum feature shift. See Fig. 14.16.

The geometric tolerance should be entered on a single line in the Tolerance Stackup report form. If the geometric tolerance zone may be modified by a material condition modifier, bonus tolerance should be entered on the next line beneath the line for the geometric tolerance. If the geometric tolerance can be related to a datum reference frame, datum feature shift should be entered on the next line beneath the line for the bonus tolerance. This approach reflects the fact that there is more than one possible contrib-

| Program: | 1(a) | | | | | Stack Information: | | | | |
|---|---|---|---|---|---|---|---|---|---|---|
| Description of Component / Assy | Part Number | Rev | Item | Description | + Dims | − Dims | Tol | Percent Contrib | Dim / Tol Source & Calcs |
| Sample Part | ABC-123 | A | 1 | Profile: Bottom Surface | | | ± 0.6000 | 55% | Profile 1.2, A |
| | | | 2 | Datum Feature Shift | | | ± 0.0000 | 0% | N/A - $DF_A$ not a Feature of Size |
| | | | 3 | Dim: Bottom Surface - Datum Feature A | 50.0000 | | ± 0.0000 | 0% | 50 Basic on dwg |
| | | | 4 | Dim: Datum Feature A - Top Surface | | 23.0000 | ± 0.0000 | 0% | 22.5 Basic on dwg (Mean Shift) |
| | | | 5 | Profile: Bottom Surface | | | ± 0.5000 | 45% | Profile 1, A (Unilateral Positive) |
| | | | 6 | Datum Feature Shift | | | ± 0.0000 | 0% | N/A - $DF_A$ not a Feature of Size |

**FIGURE 14.15** Sample Tolerance Stackup report format: Basic dimension entry, full view.

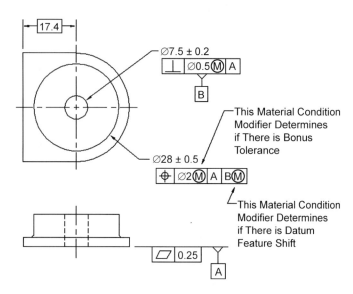

**FIGURE 14.16** Material condition modifiers in feature control frames: Bonus tolerance and datum feature shift.

utor to the Tolerance Stackup in a feature control frame. The approach highlights the contribution of each contributor by separating them onto separate lines.

By default, every geometric tolerance does not have the same number of possible contributors to a Tolerance Stackup. Every geometric tolerance has at least one contributor: the geometric tolerance itself. Any geometric tolerance that can be modified by a material condition modifier may have bonus tolerance, and any geometric tolerance that can be related to a datum reference frame may have datum feature shift.

For example, form tolerances cannot be related to a datum reference frame, so datum feature shift is not possible with form tolerances—there are only one or two special cases where a form tolerance zone can be modified with a material condition modifier, which could lead to bonus tolerance. Profile tolerances can be related to a datum reference frame, but profile tolerance zones cannot be modified by a material condition modifier, so datum feature shift is possible, but bonus tolerance is not possible with profile tolerances.

The same number of lines should be entered into the Tolerance Stackup report form for all like geometric tolerances, regardless if they are specified regardless of feature size (RFS) or with a datum reference frame that does not have datum feature shift. This means that all profile tolerances will have two lines, all positional tolerances will have three lines, etc.

This is an excellent way to ensure that no contributor to the total tolerance is overlooked. If there is no bonus tolerance for a particular geometric tolerance, simply put a zero into the Tol column and state "N/A" in the Dim/Tol Source & Calcs column for that line. If there is no datum feature shift for a particular geometric tolerance, simply put a zero into the Tol column and state "N/A" in the Dim/Tol Source & Calcs column for that line. It is a good idea to state why the bonus tolerance or datum feature shift value is not applicable (N/A), as there are different reasons why bonus tolerance and datum feature shift may not be included for a particular tolerance. Entering the contributors for each geometric tolerance onto separate lines makes it abundantly clear how much each one adds to the Tolerance Stackup. This is the best way to tell if the correct material condition modifiers have been specified.

## PROFILE TOLERANCES

ASME Y14.5M-1994 does not allow profile tolerance zones to be modified by material condition modifiers, so there is never a bonus tolerance associated with profile tolerances. However the datum feature references may be modified by material condition modifiers, so there is the possibility of datum feature shift with profile tolerances.

Profile tolerance information is entered into the Tolerance Stackup report form on two lines. The profile tolerance is entered on the first line and datum feature shift is entered on the second line. Two lines are entered into the Tolerance Stackup report form for every profile tolerance even if there is no datum feature shift for a particular profile tolerance. See Fig. 14.17.

If the profile tolerance is specified in an unequal bilateral or unilateral format, the profile tolerance and its associated basic dimensions must be converted to equal bilateral format before being entered into the Tolerance Stackup report form, as defined in Chapter 9.

Note: Profile may only be applied to basically defined features.

### Profile Tolerance Examples

*Profile Tolerance: Without Datum Feature Shift, Without a Datum Reference Frame*

The profile tolerance applied to the bottom surface in Fig. 14.17 is not related to a datum reference frame; therefore there can not be any datum feature shift. The Tolerance Stackup information for this example is shown in Fig. 14.18.

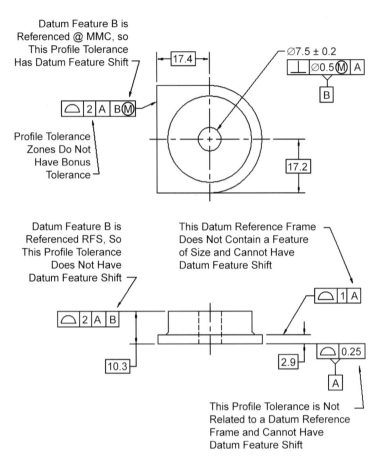

**FIGURE 14.17**   Part with various profile tolerance feature control frames.

## Profile Tolerance: Without Datum Feature Shift, Relative to a Datum Reference Frame Without Datum Features of Size

The profile tolerance applied to the surface 2.9 mm from datum feature A is not related to a datum reference frame that contains any features of size; therefore there can not be any datum feature shift. The Tolerance Stackup information for this example is shown in Fig. 14.19.

## Profile Tolerance: Without Datum Feature Shift, Relative to Datum Features of Size: RFS

The profile tolerance applied to the upper surface 10.3 mm from datum feature A is related to a datum reference frame that contains a feature of size.

| Item | Description | + Dims | − Dims | Tol | Percent Contrib | Dim / Tol Source & Calcs |
|------|-------------|--------|--------|-----|-----------------|--------------------------|
| 1 | Profile: Bottom Surface: DF$_A$ | | | ± 0.1250 | 10% | Profile 0.25 |
| 2 | Datum Feature Shift | | | ± 0.0000 | 0% | N/A - No Datum Reference Frame |
| 3 | Dim: DF$_A$ - Surface @ 2.9mm | 2.9000 | | ± 0.0000 | 0% | 2.9 Basic on dwg |
| | | | | ± | | |

**FIGURE 14.18** Sample Tolerance Stackup report format: Profile tolerance without datum feature shift and without datum reference frame.

| Description of Component / Assy | Part Number | Rev | Item | Description | + Dims | − Dims | Tol | Percent Contrib | Dim / Tol Source & Calcs |
|---|---|---|---|---|---|---|---|---|---|
| Sample Part | ABC-123 | A | 1 | Profile Bottom Surface DF$_A$ | | | ± 0.1250 | 10% | Profile 0.25 |
| | | | 2 | Datum Feature Shift | | | ± 0.0000 | 0% | N/A - No Datum Reference Frame |
| | | | 3 | Dim DF$_A$ - Surface @ 2.9mm | 2.9000 | | ± 0.0000 | 0% | 2.9 Basic on dwg |
| | | | 4 | Profile: Surface @ 2.9mm | | | ± 0.5000 | 25% | Profile 1, A |
| | | | 5 | Datum Feature Shift | | | ± 0.0000 | 0% | N/A - DF$_A$ not a Feature of Size |

**FIGURE 14.19** Sample Tolerance Stackup report format: Profile tolerance without datum feature shift and without datum features of size.

However, there is no datum feature shift because the datum feature of size is specified RFS. Another reason datum feature shift would not be included here is that the datum feature shift would act perpendicular to the Tolerance Stackup direction. The Tolerance Stackup information for this example is shown in Fig. 14.20.

*Profile Tolerance: With Datum Feature Shift, Relative to Datum Features of Size: MMC or LMC*

The profile tolerance applied to the left side surface in the top view 17.4 mm from datum feature B is related to a datum reference frame that contains a feature of size. There is datum feature shift because the datum feature of size is specified at MMC, and the datum feature shift acts in the direction of the Tolerance Stackup. The Tolerance Stackup information for this example is shown in Fig. 14.21.

## POSITIONAL TOLERANCES

Positional tolerance information is entered into the Tolerance Stackup report form on three lines. The positional tolerance is entered on the first line, bonus tolerance is entered on the second line, and datum feature shift is entered on the third line. Three lines are entered into the Tolerance Stackup report form for every positional tolerance even if there is no bonus tolerance or datum feature shift for a particular positional tolerance. See Fig. 14.22 for a sample drawing with positional tolerances. Any related basic dimension values are entered on separate lines in the Tolerance Stackup report form. See Fig. 14.23.

### Positional Tolerance Examples (see Fig. 14.22.)

*Positional Tolerance: With Bonus Tolerance, Without Datum Feature Shift, Relative to a Datum Reference Frame Without Datum Features of Size*

The Positional tolerance applied to the $2 \times \varnothing 6 \pm 0.1$ datum feature B holes is modified by an MMC material condition modifier, which means there is bonus tolerance. The specified datum reference frame does not contain any datum features of size; therefore there can not be any datum feature shift. The Tolerance Stackup information for this example is shown in Fig. 14.23.

*Positional Tolerance: Without Bonus Tolerance, with Datum Feature Shift, Relative to Datum Features of Size: MMC or LMC*

The positional tolerance applied to the $\varnothing 14 \pm 0.15$ datum feature C hole is not modified by an MMC or LMC material condition modifier, which means there is no bonus tolerance. The specified datum reference frame contains a

| Description of Component / Assy | Part Number | Rev | Item | Description | + Dims | − Dims | Tol | Percent Contrib | Dim / Tol Source & Calcs |
|---|---|---|---|---|---|---|---|---|---|
| Sample Part | ABC-123 | A | 1 | Profile: Bottom Surface, $DF_A$ | | | ± 0.1250 | 5% | Profile 0.25 |
| | | | 2 | Datum Feature Shift | | | ± 0.0000 | 0% | N/A - No Datum Reference Frame |
| | | | 3 | Dim: $DF_A$ - Upper Surface | 10.3000 | | ± 0.0000 | 0% | 10.3 Basic on dwg |
| | | | 4 | Profile: Upper Surface | | | ± 1.0000 | 40% | Profile 2, A, B |
| | | | 5 | Datum Feature Shift | | | ± 0.0000 | 0% | N/A - $DF_B$ Specified RFS |
| | | | | | | | ± | | |

**FIGURE 14.20** Sample Tolerance Stackup report format: Profile tolerance without datum feature shift and with datum features of size specified RFS.

| Description of Component / Assy | Part Number | Rev | Item | Description | + Dims | − Dims | Tol | Percent Contrib | Dim / Tol Source & Calcs |
|---|---|---|---|---|---|---|---|---|---|
| Sample Part | ABC-123 | A | | | | | | | |
| | | | 1 | Profile: Bottom Surface: DF$_A$ | | | ± 0.1250 | 5% | Profile 0.25 |
| | | | 2 | Datum Feature Shift | | | ± 0.0000 | 0% | N/A - No Datum Reference Frame |
| | | | 3 | Dim: DF$_A$ - Upper Surface | 10.3000 | | ± 0.0000 | 0% | 10.3 Basic on dwg |
| | | | 4 | Profile: Left Side Surface | | | ± 1.0000 | 30% | Profile 2, A, B @ MMC |
| | | | 5 | Datum Feature Shift | | | ± 0.4500 | 15% | = [(7.5 + 0.2) − (7.5 − 0.2 − 0.5)]/2 |

**FIGURE 14.21** Sample Tolerance Stackup report format: Profile tolerance with datum feature shift.

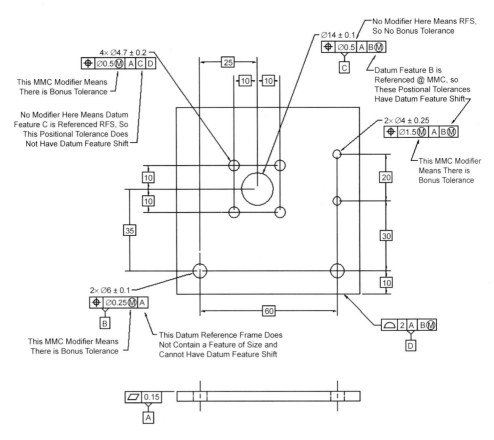

**FIGURE 14.22**   Part with various positional tolerance feature control frames.

datum feature of size referenced at MMC, so there is datum feature shift. The Tolerance Stackup information for this example is shown in Fig. 14.24.

### *Positional Tolerance: with Bonus Tolerance, Without Datum Feature Shift, Relative to Datum Feature of Size: RFS*

The positional tolerance applied to the 4× ⌀4.7 ± 0.2 holes is modified by an MMC material condition modifier, which means there is bonus tolerance. The specified datum reference frame contains a datum feature of size referenced RFS, so there is no datum feature shift. The Tolerance Stackup information for this example is shown in Fig. 14.25.

| Description of Component / Assy | Part Number | Rev | Item | Description | + Dims | − Dims | Tol | Percent Contrib | Dim / Tol Source & Calcs |
|---|---|---|---|---|---|---|---|---|---|
| Sample Part | ABC-123 | A | 1 | Position, DF $_B$ | | | ± 0.1250 | 10% | Position dia 0.25 @ MMC, A |
| | | | 2 | Bonus Tolerance | | | ± 0.1000 | 8% | = (0.1 + 0.1)/2 |
| | | | 3 | Datum Feature Shift | | | ± 0.0000 | 0% | N/A - DF$_A$ not a Feature of Size |
| | | | 4 | Dim, DF$_B$ - DF$_C$ | 35.0000 | | ± 0.0000 | 0% | 35 Basic on Dwg |

**FIGURE 14.23** Sample Tolerance Stackup report format: Positional tolerance with bonus tolerance and without datum features of size.

| Description of Component / Assy | Part Number | Rev | Item | Description | + Dims | − Dims | Tol | Percent Contrib | Dim / Tol Source & Calcs |
|---|---|---|---|---|---|---|---|---|---|
| Sample Part | ABC-123 | A | 1 | Position: $DF_B$ Hole | | | ± 0.1250 | 10% | Position dia 0.25 @ MMC, A |
| | | | 2 | Bonus Tolerance | | | ± 0.1000 | 8% | = (0.1 + 0.1)/2 |
| | | | 3 | Datum Feature Shift | | | ± 0.0000 | 0% | N/A - $DF_A$ not a Feature of Size |
| | | | 4 | Dim: $DF_B$ - $DF_C$ | 35.0000 | | ± 0.0000 | 0% | 35 Basic on Dwg |
| | | | 5 | Position: $DF_C$ Hole | | | ± 0.2500 | 20% | Position dia 0.5, A, B @ MMC |
| | | | 6 | Bonus Tolerance | | | ± 0.0000 | 0% | N/A - RFS |
| | | | 7 | Datum Feature Shift | | | ± 0.2250 | 18% | = [(6 + 0.1) − (6 − 0.1 − 0.25)]/2 |

**FIGURE 14.24**   Sample Tolerance Stackup report format: Positional tolerance without bonus tolerance and with datum feature shift.

| Description of Component / Assy | Part Number | Rev | Item | Description | + Dims | − Dims | Tol | Percent Contrib | Dim / Tol Source & Calcs |
|---|---|---|---|---|---|---|---|---|---|
| Sample Part | ABC-123 | A | 1 | Position: $DF_A$ Hole | | | ± 0.1250 | 10% | Position dia 0.25 @ MMC, A |
| | | | 2 | Bonus Tolerance | | | ± 0.1000 | 8% | = (0.1 + 0.1)/2 |
| | | | 3 | Datum Feature Shift | | | ± 0.0000 | 0% | N/A - $DF_A$ not a Feature of Size |
| | | | 4 | Dim: $DF_B$ − $DF_C$ | 35.0000 | | ± 0.0000 | 0% | 35 Basic on DWG |
| | | | 5 | Position: 4X Dia 4.7 ± 0.2 Holes | | | ± 0.2500 | 20% | Position dia 0.5, A, C, D |
| | | | 6 | Bonus Tolerance | | | ± 0.2000 | 15% | = (0.2 + 0.2)/2 |
| | | | 7 | Datum Feature Shift | | | ± 0.0000 | 0% | N/A - Datum Features Referenced RFS |

**FIGURE 14.25** Sample Tolerance Stackup report format: Positional tolerance with bonus tolerance and without datum feature shift.

*Positional Tolerance: With Bonus Tolerance, with Datum Feature*
*Shift, Relative to Datum Feature of Size: MMC or LMC*

The positional tolerance applied to the $2\times \oslash4 \pm 0.25$ holes is modified by an MMC material condition modifier, which means there is bonus tolerance. The specified datum reference frame contains datum features of size referenced at MMC; so there is datum feature shift. The Tolerance Stackup information for this example is shown in Fig. 14.26.

## ORIENTATION TOLERANCES

Orientation tolerances are commonly overlooked in Tolerance Stackups. Most part features are located by another tolerance, such as position or profile, and the orientation tolerance merely limits how much the feature may tilt. In almost all of these cases the orientation tolerance would not be included in the Tolerance Stackup, as it merely refines the orientation of the feature. The feature's location tolerance would be included in the Tolerance Stackup, as the location tolerance determines where the feature is in relation to the rest of the part.

An orientation tolerance applied to a flat surface may need to be included in a Tolerance Stackup in cases where the orientation of the surface can cause other features to tilt, reducing or increasing the gap or interference being studied. Whether an orientation tolerances applied to a flat surface is included in the Tolerance Stackup depends on the geometry of the parts being studied and how they are toleranced.

Orientation tolerance zones specified for flat surfaces or other surfaces without size may not be modified by a material condition modifier. This means there is no bonus tolerance when an orientation tolerance is applied to a flat surface or surface without size. However, there may be datum feature shift if any datum features of size are referenced at MMC or LMC in the datum reference frame.

Orientation tolerances may also be applied to features of size. When applied to the center geometry of a feature of size, orientation tolerance zones may be modified by a material condition modifier such as MMC or LMC. In these cases, the orientation tolerance may have bonus tolerance. There may also be datum feature shift if any datum features of size are referenced at MMC or LMC in the datum reference frame.

Sometimes only the orientation of a hole matters, as it may cause other features to tilt, thereby reducing or increasing a gap or interference being studied. In such a case the orientation tolerance would be included in the Tolerance Stackup. The Tolerance Analyst must recognize the relationship of each orientation tolerance to all part features, dimensions, and tolerances in

| Description of Component / Assy | Part Number | Rev | Item | Description | + Dims | − Dims | Tol | Percent Contrib | Dim / Tol Source & Calcs |
|---|---|---|---|---|---|---|---|---|---|
| Sample Part | ABC-123 | A | 1 | Position: $DF_B$ Hole | | | ± 0.1250 | 10% | Position dia 0.25 @ MMC, A |
| | | | 2 | Bonus Tolerance | | | ± 0.1000 | 8% | = f0.1 + 0.1)/2 |
| | | | 3 | Datum Feature Shift | | | ± 0.0000 | 0% | N/A − $DF_A$ not a Feature of Size |
| | | | 4 | Dim: $DF_B$ − Dia 4 Holes | 30.0000 | | ± 0.0000 | 0% | 30 Basic on Dwg |
| | | | 5 | Position: 2X Dia 4 ± 0.25 Holes | | | ± 0.7500 | 60% | Position dia 1.5, A, B @ MMC |
| | | | 6 | Bonus Tolerance | | | ± 0.2500 | 20% | = (0.25 + 0.25)/2 |
| | | | 7 | Datum Feature Shift | | | ± 0.2250 | 18% | = [(6 + 0.1) − (6 − 0.1 − 0.25)]/2 |

**FIGURE 14.26** Sample Tolerance Stackup report format: Positional tolerance with bonus tolerance and with datum feature shift.

the chain of Dimensions and Tolerances and determine if the orientation
tolerance should be included in the Tolerance Stackup. See Figs. 9.8 and 9.9
for examples.

Another fairly common case where an orientation tolerance is included
in the Tolerance Stackup is where the orientation tolerance is applied to a
secondary datum feature of size.

Where a secondary datum feature of size oriented to a primary datum
feature is in the chain of Dimensions and Tolerances, the orientation
tolerance may play a role in the Tolerance Stackup. For example, if this
datum feature of size is referenced at MMC by another feature's geometric
tolerance, the orientation tolerance would be used to calculate datum feature
shift. Figure 14.27 shows a part where an Orientation Tolerance may be
included in several Tolerance Stackups.

For example, there is only 1.5-mm nominal distance between the
cylindrical surfaces of the datum feature B hole and the $\varnothing 13 \pm 0.12$ Boss in
Fig. 14.27. It will be necessary to determine the minimum wall thickness

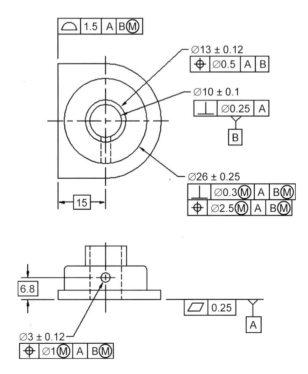

**FIGURE 14.27** Orientation tolerance on datum feature B may be part of Tolerance
Stackups.

between these features to make sure the part will not become to weak at its worst-case condition. Looking at the figure, it is clear that the orientation of datum feature B affects the wall thickness, as the wall thickness will be smaller the more datum feature B tilts. The perpendicularity tolerance applied to datum feature B must be included in the Tolerance Stackup to determine the minimum wall thickness.

Orientation tolerances applied to features of size are entered into the Tolerance Stackup using the same format as positional tolerances. Three lines are used, the first for the orientation tolerance, the second for the bonus tolerance, and the third for datum feature shift. As with positional tolerance, if there is no bonus tolerance or datum feature shift, a zero value is entered on that line and the reason for the zero is entered in the Dim/Tol Source & Calcs column.

In some cases, Orientation tolerances could have a huge effect on a Tolerance Stackup, similar to how form tolerances are described in the following section and in Chapter 20.

## FORM TOLERANCES

Form tolerances are usually not included in Tolerance Stackups. As with orientation tolerances, there are certain situations where form tolerances are included, but for most problems these situations are outside of the norm.

Although commonly overlooked in linear Tolerance Stackups, form tolerances may play a role in the effect of all interfaces between parts in a Tolerance Stackup. Depending on part geometry, the resulting form error could cause translational or rotational error elsewhere on the part.

The possible effect of a form tolerance on the Tolerance Stackup depends on several factors:

- Interface geometry
- How the interface geometry relates to the part geometry being studied
- Whether the interfacial surfaces are subject to deformation at assembly, e.g. whether the interfacial surfaces subjected to axial loading from fasteners

Whether the form error on these surfaces causes additional translational or rotational error in the Tolerance Stackup must be carefully analyzed. It is beyond the scope of this chapter to present all the considerations, rules, and case studies. Suffice to say that the Tolerance Analyst must pay careful attention to these sorts of interfaces and consider the possible consequence any form error may have on the Tolerance Stackup.

For critical Tolerance Stackups where dimensions are tight, tolerances are near process capability, and the geometry shows that the form error could cause a failure, form tolerances may need to be added to the Tolerance Stackup.

See Chapter 20 in this text for more in-depth coverage of form tolerances in Tolerance Stackups.

## RUNOUT TOLERANCES

Runout tolerance information is entered into the Tolerance Stackup report form on two lines. The runout tolerance is entered on the first line and datum feature shift is entered on the second line. A runout tolerance may only be specified RFS, so there is no bonus tolerance.

Datum features of size referenced by runout tolerances are typically specified RFS. Although it is not explicitly stated in the ASME Y14.5M-1994 standard, all of the examples show datum reference frames with datum features of size references RFS. This has led many readers to believe that runout tolerances may only be related to datum features of size referenced RFS. However, this is not true: Runout tolerances may be related to datum features of size referenced at MMC or LMC. This is not to say that is a good idea to reference the datum features of size at MMC or LMC with a runout tolerance; it merely means it is legal. In the author's experience, most datum feature references for runout are specified RFS, so this apparent confusion will most likely not impact the majority of Tolerance Stackups being done with runout.

The reader is directed to contact Advanced Dimensional Management™ for more information on this subject.

# 15

## Tolerance Stackup Direction and Tolerance Stackups with Trigonometry

The first part of this chapter discusses the direction or orientation of features and the direction of their dimensions and tolerances. Of primary importance is how the angle between part features and the direction of the Tolerance Stackup determines whether they are included in the chain of Dimensions and Tolerances.

The second part of this chapter discusses how tolerances and assembly shift may allow parts in an assembly to translate or rotate about one another, and how rotation may lead to increased variation.

The role of trigonometry in the Tolerance Stackup is discussed throughout the chapter.

### DIRECTION OF DIMENSIONS AND TOLERANCES IN THE TOLERANCE STACKUP

The direction of the Tolerance Stackup and the geometry of the parts being studied determine which dimensions and tolerances should be included in the chain of Dimensions and Tolerances. Only those dimensions and tolerances that contribute to the Tolerance Stackup are included in the chain of Dimensions and Tolerances—all other dimensions and tolerances should be excluded. It is not always easy to visualize which dimensions and tolerances affect the Tolerance Stackup. Chapter 13 discusses the importance of creating a Tolerance Stackup sketch before attempting to solve the Tolerance Stackup, as the sketch is essential for visualizing the problem and determining which tolerances should be included. Making the Tolerance Stackup sketch is the best way to make sure all the contributing dimensions, tolerances, and occurrences of assembly shift are included in the chain of Dimensions and Tolerances.

Generally speaking, only those dimensions and variables (tolerances, bonus tolerances, datum feature shift, and assembly shift) that are aligned with the direction of the Tolerance Stackup should be included in the chain of Dimensions and Tolerances. This includes dimensions and variables that can be projected or resolved into the direction of the Tolerance Stackup using trigonometry, such as the dimension and tolerance for a surface that is at a $45°$ angle with the Tolerance Stackup direction.

Usually dimensions and tolerances that are perpendicular to the Tolerance Stackup have no effect on the result and should not be included. For example, the dimensions and tolerances for horizontal surfaces rarely have an effect on a Tolerance Stackup done in the vertical direction. Figure 15.1 shows the simple part used in Chapters 7 and 8. The Tolerance Stackup direction is

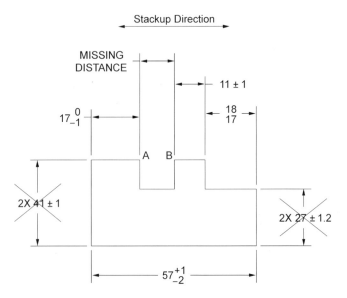

**FIGURE 15.1** Direction of dimensions in the Tolerance Stackup. Because the Tolerance Stackup direction is horizontal, only the horizontal Dimensions and Tolerances are included in the chain of Dimensions and Tolerance. The vertical Dimensions and Tolerances do not affect the horizontal Tolerance Stackup.

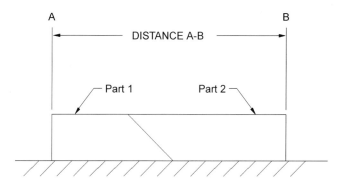

**FIGURE 15.2**   Simple Assembly with inclined surfaces.

horizontal in this example. Only the horizontal dimensions and tolerances contribute to this Tolerance Stackup. The vertical dimensions and tolerances do not play a role in the Tolerance Stackup and are therefore not included in the chain of Dimensions and Tolerances.

Sometimes one or more surfaces are at an angle to the Tolerance Stackup direction, such as 45°. The Tolerance Analyst may recognize that the angled surfaces affect the distance being studied and that the dimensions and tolerances for these surfaces must be included in the chain of Dimensions and Tolerances. In such cases the tolerances are manipulated trigonometrically and included in the chain of Dimensions and Tolerances. Depending on the alignment of the dimensions, the dimensions may also require trigonometric manipulation.

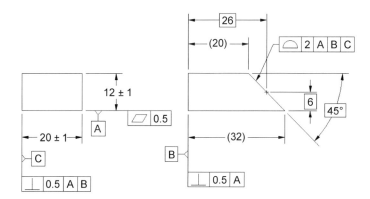

**FIGURE 15.3**   Detail drawing of part 1 with inclined surface.

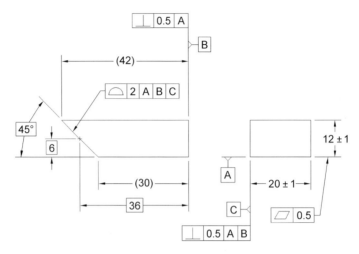

**FIGURE 15.4**   Detail drawing of part 2 with inclined surface.

Figure 15.2 shows a simple assembly of two parts. The two parts mate along inclined surfaces which are oriented 45° from the direction of the Tolerance Stackup. A Tolerance Stackup is to be performed that determines the minimum and maximum overall length for the assembly, which is shown as Distance A-B in Fig. 15.2. Detail drawings of the parts are shown in Fig. 15.3 and 15.4. The parts have been dimensioned and toleranced functionally, and an equal bilateral profile of a surface tolerance has been applied

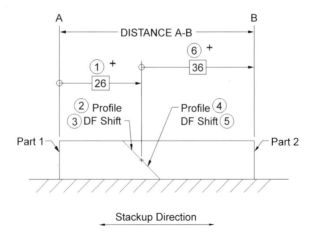

**FIGURE 15.5**   Tolerance Stackup sketch for simple assembly with inclined surfaces chain of Dimensions and Tolerances.

| Program: | Tolerance Analysis and Stackup Manual | | | | | | | | |

| Product: | Part Number ANG-5 | Rev A | Description Assembly with Inclined Surfaces |

| Problem: | Need to Determine the Overall Width of the Assembly and the Effect of the Profile Tolerances Applied to the Inclined Surfaces |

| Objective: | Determine the Overall Width of the Assembly |

Stack Information:

Stack No: Figure 15-6
Date: 07/04/02
Revision A

Direction: Horizontal
Author: BR Fischer

| Description of Component / Assy | Part Number | Rev | Item | Description | + Dims | - Dims | Tol | Percent Contrib | Dim / Tol Source & Calcs |
|---|---|---|---|---|---|---|---|---|---|
| Part 1 | ANG-5.1 | A | 1 | Dim: Datum B - Midpoint Inclined Surface | 26.0000 | | +/- 0.0000 | 0% | 26 Basic on Dwg |
| | | | 2 | Profile: Inclined Surface (See Note 1) | | | +/- 1.4142 | 50.0% | Profile 1, A, Bm: $x = 1 / \cos 45$ deg $= +/-1.4142$ |
| | | | 3 | Datum Feature Shift | | | +/- 0.0000 | 0.0% | N/A - $DF_A$ not a Feature of Size |
| Part 2 | ANG-5.2 | A | 4 | Profile: Inclined Surface (See Note 1) | | | +/- 1.4142 | 50.0% | Profile 1, A, Bm: $x = 1 / \cos 45$ deg $= +/-1.4142$ |
| | | | 5 | Datum Feature Shift | | | +/- 0.0000 | 0.0% | N/A - $DF_A$ not a Feature of Size |
| | | | 6 | Dim: Midpoint Inclined Surface - Datum B | 36.0000 | | +/- 0.0000 | 0% | 36 Basic on Dwg |

| Dimension Totals | 62.0000 | 0.0000 |
| Nominal Distance: Pos Dims - Neg Dims = | 62.0000 |

RESULTS:

| | Nom | Tol | Min | Max |
|---|---|---|---|---|
| Arithmetic Stack (Worst Case) | 62.0000 | +/- 2.8284 | 59.1716 | 64.8284 |
| Statistical Stack (RSS) | 62.0000 | +/- 2.0000 | 60.0000 | 64.0000 |
| Adjusted Statistical: 1.5*RSS | 62.0000 | +/- 3.0000 | 59.0000 | 65.0000 |

Notes:

1 - The Profile tolerance applied to the Inclined Surfaces must first be converted to an Equal-Bilateral +/- tolerance, and then projected into the direction of the Tolerance Stackup. This is done by multiplying the Equivalent Equal-Bilateral tolerance value by 1 / cosine of 45 degrees, which gives 1.4142 in this example.

2 - The Perpendicularity tolerance applied to the Datum Feature B surfaces does not contribute to the Tolerance Stackup, so it is not included in the Chain of Dimensions and Tolerances.

Assumptions:

Suggested Action:

**FIGURE 15.6** Tolerance Stackup report for simple assembly with inclined surfaces.

to the inclined surfaces of both parts. Remember, profile tolerances apply normal to the surface; so the profile tolerances values must first be converted to ± format and then must be projected in the direction of the Tolerance Stackup.

Figure 15.5 shows the Tolerance Stackup sketch for this problem. The Tolerance Stackup report is shown in Fig. 15.6. Converting the profile tolerance to ± is easy: equal bilateral profile of 2 = ± 1. The trigonometry for projecting the tolerance value in the direction of the Tolerance Stackup is shown in Figs. 15.7 and 15.8. This problem is relatively easy to solve, as there are only two mating surfaces at an angle with the Tolerance Stackup, and it is easy to recognize what must be done and do it. Depending on the trigonometric skills of the Tolerance Analyst, it may be obvious that in this case, all that is required is to multiply the equal bilateral profile tolerance value by 1/cosine 45°. Usually it is a bit more difficult to visualize the problem. With angles other than 45° it is critical to recognize that the Profile tolerance is normal to the surface, and it is the angle the tolerance vector makes with the Tolerance Stackup direction that must be solved and included in the chain of Dimensions and Tolerances. A common error is to solve for the angle the surface makes with the Tolerance Stackup direction. Review Figs. 15.7 and 15.8 carefully to make sure this point is clear.

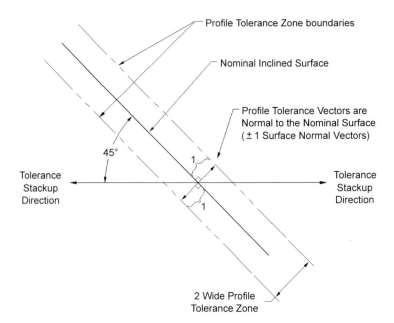

**FIGURE 15.7** Profile Tolerance zone with surface normal vectors for part with inclined surface.

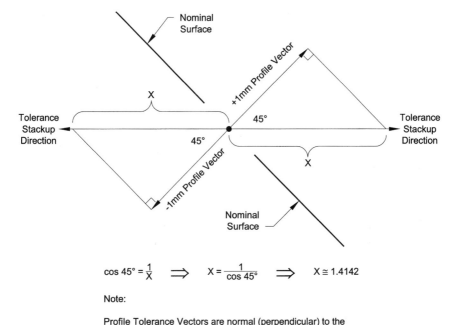

$$\cos 45° = \frac{1}{X} \implies X = \frac{1}{\cos 45°} \implies X \cong 1.4142$$

Note:

Profile Tolerance Vectors are normal (perpendicular) to the nominal surface.

Trigonometry for Converting the Profile Tolerance

**FIGURE 15.8** Trigonometry for converting the profile tolerance zone. Note: Profile tolerance vectors are normal (perpendicular) to the nominal surface.

## DIRECTION OF VARIABLES AND INCLUSION IN THE TOLERANCE STACKUP

Later in this chapter examples will be discussed that show how variables (tolerances, bonus tolerances, datum feature shift, and assembly shift) can be treated as purely translational (linear) displacements or as rotational variation that is projected as a translational displacement. In Tolerance Stackups where the variables are treated as adding translational variation only, all variables (tolerances, datum feature shift, assembly shift) that act perpendicular or normal to the Tolerance Stackup direction can be eliminated.

For datum feature shift and assembly shift to have their full effect in a Tolerance Stackup, the features they are related to or derived from must be perpendicular to the Tolerance Stackup direction. Because tolerances apply normal to the surface of a feature, a tolerance applied to a feature that is perpendicular to the Tolerance Stackup direction will be parallel to the

Tolerance Stackup direction. See Fig. 15.9. Datum feature shift and assembly shift will not contribute the Tolerance Stackup if the axes of the features that cause datum feature shift or assembly shift are parallel to the Tolerance Stackup direction. Figure 15.10 shows a detail drawing of a simple part with a datum feature of size, datum feature B. The profile tolerance applied all around the part is related to datum reference frame A, B at MMC. The profile tolerance applied to the top surface is also related to datum reference frame A, B at MMC. Figure 15.11 shows an assembly of two of the parts from Fig. 15.10 bolted together. This assembly is subject to datum feature shift and assembly shift, because of the part geometry and because of how the parts are dimensioned and toleranced. However, datum feature shift and assembly shift only act in the horizontal direction, which is perpendicular to the axis of the holes. Datum feature shift and assembly shift would be included in a Tolerance Stackup to determine the overall width of the assembly (distance A-B). Datum feature shift and assembly shift would not be included in a Tolerance Stackup to determine the overall height of the assembly (distance C-D).

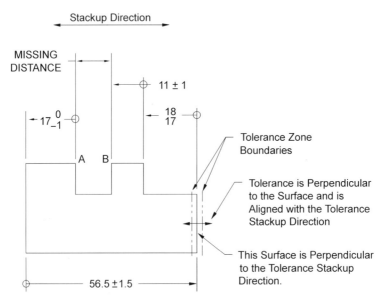

**FIGURE 15.9** Tolerances in the direction of the Tolerance Stackup: feature and zone orientation. Tolerances applied to features that are perpendicular to the Tolerance Stackup direction are parallel with the Tolerance Stackup direction. These tolerances are included in the chain of Dimensions and Tolerances without any trigonometric manipulation.

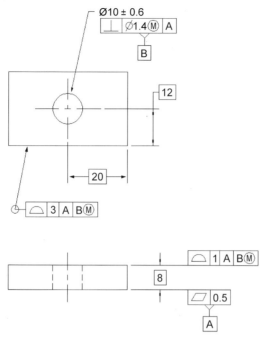

**FIGURE 15.10** Sample part for direction of datum feature shift in the Tolerance Stackup.

## RECAP OF RULES FOR DIRECTION OF DIMENSIONS AND TOLERANCES

- Only those dimensions and tolerances that are related to features that affect the Tolerance Stackup are included in the chain of Dimensions and Tolerances.

- In most cases, dimensions and tolerances that are perpendicular to the Tolerance Stackup direction are not included in the Tolerance Stackup. This includes ± tolerances, geometric tolerances, bonus tolerances, datum feature shift, and assembly shift.

- Contributing dimensions, tolerances, datum feature shift, and assembly shift that act in the direction of the Tolerance Stackup are included in the Tolerance Stackup without trigonometric manipulation.

- Contributing dimensions, tolerances, datum feature shift, and assembly shift that are at an angle other than 0°, 90°, 180°, or 270° (etc.) to the Tolerance Stackup are projected in the direction of the Tolerance Stackup using trigonometric manipulation.

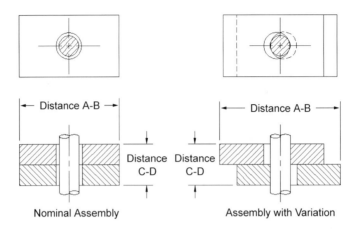

Nominal Assembly                    Assembly with Variation

In this Example, Datum Feature Shift
and Assembly Shift  act in this Direction.

Datum Feature Shift and Assembly Shift would be included in a
Tolerance Stackup to determine Horizontal Distance A-B, because
the axis of the Hole (Datum Feature B) is Perpendicular to the
Tolerance Stackup direction.  Datum Feature Shift and Assembly
Shift have an effect on a Horizontal Tolerance Stackup.

Datum Feature Shift and Assembly Shift would not be included in a
Tolerance Stackup to determine Vertical Distance C-D, because the
axis of the Hole (Datum Feature B) is Parallel to the Tolerance
Stackup direction.  Datum Feature Shift and Assembly Shift have no
effect on the Vertical Tolerance Stackup.

**FIGURE 15.11**   Datum feature shift, assembly shift and the Tolerance Stackup direction.

## CONVERTING ANGULAR DIMENSIONS AND TOLERANCES USING TRIGONOMETRY

Sometimes a Tolerance Stackup requires one or more dimensions and tolerances to be resolved into the direction of the Tolerance Stackup using trigonometry. Figure 15.12 illustrates an example where one ± dimension and tolerance are at an angle to the direction of the Tolerance Stackup (horizontal in this example). The dimension and tolerance must be converted using trigonometry to be collinear or parallel to the other dimensions. Once the trigonometry is performed, the dimension and tolerance values are multiplied by the appropriate coefficient and the results entered into the Tolerance

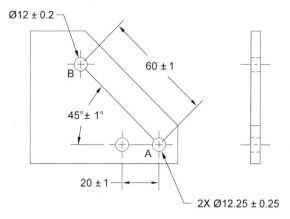

**FIGURE 15.12** Part with Dimension and Tolerance at an angle.

Stackup. Figure 15.13 shows the same part with the angular dimension and tolerance resolved into the direction of the Tolerance Stackup.

For example, an engineer from another group wants to know the minimum and maximum horizontal distance between holes A and B in Figs. 15.12 and 15.13. The only dimensions and tolerances to consider in the calculations are $60 \pm 1$ and $45° \pm 1°$. As the engineer asked for the minimum and maximum distance between the holes, both the angle and distance must be entered into the trigonometric calculations at their worst-case condition as follows:

## Converting Derived Limit Dimensions to Equal Bilateral Format

- Minimum distance: The minimum distance in the horizontal direction occurs when the angle is largest and the length of the hypotenuse is smallest.

Angle:

$$45° + 1° = 46°$$

Hypotenuse:

$$60 - 1 = 59$$

Calculation:

$$X = 59 \cos 46° = 40.98$$

This is the lower limit.

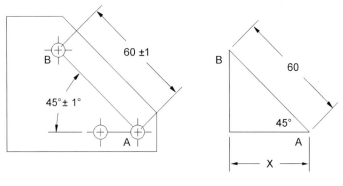

Equivalent Triangle

Nominal Calculations:

cos 45° = X/60

X = 60   cos 45°

X = 42.43

**FIGURE 15.13**   Part with Dimension and Tolerance at an angle: Angular Dimension and Tolerance resolved into horizontal direction.

- Maximum distance: The maximum distance in the horizontal direction occurs when the angle is smallest and the length of the hypotenuse is largest.

    Angle:

    $$45^\circ - 1^\circ = 44^\circ$$

    Hypotenuse:

    $$60 + 1 = 61$$

    Calculation:

    $$X = 61 \cos 44^\circ = 43.88$$

    This is the upper limit.

- Convert to equivalent equal bilateral ± tolerance:

    Upper limit (metric format) = 43.88
    Lower limit (metric format) = 40.98

- Subtract the lower limit from the upper limit to obtain the total tolerance.

  Total tolerance = 43.88 − 40.98 = 2.9

- Divide the total tolerance by 2 to obtain the equal bilateral tolerance value.

  Equal bilateral tolerance value $= \dfrac{2.9}{2} = 1.45$

- Add the equal bilateral tolerance value to the lower limit. This is the adjusted nominal value.

  Adjusted nominal value = 40.98 + 1.45 = 42.43

  (Note: The adjusted nominal value can also be obtained by subtracting the equal bilateral tolerance value from the upper limit.)

Conversion complete:

Equal bilateral equivalent = 42.43 ± 1.45

The equivalent equal bilateral dimension and tolerance value would be entered into a Tolerance Stackup to represent the variations of these two dimensions. It is likely that the distance being studied in the Tolerance Stackup would include more dimensions and tolerances, and this would be but one entry in the Tolerance Stackup.

It is important in calculations like these to remember to include the angular tolerance as well as the linear tolerance. Figure 15.14 shows the nominal, minimum, and maximum triangles with the resolved tolerance zone.

Figure 15.15 illustrates an example where a geometrically toleranced feature is at an angle to the direction of the Tolerance Stackup (horizontal in this example). The dimension must be converted using trigonometry to be colinear or parallel to the other dimensions. The positional tolerance zone for hole A is cylindrical and therefore allows the same variation or displacement in any direction normal to the axis of the hole, including horizontal—no conversion is needed for the positional tolerance zone. Once the trigonometry is performed for the basic dimension, the resolved dimension value and the converted positional tolerance value entered into the Tolerance Stackup.

Figure 15.16 shows the same part with the angular dimension and tolerance resolved into the direction of the Tolerance Stackup.

For example, an engineer from another group wants to know the minimum and maximum distance between holes A and B in Figs. 15.15 and 15.16. The only dimensions to consider in the calculations are the basic 60 and basic 45° dimensions. The horizontal equivalent dimension can be derived from these basic dimensions. The positional tolerance on hole A must also be entered into the calculations, but as mentioned above, since the tolerance zone

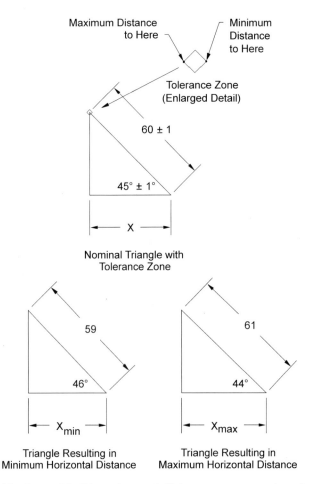

**FIGURE 15.14** Part with Dimension and Tolerance at an angle: triangles and tolerance zone.

is cylindrical, it is not necessary to include the tolerance in the trigonometric calculations. The positional tolerance on hole B is not included in the calculations as it the referenced datum feature, and since it is referenced RFS, there is no datum feature shift.

## Converting Angular Basic Dimension to Horizontal Equivalent

- Nominal distance: Since the dimensions are basic, only one triangle needs to be resolved to find the horizontal dimension: this is the nominal or basic triangle.

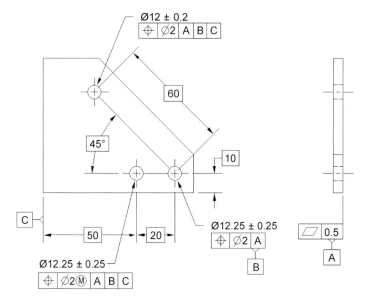

**FIGURE 15.15** Part with basic dimension and geometric tolerance at an angle.

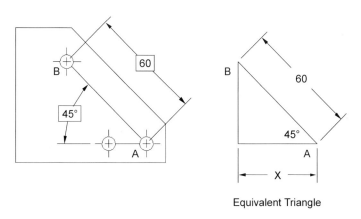

Equivalent Triangle

Nominal Calculations:

$\cos 45° = X/60$

$X = 60 \quad \cos 45°$

$X = 42.43$

**FIGURE 15.16** Part with basic dimension and geometric tolerance at an angle: Basic angular dimension resolved into the horizontal direction.

Angle:

45° basic

Distance:

60 basic

Calculation:

$X = 60 \cos 45° = 42.43$

This is the nominal horizontal dimension value.

- Convert the positional tolerance to equivalent equal bilateral ± tolerance: Divide the specified positional tolerance by 2.

Equal bilateral equivalent $= \dfrac{2}{2} = \pm 1$

Conversion complete:

Equal bilateral equivalent $= 42.43 \pm 1$

The equivalent equal bilateral dimension and tolerance value would be entered into a Tolerance Stackup to represent the variation of these two dimensions. The positional tolerance would be entered on a separate line, and it would include separate lines for bonus tolerance and datum feature shift, which in this example were both zero.

## TOLERANCE STACKUP UNITS

Tolerance Stackups may be performed in either linear or polar units. This text concentrates on Tolerance Stackups using linear units. Polar unit Tolerance Stackups are less common in most industries, but in may be very common in other industries, such as where optics are studied or perhaps in spacecraft flight path calculations. Several Tolerance Stackups done in this section of the text take rotation into account; projecting the angle one or more parts may rotate into the direction of the Tolerance Stackup. Ultimately, this rotation is translated into linear units so it is compatible with the rest of the linear variation in the Tolerance Stackup.

A Tolerance Stackup cannot combine units; that is, the summations done in a Tolerance Stackup must be done using only one type of units. For example, line items 1, 2, and 3 can't be reported in linear units and lines 4, 5, and 6 be reported in angular units. If the goal of a Tolerance Stackup is to determine a minimum or maximum distance, then the Tolerance Stackup should be reported in linear units. If the goal of a Tolerance Stackup is to determine a minimum or maximum angle, then the Tolerance Stackup should be reported in angular units.

It is appropriate, however, to derive some or all of the values in a linear Tolerance Stackup from angular relationships and units, if needed. It is also appropriate to derive some or all of the values in an angular Tolerance Stackup from linear relationships and units, if needed. In fact, the entire Tolerance Stackup could be done using one set of units and reported in another, say where all the tolerances are treated and summed as linear displacements, but the final results are converted into angular units and reported as such. The approach used to solve each Tolerance Stackup problem must be carefully considered before proceeding.

It is common to proceed down a path only to find a different path is necessary. This is okay. The original Tolerance Stackup should be saved and possibly copied for use in the revised approach.

The units to report in the Tolerance Stackup are determined by the goal of the Tolerance Stackup, often by the initial question asked by the person requesting the study, such as

- What is the minimum gap possible between the flanges on these two parts? This question leads the analyst to understand that the requestor wants a linear Tolerance Stackup (gap), reported in linear units. Other keywords for a linear Tolerance Stackup are *distance*, *space*, *overlap*, and *displacement*.
- How much can the machined face on this part tilt relative to the slot on the mating part? This question leads the analyst to understand that the requestor wants an angular Tolerance Stackup (tilt), reported in angular units. Other keywords for an angular Tolerance Stackup are *angle*, *rotation*, and *inclination*.

Anytime someone asks for a Tolerance Stackup to be performed, it is important to ask specific questions to make sure their request is clearly understood. Keywords such as *clearance*, *interference*, *dimension*, *relationship*, and others should flag the analyst to ask more questions to make sure the goal of the Tolerance Stackup is understood.

## ROTATION OF PARTS WITHIN A LINEAR TOLERANCE STACKUP

Another common situation is where one or more components in an assembly may rotate or tilt within a Tolerance Stackup. Parts may rotate because of tolerances specified on surfaces, or application of forces that deform part features, or they may rotate by their holes or slots shifting about fasteners, pins, shafts, keys, tabs, etc. The last factor in this list is a form of assembly shift, and is referred to as *rotational assembly shift*.

In cases such as these it may be necessary to solve the Tolerance Stackup twice using two methods to determine which leads to the greatest possible variation: first treating all the variation as if it was purely translation along the direction of the Tolerance Stackup, and second by treating the variation as a combination of rotation and translation and using trigonometry to resolve the effects back into the direction of the Tolerance Stackup. Often an educated guess is needed to determine which is more likely, translation or rotation. Assembly personnel may be confident that they can install a part in the horizontal position, eliminating the possibility of rotation for that com-

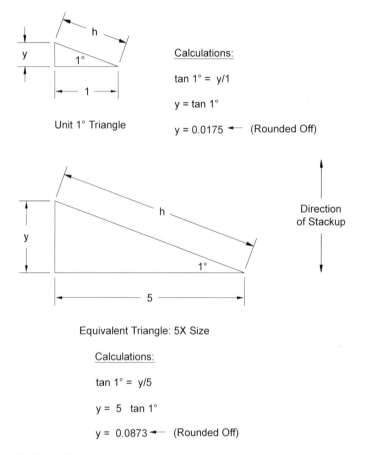

Unit 1° Triangle

Calculations:

$\tan 1° = y/1$

$y = \tan 1°$

$y = 0.0175$ ◂— (Rounded Off)

Direction of Stackup

Equivalent Triangle: 5X Size

Calculations:

$\tan 1° = y/5$

$y = 5 \ \tan 1°$

$y = 0.0873$ ◂— (Rounded Off)

**Figure 15.17**   Like triangles and resulting linear displacement along Tolerance Stackup direction.

ponent, but they may not be confident about where the component is placed, which leads to including the translational tolerance in the Tolerance Stackup. Each case is different and must be considered carefully.

The possible effects of rotation are very important. It cannot be overstated how important they may be. Part rotation can be a hidden source of large amounts of possible variation. Without recognizing where it may occur, determining the likelihood of its occurrence and analyzing its effect on the assembly, the designer may believe that the translational Tolerance Stackup performed is an adequate model of the possible variation.

Rotational variation is greatest when it is projected a large distance. This is a simple function of angular relationships, or *like triangles*. For example, if a surface has $\pm 1°$ variation from nominal, and the surface is 1 unit long, the maximum equivalent linear displacement due to the angle is 0.0175 linear units. If that same surface is 5 units long, the maximum equivalent linear displacement due to the angle increases to $5(0.0175) = \sim 0.0873$ linear units (see Fig. 15.17).

The following examples compare the effects of rotation and translation in Tolerance Stackups done for a simple assembly with $\pm$ dimensions and tolerances. Figures 15.18 through 15.29 show the effects where locating features are far apart in the assembly, and Figs. 15.30 through 15.41 show the effects where the same locating features are closer together.

## Rotation With Part Features Farther Apart

Figure 15.18 shows a simple assembly consisting of two parts: a plate with pins and a bar with mating clearance holes. Notice that the clearance holes and pins are spaced as far apart as practical in the assembly. A customer wants to know the maximum and minimum vertical distance between points A and B on the assembled parts as shown.

Figures 15.19 and Fig. 15.20 show the drawings for each part with $\pm$ dimensions and tolerances. The plate is detailed in Fig. 15.19 and the bar is detailed in Fig. 15.20.

In Fig. 15.21 the worst-case assembly is shown, the variation is assumed to be purely translational (linear) and the parts are shown translated. Rotation is not considered in this Tolerance Stackup. The holes in the bar are biased downward within their tolerance zones, and the pins in the plate are biased upward within their tolerance zones. The height of the plate is smallest, and the height of the Bar is largest.

The Tolerance Stackup in Fig. 15.22 represents the assembly shown in Fig. 15.21, calculates the effects of the dimensions and tolerances as linear variation alone, and shows the worst-case (smallest) distance between points A and B is 0.090.

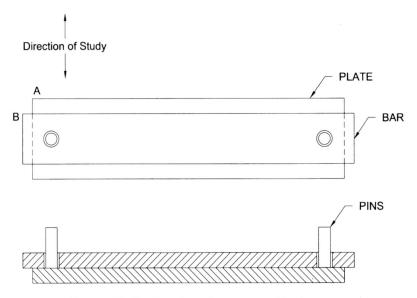

**FIGURE 15.18**   Rotation of parts: assembly, far apart.

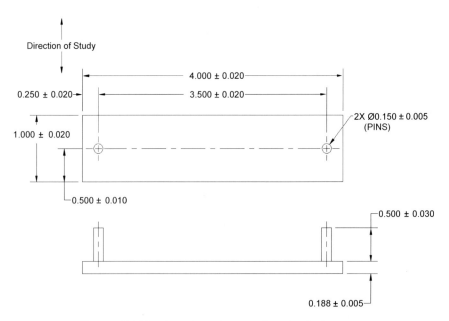

**FIGURE 15.19**   Rotation of parts: plate detail, far apart.

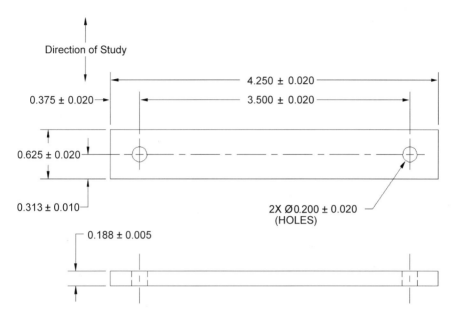

**FIGURE 15.20** Rotation of parts: bar detail, far apart.

The assembly in Fig. 15.23 shows the parts with a combination of linear and rotational displacement leading to the minimum distance A − B. The pins on the plate are at their LMC (smallest) size and are displaced to facilitate the maximum rotational effect on the bar: the pin on the left is translated upward within its tolerance zone, and the pin on the right is translated downward within its tolerance zone. The holes in the bar are at their LMC (largest) size and are also displaced and sized to facilitate the maximum rotational

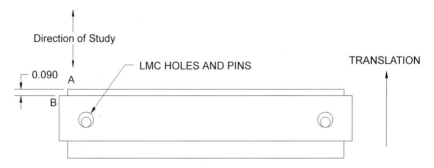

**FIGURE 15.21** Rotation of parts: worst-case assembly, translation only, far apart.

Tolerance Stack　　　　　　　　　　　　　　　　　　　　　　　　　　　Release 1.2a

| Program: | Training Manual Exercises | | |
|---|---|---|---|
| Product: | Part Number | Rev | Description |
| | 43-001 | A | Bar and Plate Assembly, Holes and Pins Spaced Far Apart |
| Problem: | Need to Know the Minimum Distance Between Points A & B in the Y Direction | | |
| Objective: | Determine the Minimum and Maximum Distance in the Y Direction: Translation Only | | |

Stack Information:

| | |
|---|---|
| Stack No: | Bar and Plate Assembly - Far Apart |
| Date: | 07/04/02 |
| Revision: | A |
| Direction: | XY Plane |
| Author: | BR Fischer |

| Description of Component / Assy | Part Number | Rev | Item | Description | + Dims | − Dims | Tol | Percent Contrib | Dim / Tol Source & Calcs |
|---|---|---|---|---|---|---|---|---|---|
| Plate | 43-002 | A | 1 | Dim Upper Surface - Lower Surface | 1.0000 | | ± 0.0200 | 21% | 1.000 ± 0.020 on Dwg |
| | | | 2 | Dim Lower Edge - CL Pins | | 0.5000 | ± 0.0100 | 10% | 0.500 ± 0.010 on Dwg |
| | | | 3 | Assembly Shift (Mounting Holes$_{LMC}$ - Pins$_{LMC}$)/2 | | | ± 0.0375 | 38% | = [(0.2 ± 0.02) − (0.15 − 0.005)] /2 |
| Bar | 43-003 | A | 4 | Dim CL Holes - Lower Edge | 0.3130 | | ± 0.0100 | 10% | 0.313 ± 0.010 on Dwg |
| | | | 5 | Dim Lower Edge - Upper Surface | | 0.6250 | ± 0.0200 | 21% | 0.625 ± 0.020 on Dwg |
| | | | | Dimension Totals | 1.3130 | 1.1250 | | | |
| | | | | Nominal Distance: Pos Dims − Neg Dims = | 0.1880 | | | | |

RESULTS:

| | Norm | Tol | Min | Max |
|---|---|---|---|---|
| Arithmetic Stack (Worst Case) | 0.1880 | ± 0.0975 | 0.0905 | 0.2855 |
| Statistical Stack (RSS) | 0.1880 | ± 0.0491 | 0.1389 | 0.2371 |
| Adjusted Statistical 1.5*RSS | 0.1880 | ± 0.0736 | 0.1144 | 0.2616 |

**FIGURE 15.22** Rotation of parts: Tolerance Stackup report, translation only, far apart.

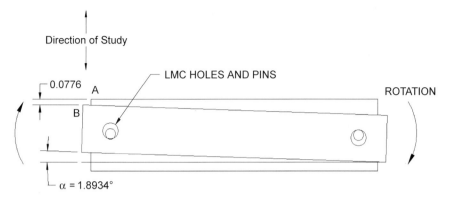

**FIGURE 15.23** Rotation of parts: worst-case assembly, translation and rotation, far apart.

effect: the hole on the left is translated downward within its tolerance zone, and the hole on the right is translated upward within its tolerance zone. Detailed drawings of these worst-case parts can be seen in Fig. 15.24. The holes in the bar are allowed to rotate about the pins. Notice in this example the locational tolerances of the holes and pins and the assembly shift between the holes and pins are treated as rotational assembly shift in the Tolerance Stackup—not only can the parts shift linearly, but they can rotate as well. All remaining tolerances are treated as purely translational displacements. Obviously the point of these examples is to show that the effect of the variation may be much greater when it is considered as rotation as opposed to treating it as purely translational.

The Tolerance Stackup in Fig. 15.25 represents the assembly shown in Fig. 15.23, calculates the effects of the dimensions and tolerances as rotational variation and linear variation, and shows the worst-case distance between points A and B is 0.0776. Comparing the results in the two Tolerance Stackups shown in Figs. 15.22 and 15.25, the combination of rotational (or angular) variation and linear variation is greater than treating the possible variation as purely translational.

Once the effects of the rotational variation are converted into linear units, the effects can be entered into the Tolerance Stackup. After the equivalent linear displacement is entered into the Tolerance Stackup, the effect on the geometry being studied can be clearly seen and analyzed. If there is a problem, the dimensioning, tolerancing, part geometry, or assembly procedure can be changed to minimize or eliminate the effect of rotation.

The method for calculating the effect of the rotational variation follows. The steps below are for parts toleranced using plus/minus. The steps would be slightly different for parts toleranced using GD&T for several reasons. First,

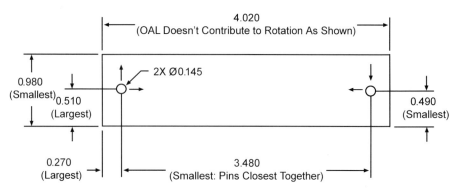

Plate: Worst-Case Dimensions for Rotational Shift

To obtain the worst-case rotational shift, the Pins are biased inwards within their tolerance zones. The Pin on the left side is biased upward and the Pin on the right side is biased downward within each respective tolerance zone. The overall height of the plate is smallest; the overall length doesn't contribute to tolerance stackup in this example.

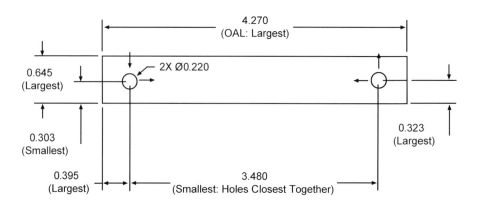

Bar: Worst-Case Dimensions for Rotational Shift

To obtain the worst-case rotational shift, the Holes are biased inwards within their tolerance zones. The Hole on the left side is biased downward and the Hole on the right side is biased upward within each respective tolerance zone. The overall height of the Bar is largest; the overall length contributes to tolerance stackup and is largest in this example.

**FIGURE 15.24**  Rotation of parts: worst-case part dimensions, far apart.

Tolerance Stack   Release 1.2a

| Program: | Training Manual Exercises | | |
|---|---|---|---|
| Product: | Part Number | Rev | Description |
| | 43-001 | A | Bar and Plate Assembly. Holes and Pins Spaced Far Apart |
| Problem: | Need to Know the Minimum Distance Between Points A & B in the Y Direction | | |
| Objective: | Determine the Minimum and Maximum Distance in the Y Direction. Translation and Rotation | | |

**Stack Information**

| | |
|---|---|
| Stack No | Bar and Plate Assembly - Far Apart |
| Date | 07/04/02 |
| Revision | A |
| Direction: | XY Plane |
| Author: | BR Fischer |

| Description of Component / Assy | Part Number | Rev | Item | Description | + Dims | − Dims | Tol | Percent Contrib | Dim / Tol Source & Calcs |
|---|---|---|---|---|---|---|---|---|---|
| Plate | 43-002 | A | 1 | Dim: Upper Surface - Lower Surface | 1.0000 | | ± 0.0200 | 18% | 1.000 ± .020 on Dwg |
| | | | 2 | Dim: Lower Edge - CL Pins | | 0.5000 | ± 0.0000 | 0% | 0.500 ± .010 on Dwg |
| | | | 3 | Rotational Assy Shift: Rot'n About (Mtg Hole$_{SMC}$ − Pin$_{SMC}$) / 2 | | | ± 0.0706 | 64% | See Figure 46(i) |
| Bar | 43-003 | A | 4 | Dim: CL Holes - Lower Edge | 0.3128 | | ± 0.0000 | 0% | 0.313 ± .010 on Dwg,  y = dist cos 1.8934° |
| | | | 5 | Dim: Lower Edge - Upper Surface | | 0.6247 | ± 0.0200 | 18% | 0.625 ± .020 on Dwg,  y = 0.020 cos 1.8934° |
| | | | | Dimension Totals | 1.3128 | 1.1247 | | | |
| | | | | Nominal Distance: Pos Dims − Neg Dims = | 0.1882 | | | | |

**RESULTS:**

| | Nom | Tol | Min | Max |
|---|---|---|---|---|
| Arithmetic Stack (Worst Case) | 0.1882 | ± 0.1105 | 0.0776 | 0.2987 |
| Statistical Stack (RSS) | 0.1882 | ± 0.0760 | 0.1122 | 0.2642 |
| Adjusted Statistical 1.5*RSS | 0.1882 | ± 0.1140 | 0.0742 | 0.3022 |

**FIGURE 15.25** Rotation of parts: Tolerance Stackup report, translation and rotation, far apart.

cylindrical Positional tolerance zones may be specified with GD&T—such tolerance zones do not allow the pins or holes to be fully biased in two directions at once. That is, say for a diameter 1 positional tolerance zone, the holes could not be displaced ±0.5 vertically from nominal and ±0.5 horizontally from nominal at the same time. The Tolerance Analyst may decide to calculate both conditions separately to determine which yields the greatest rotational variation. Second, MMC or LMC material condition modifiers

Worst-Case rotation occurs when the holes and pins are at the extreme locations discussed in step 1. Their worst-case center-to-center distance ($d_1$) lies along the hypotenuse of the triangles below.

Use the Pythagorean Theorem to calculate center-to-center distance $d_1$:

$$d_1 = \sqrt{(\text{Horizontal Distance } x_1)^2 + (\text{Vertical Distance } y_1)^2}$$

$$= 3.48005747$$

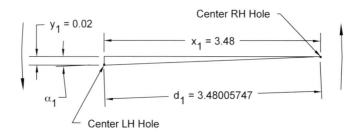

Triangle 1: Worst-case Center-to-Center Distance Between Holes

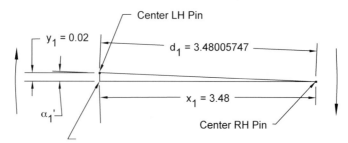

Triangle 2: Worst-case Center-to-Center Distance Between Pins

**FIGURE 15.26**   Rotation of parts: center-to-center distance, far apart.

may be applied to positional tolerance zones, which allow the tolerance zones to increase in size. The effect of the material condition modifiers would also have to be included when calculating the rotational variation.

The total possible worst-case angle of rotation ($\alpha$) is the sum of two angles ($\alpha_1$ and $\alpha_2$) projected over a distance. $\alpha_1$ and $\alpha_2$ are defined below.

### Steps to Calculate Worst-Case Rotational Shift for Parts Toleranced Using Plus/Minus:

1.  Determine the worst-case part geometry that leads to the greatest possible rotational variation. The worst case occurs when the hole size is largest (LMC) and the pin size is smallest (LMC), the holes and pins are closest together within their tolerance zones, and the holes and pins are biased vertically in opposite directions as seen in Fig. 15.24. In assemblies consisting of two parts with clearance holes that share common fasteners, the holes in both parts would be at their largest (LMC) size.

2.  The hole and pin center-to-center distance ($d_1$) must be calculated for use in the formulas in Fig. 15.26. The math and calculations can also be seen in Fig. 15.26. (Note that the distance should be the same for both mating parts.)

3.  Calculate the angle of rotation contributed by the $\pm$ location tolerances on the mating part features ($\alpha_1$). The math and calculations can be seen in Fig. 15.27.

4.  Calculate the angle of rotation contributed by the assembly shift ($\alpha_2$). The math and calculations can be seen in Fig. 15.28. For this example, the assembly shift only occurs once, as the holes in the bar may shift about the pins in the plate. For assemblies consisting of two parts with clearance holes that share common fasteners the rotational effect of the assembly shift must be calculated and added twice, once for each part about the fasteners.

5.  Determine the total angle of rotation ($\alpha$), which is the sum of $\alpha_1$ and $\alpha_2$. For this example, $\alpha_1 = 0.6586°$ and $\alpha_2 = 1.2348°$.

    $$\alpha = \alpha_1 + \alpha_2 = 0.6586° + 1.2348° = 1.8934°$$

    This is the total angle of rotation.

6.  Project the total angle of rotation to one of the points under consideration. In this example, the angle is projected out to the corners of the bar at point B. Convert the projected angle to linear units in the direction of the Tolerance Stackup as shown in Fig. 15.29. This linear displacement is added to the Tolerance Stackup report as

Solve for $\alpha_1'$:    $\tan \alpha_1' = \dfrac{0.02}{3.48}$    $\Rightarrow$    $\alpha_1' = 0.3293°$

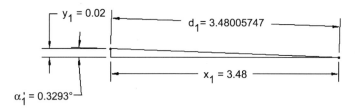

Worst-case Angle from the ± Location Tolerances on the Holes

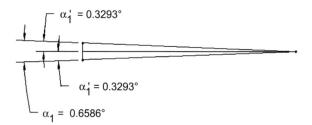

Worst-case Angle from the ± Location Tolerances on the Pins

Solve for $\alpha_1$:    $\alpha_1 = \alpha_1' + \alpha_1' = 0.3293° + 0.3293° = 0.6586°$

Worst-case Angle from the ± Location Tolerances on Holes and Pins

**FIGURE 15.27**    Rotation of parts: worst-case angle from ± location tolerances, far apart.

Solve for Worst-case Angle from Assembly Shift $\alpha_2$:

Given:

$r_2$ = largest hole radius = $\dfrac{.200 + .020}{2} = \dfrac{.22}{2} = .11$

$r_1$ = smallest pin radius = $\dfrac{.150 - .005}{2} = \dfrac{.145}{2} = .0725$

$d_1$ = center-to-center distance = 3.48005747 (from Step 1)

$r_2 - r_1$ = the radial clearance between the largest hole (LMC) and the smallest (LMC) pin

$$\alpha_2 = 2 \sin^{-1}\left(\frac{r_2 - r_1}{d_1}\right) =$$

$$\alpha_2 = 2 \sin^{-1}\left(\frac{.11 - .0725}{3.48005747}\right) =$$

$$\alpha_2 = 2 \sin^{-1}\left(\frac{.0375}{3.48005747}\right) =$$

$$\alpha_2 = 2 \sin^{-1}(0.01077568) =$$

$$\alpha_2 = 1.2348°$$

**FIGURE 15.28**  Rotation of parts: worst-case angle from assembly shift, far apart.

rotational assembly shift. Because the locational ± tolerances for the holes and pins are included in the calculations above, these tolerances are not included with the dimensions on the Tolerance Stackup report form. Notice on lines 2 and 4 the tolerance values are zero.

The dimension values for the bar entered into lines 4 and 5 of the Tolerance Stackup report in Fig. 15.25 have been trigonometrically manipulated. The tolerance value for the height of the bar on line 5 has also been trigonometrically manipulated. This is required because the bar has rotated,

Calculate the Projected Linear Displacement:

$\alpha$ = 1.8934°    (Angle of Rotation)

$d_2$ = 4.27    (Longest Bar)

$y_2$ = Projected Linear Displacement

Solve for $y_2$:

$y_2$ = $d_2$ sin $\alpha$ = 4.27 sin 1.8934°

$y_2$ = .1411

± Equivalent: .1411 / 2 = ±.0706

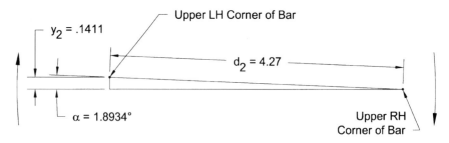

**FIGURE 15.29**  Rotation of parts: projected linear displacement, far apart.

and these dimension and tolerance values are no longer directly aligned with the Tolerance Stackup direction.

(Note: I must gratefully acknowledge Eric Schulz, Mathematics Professor at Walla Walla Community College for his help in correctly visualizing and solving this problem. Thank you Eric!)

## Rotation With Part Features Closer Together

The point was made prior to this exercise that the effect of rotation is greater when it is projected over a longer distance. In this example, the same assembly as above is considered, except the holes and pins are closer together—every other dimension and tolerance is the same. As will be shown, the effect of the rotation is greater where the locating features are closer together, and it is projected over the same distance. Notice that the resulting angle of rotation in this example is far greater than in the above example.

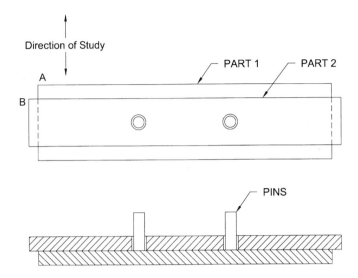

**FIGURE 15.30**  Rotation of parts: assembly, close together.

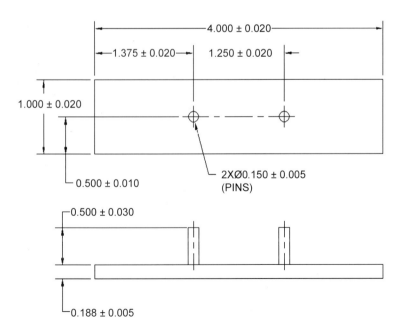

**FIGURE 15.31**  Rotation of parts: plate detail, close together.

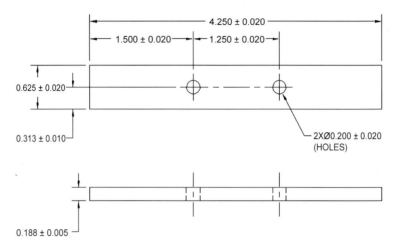

**FIGURE 15.32**   Rotation of parts: bar detail, close together.

Figure 15.30 shows a simple assembly consisting of two parts: a plate with pins and a bar with clearance holes. Notice that the clearance holes and pins are spaced closer together in this assembly. As before, a customer wants to know the maximum and minimum vertical distance between points A and B on the parts as shown.

Figures 15.31 and 15.32 show the drawings for each part with ± dimensions and tolerances. The plate is detailed in Fig. 15.31, and the bar is detailed in Fig. 15.32.

In Fig. 15.33 the worst-case assembly is shown, the variation is assumed to be purely translational (linear), and the parts are shown translated.

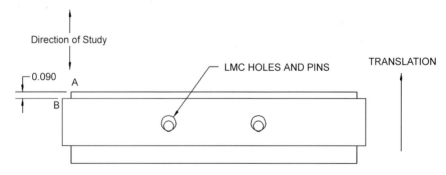

**FIGURE 15.33**   Rotation of parts: worst-case assembly, translation only, close together.

Tolerance Stack

Release 1.2a

| Program: | Training Manual Exercises | | |
|---|---|---|---|
| Product: | Part Number | Rev | Description |
| | 44-001 | A | Bar and Plate Assembly, Holes and Pins Close Together |
| Problem: | Need to Know the Minimum Distance Between Points A & B in the Y Direction | | |
| Objective: | Determine the Minimum and Maximum Distance in the Y Direction, Translation Only | | |

**Stack Information**

| Stack No: | Bar and Plate Assembly - Close Together |
|---|---|
| Date: | 07/04/02 |
| Revision | A |
| Direction: | XY Plane |
| Author: | BR Fischer |

| Description of Component / Assy | Part Number | Rev | Item | Description | + Dims | − Dims | Tol | Percent Contrib | Dim / Tol Source & Calcs |
|---|---|---|---|---|---|---|---|---|---|
| Plate | 44-002 | A | 1 | Dim Upper Surface - Lower Surface | 1.0000 | | ± 0.0200 | 21% | 1.000 ± 0.020 on Dwg |
| | | | 2 | Dim Lower Edge - CL Pins | | 0.5000 | ± 0.0100 | 10% | 0.500 ± 0.010 on Dwg |
| | | | 3 | Assembly Shift (Mounting Holes$_{LMC}$ - Pins$_{LMC}$)/2 | | | ± 0.0375 | 38% | = [(0.2 + 0.02) − (0.15 − 0.005)]/2 |
| Bar | 44-003 | A | 4 | Dim CL Holes - Lower Edge | 0.3130 | | ± 0.0100 | 10% | 0.313 ± 0.010 on Dwg |
| | | | 5 | Dim Lower Edge - Upper Surface | | 0.6250 | ± 0.0200 | 21% | 0.625 ± 0.020 on Dwg |
| | | | | Dimension Totals | 1.3130 | 1.1250 | | | |
| | | | | Nominal Distance - Pos Dims − Neg Dims = | 0.1880 | | | | |

**RESULTS:**

| | Nom | Tol | Min | Max |
|---|---|---|---|---|
| Arithmetic Stack (Worst Case) | 0.1880 | ± 0.0975 | 0.0905 | 0.2855 |
| Statistical Stack (RSS) | 0.1880 | ± 0.0491 | 0.1389 | 0.2371 |
| Adjusted Statistical 1.5*RSS | 0.1880 | ± 0.0736 | 0.1144 | 0.2616 |

**FIGURE 15.34**  Rotation of parts: Tolerance Stackup report, translation only, close together.

Rotation is not considered in the Tolerance Stackup. The holes in the bar are biased downward within their tolerance zones and the pins in the plate are biased upward within their tolerance zones. The height of the plate is smallest, and the height of the bar is largest.

When the variation is considered as translation only, the horizontal spacing of the holes and pins has no effect on the vertical distance between points A and B. The translational Tolerance Stackup is shown in Fig. 15.34.

Comparing the results in the two Tolerance Stackups shown in Figs. 15.22 and 15.34, where only translation was considered, the results are exactly the same.

The assembly in Fig. 15.35 shows the parts with a combination of linear and rotational displacements leading to the minimum distance A − B. The pins on the plate are at their LMC (smallest) size and are displaced to facilitate the maximum rotational effect on the bar: the pin on the left is translated upward within its tolerance zone, and the pin on the right is translated downward within its tolerance zone. The holes in the bar are at their LMC (largest) size and are also displaced and sized to facilitate the maximum rotational effect: the hole on the left is translated downward within its tolerance zone, and the hole on the right is translated upward within its tolerance zone. Detailed drawings of these worst-case parts can be seen in Fig. 15.36. The holes in the bar are allowed to rotate about the pins. Notice in this example the locational tolerances of the holes and pins, and the assembly shift between the holes and pins are treated as rotational assembly shift in the Tolerance Stackup—not only can the parts shift linearly, but they can rotate as well. All remaining tolerances are treated as purely translational displacements.

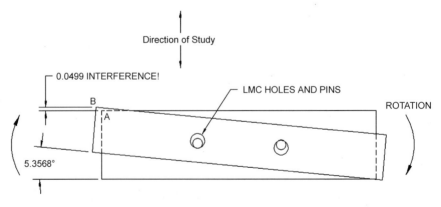

**FIGURE 15.35** Rotation of parts: worst-case assembly translation and rotation, close together.

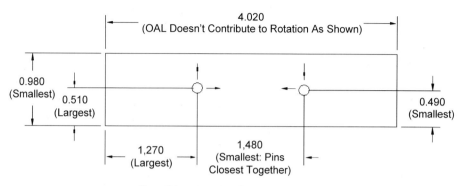

Plate: Worst-Case Dimensions for Rotational Shift

To obtain the worst-case rotational shift, the Pins are biased inwards within their tolerance zones. The Pin on the left side is biased upward and the Pin on the right side is biased downward within each respective tolerance zone. The overall height of the plate is smallest; the overall length doesn't contribute to tolerance stackup in this example.

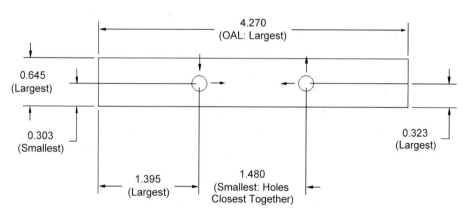

Bar: Worst-Case Dimensions for Rotational Shift

To obtain the worst-case rotational shift, the Holes are biased inwards within their tolerance zones. The Hole on the left side is biased downward and the Hole on the right side is biased upward within each respective tolerance zone. The overall height of the Bar is largest; the overall length contributes to tolerance stackup and is largest in this example.

**FIGURE 15.36** Rotation of parts: worst-case part dimensions, close together.

Tolerance Stack                                                                           Release 1.2a

| Program: | Training Manual Exercises | | | | | | | | Stack Information |
|---|---|---|---|---|---|---|---|---|---|

Product: Part Number 44-001  Rev A  Description: Bar and Plate Assembly: Holes and Pins Close Together

Problem: Need to Know the Minimum Distance Between Points A & B in the Y Direction

Objective: Determine the Minimum and Maximum Distance in the Y Direction: Translation and Rotation

Stack Information

Stack No: Bar and Plate Assembly - Close Together
Date: 07/04/02
Revision: A
Direction: XY Plane
Author: BR Fischer

| Description of Component/Assy | Part Number | Rev | Item | Description | + Dims | – Dims | Tol | Percent Contrib | Dim / Tol Source & Calcs |
|---|---|---|---|---|---|---|---|---|---|
| Plate | 44-002 | A | 1 | Dim: Upper Surface - Lower Surface | 1.0000 | | ± 0.0200 | 8% | 1.000 ± 0.020 on Dwg |
| | | | 2 | Dim: Lower Edge - CL Pins | | 0.5000 | ± 0.0000 | 0% | 0.500 ± 0.010 on Dwg |
| | | | 3 | Rotational Assy Shift: Rotation About (Mtg Holes$_{LMC}$ - Pins$_{LMC}$) / 2 | | | ± 0.1993 | 83% | See Figure 47(i) |
| Bar | 44-003 | A | 4 | Dim: CL Holes - Lower Edge | 0.3116 | | ± 0.0000 | 0% | 0.313 ± 0.010 on Dwg,  y = dist cos 5.3568° |
| | | | 5 | Dim: Lower Edge - Upper Surface | | 0.6223 | ± 0.0199 | 8% | 0.625 ± 0.020 on Dwg,  y = dist cos 5.3568° |

Dimension Totals: 1.3116   1.1223

Nominal Distance: Pos Dims – Neg Dims = 0.1894

RESULTS:

| | Nom | Tol | Min | Max |
|---|---|---|---|---|
| Arithmetic Stack (Worst Case) | 0.1894 | ± 0.2392 | –0.0499 | 0.4286 |
| Statistical Stack (RSS) | 0.1894 | ± 0.2013 | –0.0119 | 0.3907 |
| Adjusted Statistical 1.5*RSS | 0.1894 | ± 0.3019 | –0.1126 | 0.4913 |

**FIGURE 15.37**  Rotation of parts: Tolerance Stackup report, translation and rotation, close together.

The Tolerance Stackup in Fig. 15.37 represents the assembly shown in Fig. 15.35, calculates the effects of the dimensions and tolerances as rotational variation and linear variation and shows the worst-case distance between points A and B is 0.0499 interference! As in the previous example, comparing the results in the two Tolerance Stackups shown in Figs. 15.34 and 15.37, the combination of rotational (or angular) variation and linear variation is greater than treating the possible variation as purely translational.

Comparing the results in the two Tolerance Stackups shown in Figs. 15.25 and 15.37, where translation *and* rotation were considered, and the pins and holes are spaced, respectively, farther apart and closer together, the variation is *much* greater when the holes and pins are closer together. The tolerances are exactly the same in both examples, but the resulting angle is nearly three times greater when the pins and holes are closer together. Remember, the only difference between these examples is the center-to-center distance between the holes and pins.

The total possible worst-case angle of rotation ($\alpha$) is the sum of two angles ($\alpha_1$ and $\alpha_2$) projected over a distance. $\alpha_1$ and $\alpha_2$ are defined below.

## Steps to Calculate Worst-Case Rotational Shift
## For Parts Toleranced Using Plus/Minus:

1. Determine the worst-case part geometry that leads to the greatest possible rotational variation. The worst case occurs when the hole size is largest (LMC) and the pin size is smallest (LMC), the holes and pins are closest together within their tolerance zones, and the holes and pins are biased vertically in opposite directions as seen in Fig. 15.36. In assemblies consisting of two parts with clearance holes that share common fasteners, the holes in both parts would be at their largest (LMC) size.

2. The hole and pin center-to-center distance ($d_1$) must be calculated for use in the formulas that follow. The math and calculations can be seen in Fig. 15.38. (Note that the distance should be the same for both mating parts.)

3. Calculate the angle of rotation contributed by the $\pm$ location tolerances on the mating part features ($\alpha_1$). The math and calculations can be seen in Fig. 15.39.

4. Calculate the angle of rotation contributed by the assembly shift ($\alpha_2$). The math and calculations can be seen in Fig. 15.40. For this example, the assembly shift only occurs once, as the holes in the bar may shift about the pins in the plate. For assemblies consisting of two parts with clearance holes that share common fasteners the

Worst-Case rotation occurs when the holes and pins are at the extreme locations discussed in step 1. Their worst-case center-to-center distance ($d_1$) lies along the hypotenuse of the triangles below.

Use the Pythagorean Theorem to calculate center-to-center distance $d_1$:

$$d_1 = \sqrt{(\text{Horizontal Distance } x_1)^2 + (\text{Vertical Distance } y_1)^2}$$

$$= 1.23016259$$

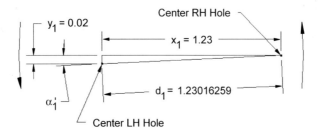

Triangle 1: Worst-case Center-to-Center Distance Between Holes

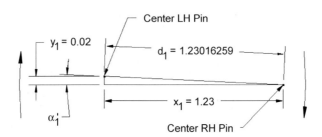

Triangle 2: Worst-case Center-to-Center Distance Between Pins

**FIGURE 15.38**   Rotation of parts: center-to-center distance, close together.

rotational effect of the assembly shift must be calculated and added twice, once for each part about the fasteners.

5.  Determine the total angle of rotation ($\alpha$), which is the sum of $\alpha_1$ and $\alpha_2$. For this example, $\alpha_1 = 1.8631°$ and $\alpha_2 = 3.4937°$.

$$\alpha = \alpha_1 + \alpha_2 = 1.8631° + 3.4937° = 5.3568°$$

This is the total angle of rotation.

Solve for $\alpha_1'$: $\quad \tan \alpha_1' = \dfrac{0.02}{1.23} \quad \Rightarrow \quad \alpha_1' = 0.9316°$

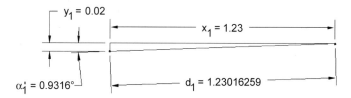

Worst-case Angle from the ± Location Tolerances on the Holes

Worst-case Angle from the ± Location Tolerances on the Pins

Solve for $\alpha_1$: $\quad \alpha_1 = \alpha_1' + \alpha_1' = 0.9316° + 0.9316° = 1.8631°$
(Reflects Rounding Error)

$\alpha_1' = 0.9316°$

$\alpha_1' = 0.9316°$

$\alpha_1 = 1.8631°$

Worst-case Angle from the ± Location Tolerances on Holes and Pins

**FIGURE 15.39**  Rotation of parts: worst-case angle from ± location tolerances, close together.

Solve for Worst-case Angle from Assembly Shift $\alpha_2$:

Given:

$r_2$ = largest hole radius = $\dfrac{.200 + .020}{2} = \dfrac{.22}{2} = .11$

$r_1$ = smallest pin radius = $\dfrac{.150 - .005}{2} = \dfrac{.145}{2} = .0725$

$d_1$ = center-to-center distance = 1.23016259  (from Step 1)

$r_2 - r_1$ = the radial clearance between the largest hole (LMC) and the smallest (LMC) pin

$$\alpha_2 = 2 \sin^{-1}\left(\frac{r_2 - r_1}{d_1}\right) =$$

$$\alpha_2 = 2 \sin^{-1}\left(\frac{.11 - .0725}{1.23016259}\right) =$$

$$\alpha_2 = 2 \sin^{-1}\left(\frac{.0375}{1.23016259}\right) =$$

$$\alpha_2 = 2 \sin^{-1}(0.03048378) =$$

$$\alpha_2 = 3.4937°$$

**FIGURE 15.40** Rotation of parts: worst-case angle from assembly shift, close together.

6. Project the total angle of rotation to one of the points under consideration. In this example, the angle is projected out to the corners of the bar at point B. Convert the projected angle to linear units in the direction of the Tolerance Stackup as shown in Fig. 15.41. This linear displacement is added to the Tolerance Stackup report as rotational assembly shift. Because the locational ± tolerances for the holes and pins are included in the calculations above, these tolerances are not included with the dimensions on the Tolerance Stackup report form. Notice on lines 2 and 4 the tolerance values are zero.

Calculate the Projected Linear Displacement:

$\alpha$ = 5.3568°   (Angle of Rotation)

$d_2$ = 4.27   (Longest Bar)

$y_2$ = Projected Linear Displacement

Solve for $y_2$:

$y_2 = d_2 \sin \alpha$ = 4.27 sin 5.3568°

$y_2$ = .3986

± Equivalent: .3986 / 2 = ±.1993

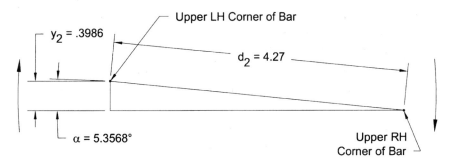

**FIGURE 15.41** Rotation of parts: projected linear displacement, close together.

The dimension values for the bar entered into lines 4 and 5 of the Tolerance Stackup report in Fig. 15.37 have been trigonometrically manipulated. The tolerance value for the height of the bar on line 5 has also been trigonometrically manipulated. This is required because the bar has rotated, and these dimension and tolerance values are no longer directly aligned with the Tolerance Stackup direction.

# 16

## Putting It All Together: Tolerance Stackups with GD&T Solved Using the Advanced Dimensional Management Method

This chapter presents a series of seven Tolerance Stackup examples based on an assembly of parts mainly dimensioned and toleranced using GD&T. Plus/ minus dimensions and tolerances are only used to define Features of Size and for simple thicknesses. The Tolerance Stackups are solved using Advanced Dimensional Management's ™ Tolerance Stackup sketch techniques and Tolerance Stackup reporting techniques described in Chapters 13 and 14. All of the tools and techniques learned up to this point are included in these Tolerance Stackups.

The drawings that follow are to be used with Examples 16.1 through 16.7. These problems are based on an assembly where a ground plate is mounted inside an enclosure. Assembly drawings and detail drawings of each part are included. There are three optional drawings for the ground plate and three corresponding optional drawings for the enclosure, labeled Options 1, 2, and 3. The Option 1 ground plate is to be used with the Option 1 enclosure, the Option 2 ground plate is to be used with the Option 2 enclosure, and the Option 3 ground plate is to be used with the Option 3 enclosure. The tolerancing schemes for each pair of drawings are coordinated, and each scheme is slightly different—the main difference between the schemes is in the datum reference frame. Tolerance Stackup Examples 16.5 to 16.7 compare the effects of using these various dimensioning and tolerancing schemes.

The Tolerance Stackups presented in Examples 16.1 through 16.7 represent some of the more important Tolerance Stackups that would be performed on such an assembly. These Tolerance Stackups would be done as part of the design process, to verify that the part and assembly geometry satisfies the functional requirements, to verify that the dimensioning and tolerancing schemes satisfy the functional requirements, to verify that the dimension and tolerance values satisfy their functional requirements, and to verify the assembly procedure satisfies the functional requirements.

Probably the first and most important Tolerance Stackup required for these parts is a fixed-fastener calculation. Fixed-fastener calculations are described in Chapter 18. The fixed-fastener formula would be used to

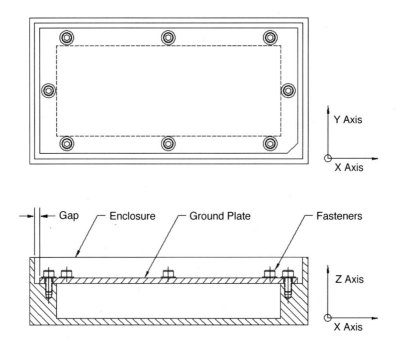

The Ground Plate is mounted in the Enclosure. The design criteria dictates that a gap must be maintained between the edges of the Ground Plate and the inside edges of the Enclosure—the edges of the Ground Plate must not touch the inside edges of the Enclosure.

**FIGURE 16.1**  Enclosure assembly for 16–18.

determine the required size of the clearance holes in the ground plate—the formula may also used to determine the allowable positional tolerance values for the holes in the ground plate and the enclosure. The fixed-fastener calculation determines the smallest allowable size for the clearance holes based on their positional tolerance, the positional tolerance on the mating threaded holes, and the maximum outer diameter (OD) of the fasteners. Although important, this calculation is not included in this

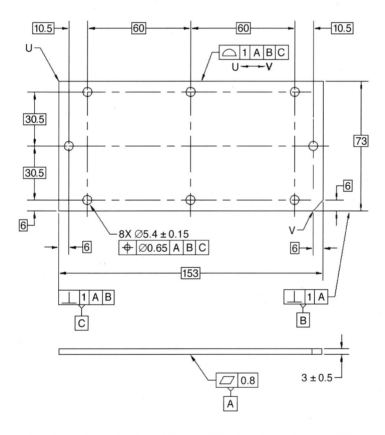

The Ground Plate has been toleranced simply, using surfaces as Datum Features. The surface that mates with the Enclosure is Datum Feature A, Datum Features B and C are surfaces along the edges of the Plate.

**FIGURE 16.2** Ground plate for 16–18: Option 1.

chapter, because the theory and techniques have not yet been covered in the text.

## ASSEMBLY DRAWINGS AND DETAIL DRAWINGS FOR EXAMPLES 16.1 THROUGH 16.8

The enclosure assembly drawing shown in Fig. 16.1 defines three axes of a Cartesian coordinate system: the $X$ axis, the $Y$ axis, and the $Z$ axis. The direction of the following Tolerance Stackups will be described in terms of these axes. As stated in Chapters 13 and 14, it is very important to describe and

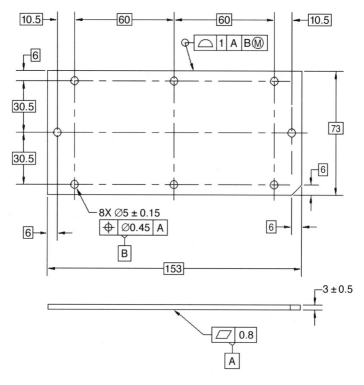

The Ground Plate has been toleranced to satisfy its functional requirements. The Ground Plate mates on the flat surface (Datum Feature A), and is located by the 8 clearance holes (Datum Feature B).

Because Datum Feature B is a pattern of holes, it should only be referenced at MMC or LMC; it should not be referenced RFS.

**FIGURE 16.3** Ground plate for 16–18: Option 2.

label the direction of the Tolerance Stackup. The nominal gap between the ground plate and the enclosure is also highlighted in this figure. The Tolerance Stackups in Examples 16.5 through 16.7 determine if a gap remains after assembly using the three optional dimensioning and tolerancing schemes.

Figures 16.2 to 16.4 show the three optional ground plate drawings. These drawings only differ in their dimensioning and tolerancing scheme.

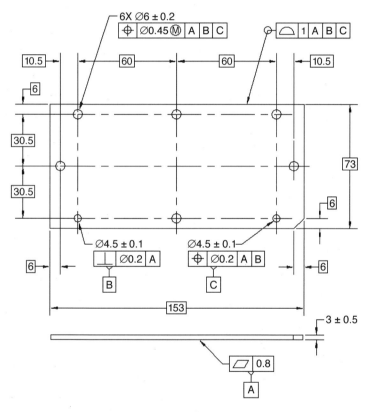

The Ground Plate has been functionally toleranced to minimize tolerance accumulation. The Ground Plate mates on the flat surface (Datum Feature A), and is located by the lower left clearance hole (Datum Feature B) and lower right clearance hole (Datum Feature C).

The other 6 clearance holes do not locate the part in the assembly, and are therefore free to have larger size and location tolerance, if desired. The assembly procedure must adhere to the GD&T: fasteners through Datum Features B & C first, then through the remaining 6 holes.

**FIGURE 16.4** Ground plate for 16–18: Option 3.

The most important difference between the three options is the datum reference frame used.

Figures 16.5 to 16.7 show the three optional enclosure drawings. Like the ground plate drawings, these drawings only differ in their dimensioning and tolerancing scheme.

Figure 16.8 shows the enclosure assembly with its cover. The cover must fit within the enclosure. Figure 16.9 is a drawing of the cover.

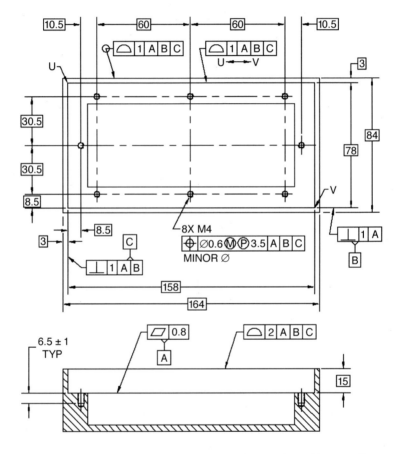

The Enclosure has been toleranced simply using surfaces as Datum Features. The upper inside surface that mates with the Ground Plate is Datum Feature A, Datum Features B and C are inside surfaces.

The geometric tolerances were applied to the minor diameter of the tapped holes to facilitate inspection using a functional gage rather than for functional reasons.

**FIGURE 16.5**   Enclosure for 16–18: Option 1.

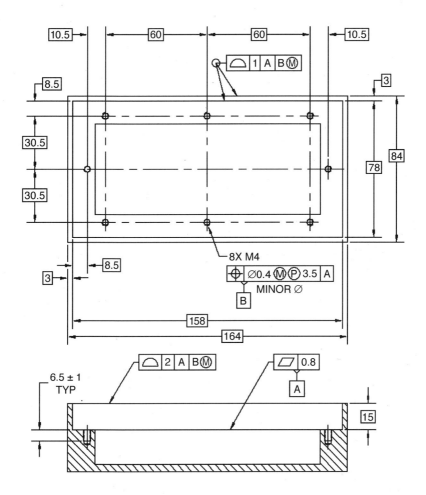

**FIGURE 16.6** Enclosure for 16–18: Option 2.

The Enclosure has been toleranced to satisfy its functional requirements. The upper inside surface that mates with the Ground Plate is Datum Feature A, and Datum Feature B is the 8 tapped holes.

Datum Feature B is a pattern of holes and should be referenced at MMC or LMC; it should not be referenced RFS. The geometric tolerances and datum references specified for the tapped holes apply to their minor diameter. This was done to facilitate inspection using a functional gage rather than for functional reasons.

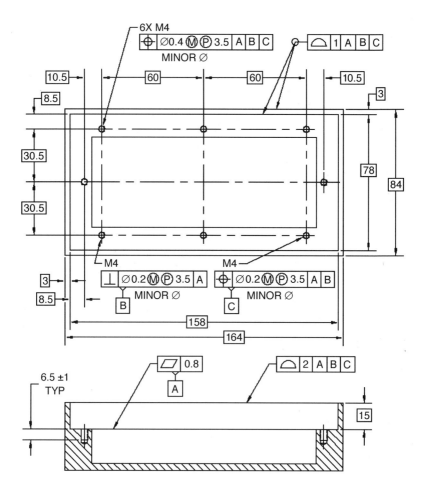

The Enclosure has been functionally toleranced to minimize tolerance accumulation. The surface that mates with the Ground Plate is Datum Feature A, Datum Features B and C are the minor diameters of the lower left and lower right tapped holes.

The other 6 tapped holes do not locate the part in the assembly, and may have a larger location tolerance. The geometric tolerances were applied to the minor diameter of the tapped holes to facilitate inspection using a functional gage rather than for functional reasons. The assembly procedure must adhere to the GD&T: fasteners are started in Datum Features B & C first, then into the remaining 6 holes.

**FIGURE 16.7**   Enclosure for 16–18: Option 3.

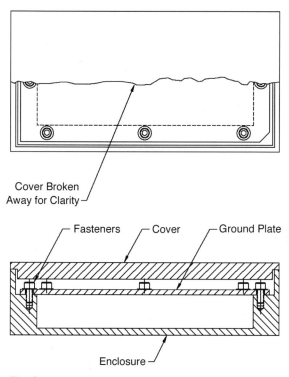

Cover Broken
Away for Clarity

Fasteners — Cover — Ground Plate

Enclosure

The Cover must fit within the Enclosure and it must not
contact the fasteners.

**FIGURE 16.8** Enclosure assembly with cover.

## EXAMPLES

*Example 6.1. Screw Thread Depth Tolerance Stackup*

Determine if the M4 screws bottom out in the threaded holes.
  Extra data:

- M4 washer thickness = 1.2 ± 0.1.
- M4 × 8 socket head cap screw length = 8 ± 0.2 (from vendor drawing). Length is from bottom of head to end of screw.

Solve Example 16.1 as follows:

- Stackup direction is along the $Z$ axis (see Fig. 16.10).
- Use Option 1 parts for this example.

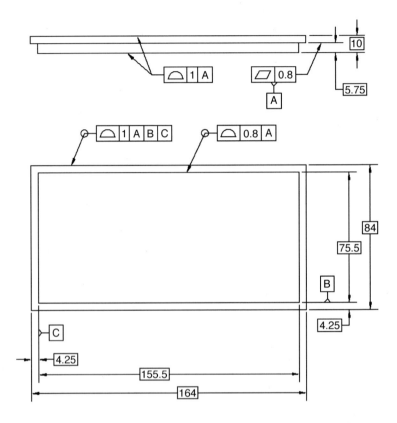

The Cover has been toleranced to satisfy its functional requirements. The surface that mates with the Enclosure is Datum Feature A, Datum Feature B is the longer inside surface, and Datum Feature C is the shorter inside surface.

The form and orientation of Datum Features B & C are controlled by the all-around 0.8 Profile of a Surface tolerance back to Datum A.

**FIGURE 16.9** Cover for 16–18.

The Tolerance Stackup in Example 16.1 determines if the M4 screws bottom out in the threaded holes in the enclosure. The Tolerance Stackup report is shown in Fig. 16.11 and the Tolerance Stackup sketch is shown in Fig. 16.12. The Tolerance Stackup sketch is included as page 2 of the Tolerance Stackup report.

*Results.* The worst-case Tolerance Stackup result shows that the threaded holes extend 0.9 mm beyond the ends of the screws—the screws do not bottom out in the threaded holes.

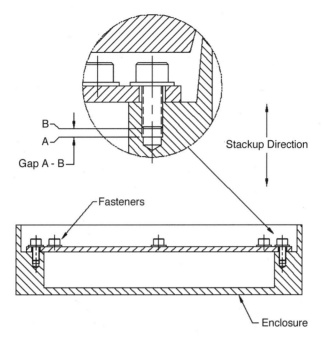

B

A

Gap A - B

Stackup Direction

Fasteners

Enclosure

Enclosure Assembly with Cover: Does Screw
Bottom Out Before Seating?

**FIGURE 16.10**   Example 16.1: Screw thread depth.

## *Example 16.2.   Cover Fit Tolerance Stackup Along X Axis*

Determine if the cover fits within the enclosure along $X$ axis.
Solve Example 16.2 as follows:

- Stackup direction is along the $X$ axis (see Fig. 16.13).
- Use Option 2 parts for the example.

The Tolerance Stackup in Example 16.2 determines if the cover fits within the enclosure. The Tolerance Stackup report is shown in Fig. 16.14, and the Tolerance Stackup sketch is shown in Fig. 16.15. The detail in Fig. 16.15 shows that the top surface of the enclosure may tilt relative to the inside surfaces of the enclosure. This occurs because the top surface of the enclosure is not the primary datum feature for the profile tolerance on the inside surfaces of the enclosure. This tilting can lead to an apparent foreshortening of the opening for the cover. Line item 6 in the Tolerance Stackup report includes the linear equivalent for the angle between the top surface and the

Program: Electronics Packaging Program AV-11

Product:

| Part Number | Rev | Description |
|---|---|---|
| 123456678-001 | A | Ground Plate Enclosure Assembly |

Problem: Screws Must Not Bottom Out in Tapped Holes

Objective: Determine if the M4 Holes in the Enclosure are Deep Enough

Stack Information:

Stack No: Example 16-1
Date: 07/04/02
Revision: A

Direction: Z Axis
Author: BR Fischer

| Description of Component / Assy | Part Number | Rev | Item | Description | + Dims | - Dims | Tol | Percent Contrib | Dim / Tol Source & Calcs |
|---|---|---|---|---|---|---|---|---|---|
| Enclosure | 123456678-002 | A | 1 | Dim: Bottom M4 Tapped Hole - DF$_A$ | 6.5000 | | +/- 1.0000 | 56% | 6.5 +/-1 on Dwg |
| Ground Plate | 123456678-004 | A | 2 | Dim: Bottom Surface - Top Surface | 3.0000 | | +/- 0.5000 | 28% | 3 +/-0.5 on Dwg |
| M4 Washer | | | 3 | Dim: Bottom Surface - Top Surface | 1.2000 | | +/- 0.1000 | 6% | 1.2 +/- 0.1 fm Machinery's Hdbk 23rd Ed. |
| M4 X 8 SHCS | | | 4 | Dim: Underside of Head - End of Screw | | 8.0000 | +/- 0.2000 | 11% | 8 +/-0.2 fm Vendor Dwg |
| | | | | Dimension Totals | 10.7000 | 8.0000 | | | |
| | | | | Nominal Distance: Pos Dims - Neg Dims = | 2.7000 | | | | |

RESULTS:

| | Nom | Tol | Min | Max |
|---|---|---|---|---|
| Arithmetic Stack (Worst Case) | 2.7000 | +/- 1.8000 | 0.9000 | 4.5000 |
| Statistical Stack (RSS) | 2.7000 | +/- 1.1402 | 1.5598 | 3.8402 |
| Adjusted Statistical: 1.5*RSS | 2.7000 | +/- 1.7103 | 0.9897 | 4.4103 |

Notes:

Assumptions: - Used Enclosure and Ground Plate Option 1 for this study.

Suggested Action:

**FIGURE 16.11** Tolerance Stackup report for Example 16.1.

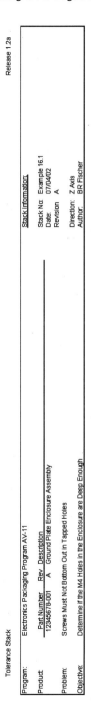

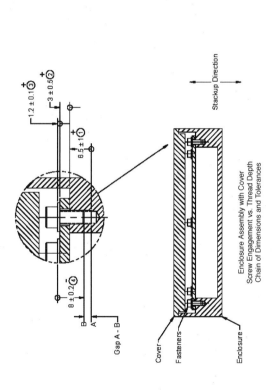

**FIGURE 16.12** Tolerance Stackup sketch for Example 16.1.

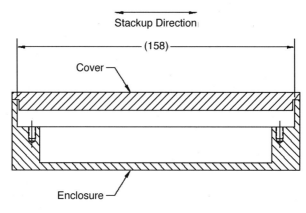

**FIGURE 16.13**   Example 16.2: Cover fit stack, along $X$ axis.

inside surfaces of the enclosure. This is the distance projected along the maximum depth that the cover protrudes into the enclosure. The calculations can be seen in Fig. 16.16. The Tolerance Stackup sketch and the detail are included as page 2 and the calculations are included as page 3 of the Tolerance Stackup report.

Although this Tolerance Stackup appears to be fairly simple at first glance, it becomes more complex once the tilting allowed by the enclosure's datum reference frame is taken into account.

*Results.*   The worst-case Tolerance Stackup result shows that there is 0.6204-mm clearance between the cover and the enclosure—the cover fits inside the enclosure along the $X$ axis.

### Example 16.3.   Cover Fit Tolerance Stackup Along the Y Axis

Determine if the cover fits within the enclosure along $Y$ axis.
Solve Example 16.3 as follows:

- Stackup direction is along $Y$ axis (see Fig. 16.17).
- Use Option 2 parts for this example.

This Tolerance Stackup is the same as the previous Tolerance Stackup except it is done in the $Y$ axis direction. The potential for variation along the $Y$ axis is greater than along the $X$ axis, as the enclosure and cover are shorter in this direction, which leads to a greater angle of foreshortening. The Tolerance Stackup in Example 16.3 determines if the cover fits within the enclosure. The

Program:   Electronics Packaging Program AV-11

Product:
| Part Number | Rev | Description |
|---|---|---|
| 12345678-001 | A | Ground Plate Enclosure Assembly |

Problem:   Cover Must Fit into the Enclosure

Objective:   Determine if the Largest Cover Fits Inside the Enclosure Along the X Axis

Stack Information:
Stack No:   Example 16-2
Date:   07/04/02
Revision:   A
Direction:   X Axis
Author:   BR Fischer

| Description of Component / Assy | Part Number | Rev | Item | Description | + Dims | - Dims | Tol | Percent Contrib | Dim / Tol Source & Calcs |
|---|---|---|---|---|---|---|---|---|---|
| Enclosure | 12345678-002 | A | 1 | Profile: LH Inside Surface | | | +/- 0.5000 | 27% | Profile 1, A, Bm |
| | | | 2 | Datum Feature Shift | | | +/- 0.0000 | 0% | N/A- Sim Reqts |
| | | | 3 | Dim: LH Inside Surface - RH Inside Surface of Enclosure | 158.0000 | | +/- 0.0000 | 0% | 158 Basic on Dwg |
| | | | 4 | Profile: RH Inside Surface | | | +/- 0.5000 | 27% | Profile 1, A, Bm |
| | | | 5 | Datum Feature Shift | | | +/- 0.0000 | 0% | N/A- Sim Reqts |
| | | | 6 | Trig Effect of Angle Between Top Surface and Sides of Enclosure | 0.0398 | | +/- 0.0398 | 2% | = (( 2 * 6.25) / 157) / 2 (Like Triangles) w/ (Mean Shift) |
| Cover | 12345678-003 | A | 7 | Profile: RH Surface | | | +/- 0.4000 | 22% | Profile 0.8, A All-Around |
| | | | 8 | Datum Feature Shift | | | +/- 0.0000 | 0% | N/A - DF$_A$ not a Feature of Size |
| | | | 9 | Dim: RH Surface - LH Surface of Cover | | 155.5000 | +/- 0.0000 | 0% | 155.5 Basic on Dwg |
| | | | 10 | Profile: LH Surface | | | +/- 0.4000 | 22% | Profile 0.8, A |
| | | | 11 | Datum Feature Shift | | | +/- 0.0000 | 0% | N/A - DF$_A$ not a Feature of Size & Sim Reqts |

Dimension Totals:   + Dims 158.0000   - Dims 155.5398

Nominal Distance: Pos Dims - Neg Dims = 2.4602

RESULTS:

| | Nom | Tol | Min | Max |
|---|---|---|---|---|
| Arithmetic Stack (Worst Case) | 2.4602 | +/- 1.8398 | 0.6204 | 4.3000 |
| Statistical Stack (RSS) | 2.4602 | +/- 0.9064 | 1.5538 | 3.3666 |
| Adjusted Statistical: 1.5*RSS | 2.4602 | +/- 1.3596 | 1.1006 | 3.8198 |

Notes:   - 'Item 6 "Trig Effect..." is included because the top surface of the Enclosure (which the Cover sits on) is not the Datum Feature for the inside surfaces. Both are related to Datum Feature A, which is the Ground Plate mounting surface. Consequently, the top and the inside surfaces can tilt relative to each other, foreshortening the apparent width that the Cover fits into, because the Cover will orient to the top surface of the Enclosure. The angle is projected over the maximum height of the vertical surfaces of the Cover, which is 6.25mm – this represents the worst-case.
- The angle between the top surface and the inside surfaces may only decrease the width of the opening – it cannot increase the width of the opening. The foreshortening value is divided by 2, giving the correct minimum and maximum limits due to foreshortening. Therefore a Mean Shift of 1/2 the foreshortening value must be included in the negative column on the same line as the trig effect.
- Datum Feature Shift on Line 5 does not contribute to the Tolerance Stackup because the Profile Tolerance is specified all-around for the Enclosure opening. Both the Left and Right Surfaces are to be inspected at the same time (in the same setup), as they are controlled by the same tolerance. Consequently the Datum Feature Shift does not affect the possible distance between these surfaces.

Assumptions:   - Used Enclosure Option 2 for this study.

Suggested Action:   None. Even with the foreshortening of the opening there is still ~0.6 clearance between the Cover and the Enclosure.

**FIGURE 16.14**   Tolerance Stackup report for Example 16.2.

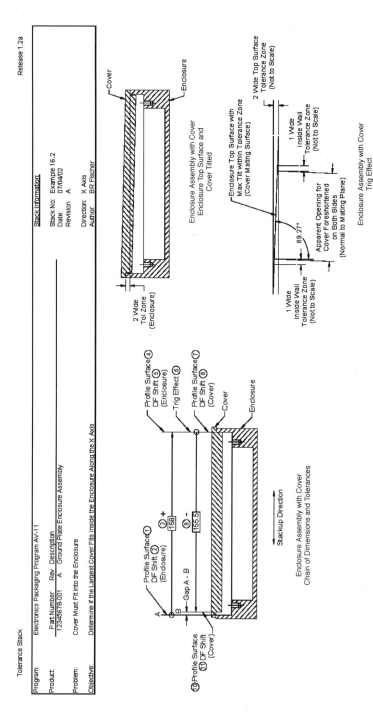

**FIGURE 16.15** Tolerance Stackup sketch and detail for Example 16.2.

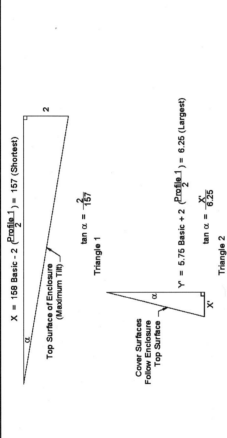

Release 1.2a

Tolerance Stack

| Program: | Electronics Packaging Program AV-11 | | |
|---|---|---|---|
| Product | Part Number | Rev | Description |
| | 12345678-001 | A | Ground Plate Enclosure Assembly |
| Problem: | Cover Must Fit Into the Enclosure | | |
| Objective: | Determine If the Largest Cover Fits Inside the Enclosure Along the X Axis | | |

Stack Information
Stack No: Example 16.2
Date: 07/04/02
Revision A

Direction: X Axis
Author: BR Fischer

$X = 158 \text{ Basic} - 2 \left(\dfrac{\text{Profile 1}}{2}\right) = 157 \text{ (Shortest)}$

Top Surface of Enclosure
(Maximum Tilt)

$\tan\alpha = \dfrac{2}{157}$

**Triangle 1**

Cover Surfaces
Follow Enclosure
Top Surface

$Y = 5.75 \text{ Basic} + 2 \left(\dfrac{\text{Profile 1}}{2}\right) = 6.25 \text{ (Largest)}$

$\tan\alpha = \dfrac{X'}{6.25}$

**Triangle 2**

**Like Triangles:**

**Solve For X':**

$\dfrac{X'}{6.25} = \dfrac{2}{157} \implies X' = \dfrac{2\,(6.25)}{157} = 0.0796$

**Trig Effect:**

Foreshortening of apparent cover opening in enclosure = 0.0796
(Line Item 6 in Report)

Like Triangles:  Cover Fit Tolerance Stackup
Trigonometry,  Along X Axis

**FIGURE 16.16** Tolerance Stackup calculations for Example 16.2.

Stackup Direction

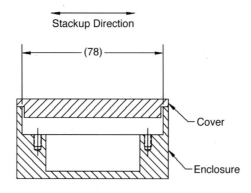

Does Cover Fit within Enclosure?

**FIGURE 16.17**   Example 16.3: Cover fit stack, along $Y$ axis.

Tolerance Stackup report is shown in Fig. 16.18 and the Tolerance Stackup sketch is shown in Fig. 16.19. The detail in Fig. 16.19 shows that the top surface of the enclosure may tilt relative to the inside surfaces of the enclosure. This occurs because the top surface of the enclosure is not the primary datum feature for the profile tolerance on the inside surfaces of the enclosure. This tilting can lead to an apparent foreshortening of the opening for the cover. Line item 6 in the Tolerance Stackup report includes the linear equivalent for the angle between the top surface and the inside surfaces of the enclosure. This is the distance projected along the maximum depth that the cover protrudes into the enclosure. The calculations can be seen in Fig. 16.20. The Tolerance Stackup sketch and the detail are included as page 2 and the calculations are included as page 3 of the Tolerance Stackup report.

Although this Tolerance Stackup appears to be fairly simple at first glance, it becomes more complex once the tilting allowed by the enclosure's datum reference frame is taken into account.

*Results.*   The worst-case Tolerance Stackup results shows that there is 0.5377-mm clearance between the cover and the enclosure—the cover fits inside the enclosure along the $Y$ axis, albeit with slightly less clearance than along the $X$ axis.

### Example 16.4.   *Screw Head Clearance Tolerance Stackup*

Determine if the cover contacts the M4 screw heads.
Extra data:

- M4 washer thickness $= 1.2 \pm 0.1$.
- M4 $\times$ 8 screw head height $= {4 \atop 3.82}$

| Program: | Electronics Packaging Program AV-11 | | | | | Stack Information: | | | |
|---|---|---|---|---|---|---|---|---|---|

| Product: | Part Number | Rev | Description | | | Stack No: | Example 16-3 |
|---|---|---|---|---|---|---|---|
| | 12345678-001 | A | Ground Plate Enclosure Assembly | | | Date: | 07/04/02 |
| | | | | | | Revision: | A |

Problem: Cover Must Fit into the Enclosure

Objective: Determine if the Largest Cover Fits Inside the Enclosure Along the Y Axis

Direction: Y Axis
Author: BR Fischer

| Description of Component / Assy | Part Number | Rev | Item | Description | + Dims | - Dims | Tol | Percent Contrib | Dim / Tol Source & Calcs |
|---|---|---|---|---|---|---|---|---|---|
| Enclosure | 12345678-002 | A | 1 | Profile LH Inside Surface | | | +/- 0.5000 | 27% | Profile 1, A, Bm |
| | | | 2 | Datum Feature Shift | | | +/- 0.0000 | 0% | N/A- Sim Reqts |
| | | | 3 | Dim: LH Inside Surface - RH Inside Surface of Enclosure | 158.0000 | | +/- 0.0000 | 0% | 158 Basic on Dwg |
| | | | 4 | Profile RH Inside Surface | | | +/- 0.5000 | 27% | Profile 1, A, Bm |
| | | | 5 | Datum Feature Shift | | | +/- 0.0000 | 0% | N/A- Sim Reqts |
| | | | 6 | Trig Effect of Angle Between Top Surface and Sides of Enclosure | | 0.0812 | +/- 0.0812 | 4% | = (( 2 * 6.25) / 77) / 2 ( Like Triangles) w/ (Mean Shift) |
| Cover | 12345678-003 | A | 7 | Profile RH Surface | | | +/- 0.4000 | 21% | Profile 0.8, A, All-Around |
| | | | 8 | Datum Feature Shift | | | +/- 0.0000 | 0% | N/A - DF, not a Feature of Size |
| | | | 9 | Dim: RH Surface - LH Surface of Cover | | 155.5000 | +/- 0.0000 | 0% | 155.5 Basic on Dwg |
| | | | 10 | Profile LH Surface | | | +/- 0.4000 | 21% | Profile 0.8, A |
| | | | 11 | Datum Feature Shift | | | +/- 0.0000 | 0% | N/A - DF, not a Feature of Size & Sim Reqts |

Dimension Totals 158.0000 | 155.5812
Nominal Distance: Pos Dims - Neg Dims = 2.4188

**RESULTS:**

| | Nom | Tol | Min | Max |
|---|---|---|---|---|
| Arithmetic Stack (Worst Case) | 2.4188 | +/- 1.8812 | 0.5377 | 4.3000 |
| Statistical Stack (RSS) | 2.4188 | +/- 0.9092 | 1.5097 | 3.3280 |
| Adjusted Statistical: 1.5*RSS | 2.4188 | +/- 1.3638 | 1.0551 | 3.7826 |

Notes: - "Item 6 "Trig Effect..." is included because the top surface of the Enclosure (which the Cover sits on) is not the Datum Feature for the inside surfaces. Both are related to Datum Feature A, which is the Ground Plate mounting surface. Consequently, the top and the inside surfaces can tilt relative to each other, foreshortening the apparent width that the Cover fits into, because the Cover will orient to the top surface of the Enclosure. The angle is projected over the maximum height of the Enclosure, which is 6.25mm - this represents the worst-case.
- The angle between the top surface and the inside surfaces may only decrease the width of the opening - it cannot increase the width of the opening. The foreshortening value is divided by 2, giving the correct minimum and maximum limits due to foreshortening. Therefore a Mean Shift of 1/2 the foreshortening value must be included in the negative column on the same line as the trig effect. The foreshortening value is divided by 2, giving the correct minimum and maximum limits due to foreshortening.
- Datum Feature Shift on Line 5 does not contribute to the Tolerance Stackup because the Profile Tolerance is specified all-around for the Enclosure opening. Both the Left and Right Surfaces are to be inspected at the same time (in the same setup), as they are controlled by the same tolerance. Consequently the Datum Feature Shift does not affect the possible distance between these surfaces.

Assumptions: - Used Enclosure Option 2 for this study.

Suggested Action: None. Even with the foreshortening of the opening there is still -0.5 clearance between the Cover and the Enclosure.
Notice that the Trig Effect is line 6 is greater in Y direction than in the X direction. This is because the 2mm Profile tolerance applies along a shorter distance in this direction.

**FIGURE 16.18** Tolerance Stackup report for Example 16.3.

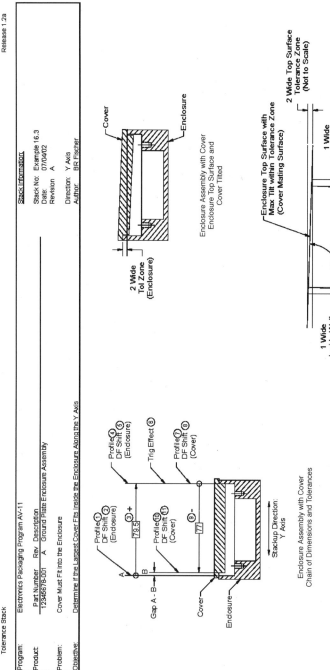

**FIGURE 16.19**  Tolerance Stackup sketch and detail for Example 16.3.

Tolerance Stack

Release 1.2a

| Program: | Electronics Packaging Program AV-11 | | | **Stack Information:** | |
|---|---|---|---|---|---|
| Product: | Part Number | Rev | Description | Stack No: | Example 16.3 |
| | 12345678-001 | A | Ground Plate Enclosure Assembly | Date: | 07/04/02 |
| | | | | Revision: | A |
| Problem: | Cover Must Fit into the Enclosure | | | Direction: | Y Axis |
| Objective: | Determine If the Largest Cover Fits Inside the Enclosure Along the Y Axis | | | Author: | BR Fischer |

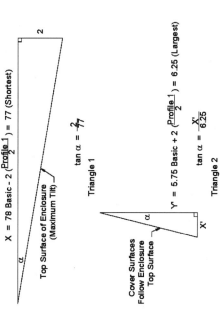

X = 78 Basic − 2 ($\frac{Profile\ 1}{2}$) = 77 (Shortest)

Top Surface of Enclosure
(Maximum Tilt)

$\tan \alpha = \frac{2}{77}$

Triangle 1

Cover Surfaces
Follow Enclosure
Top Surface

Y = 5.75 Basic + 2 ($\frac{Profile\ 1}{2}$) = 6.25 (Largest)

$\tan \alpha = \frac{X'}{6.25}$

Triangle 2

**Like Triangles:**

**Solve For X':**

$\frac{X'}{6.25} = \frac{2}{77} \Rightarrow X' = 2\frac{6.25}{77} = 0.1623$

**Trig Effect:**

Foreshortening of apparent cover opening in enclosure = 0.1623
(Line Item 6 in Report)

Like Triangles: Cover Fit Tolerance Stackup
Trigonometry, Along Y Axis

**FIGURE 16.20** Tolerance Stackup calculations for Example 16.3.

Solve Example 16.4 as follows:

- Stackup direction is along the *Z* axis (see Fig. 16.21).
- Use Option 1 parts for this example.

The Tolerance Stackup in Example 16.4 determines if the screw heads contact the bottom of the cover. The Tolerance Stackup report is shown in Fig. 16.22, and the Tolerance Stackup sketch is shown in Fig. 16.23.

*Results.*   The worst-case Tolerance Stackup results shows that there is a potential interference of 1.05 mm between the screw heads and the cover— the screw heads contact the bottom of the cover. The result indicates interference because it is a negative number—a positive result indicates clearance in this example. Possible solutions are to decrease the profile tolerance on the top surface of the enclosure, increase the nominal distance

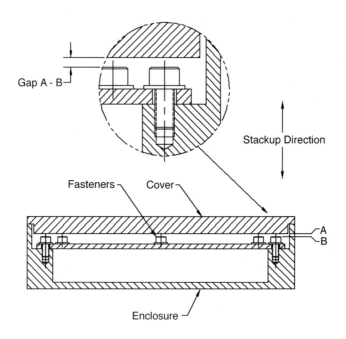

Do Screw Heads Contact the Cover?

**FIGURE 16.21**   Example 16.4: Screw heads vs. cover.

Tolerance Stack

Release 1.2a

**Program:** Electronics Packaging Program AV-11

**Product:**

| Part Number | Rev | Description |
|---|---|---|
| 123456678-001 | A | Ground Plate Enclosure Assembly |

**Problem:** Cover Must Not Contact Heads of Screws

**Objective:** Determine If the Cover Hits the Screws

**Stack Information:**

Stack No: Example 16.4
Date: 07/04/02
Revision: A

Direction: Z Axis
Author: BR Fischer

| Description of Component / Assy | Part Number | Rev | Item | Description | + Dims | - Dims | Tol | Percent Contrib | Dim / Tol Source & Calcs |
|---|---|---|---|---|---|---|---|---|---|
| Cover | 123456678-003 | A | 1 | Profile: Bottom Surface | | | ± 0.5000 | 23% | Profile 1, A |
| | | | 2 | Datum Feature Shift | | | ± 0.0000 | 0% | N/A - DF$_A$ not a Feature of Size |
| | | | 3 | Dim: Bottom Surface - DF$_A$ | | 5.7500 | ± 0.0000 | 0% | 5.75 Basic on Dwg |
| Enclosure | 123456678-002 | A | 4 | Profile: Top Surface | | | ± 1.0000 | 46% | Profile 2, A, B$_m$ |
| | | | 5 | Datum Feature Shift (DF$_B$ @ umc - DFS$_B$) / 2 | | | ± 0.0000 | 0% | N/A - DF$_B$ Shift is perpendicular to Stack - No Effect |
| | | | 6 | Dim: Top Surface - DF$_A$ | 15.0000 | | ± 0.0000 | 0% | 15 Basic on Dwg |
| Ground Plate | 123456678-004 | A | 7 | Dim: Bottom Surface - Top Surface | | 3.0000 | ± 0.5000 | 23% | 3 ± 0.5 on Dwg |
| M4 Washer | | | 8 | Dim: Bottom Surface - Top Surface | | 1.2000 | ± 0.1000 | 5% | 1.2 ± 0.1 fm Machinery's Hdbk 23rd Ed. |
| M4 X 8 SHCS | | | 9 | Dim: Bottom Surface - Top Surface | | 3.9100 | ± 0.0900 | 4% | 3.82 - 4 fm Machinery's Hdbk 23rd Ed. |
| | | | | Dimension Totals | 15.0000 | 13.8600 | | | |
| | | | | Nominal Distance Pos Dims - Neg Dims = | | 1.1400 | | | |

**RESULTS:**

| | Nom | Tol | Min | Max |
|---|---|---|---|---|
| Arithmetic Stack (Worst Case) | 1.1400 | ± 2.1900 | -1.0500 | 3.3300 |
| Statistical Stack (RSS) | 1.1400 | ± 1.2321 | -0.0921 | 2.3721 |
| Adjusted Statistical 1.5*RSS | 1.1400 | ± 1.8482 | -0.7082 | 2.9882 |

**Notes:**

**Assumptions:** - Used Enclosure Option 1 for this study.

**Suggested Action:**
- Decrease Profile tolerance on Top Surface of Enclosure to 0.75
- Or increase basic 15 dimension (DF$_A$ to Top Surface of Enclosure) to 15.75 basic on Enclosure.

**FIGURE 16.22** Tolerance Stackup report for Example 16.4.

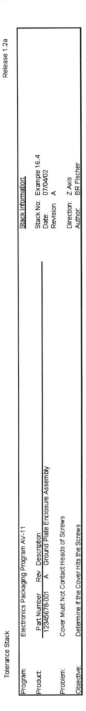

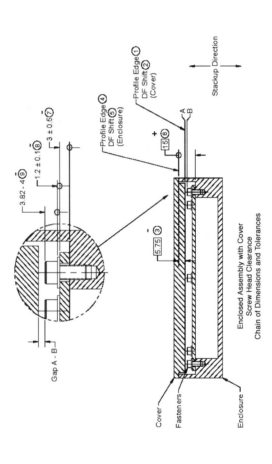

**FIGURE 16.23**  Tolerance Stackup sketch for Example 16.4.

from datum feature A to the top surface of the enclosure, or decrease the depth of the cover, to name a few.

### Example 16.5. Ground Plate to Enclosure Gap Study: Option 1 Parts

Determine if the ground plate contacts the inside walls of the enclosure.
   Extra data:

- M4 tapped hole Dims: Minor diameter = 3.242–3.422.
- M4 × 8 socket head cap screw Dims: Major diameter = 3.82–4.

Solve Example 16.5 as follows:

- Stackup direction is along the *Y* axis (see Fig. 16.24).
- Use Option 1 parts for this example.
- Use minimum and maximum minor diameter for bonus tolerance calculations on the M4 tapped holes in the enclosure. (This is because the positional tolerance is specified on the minor diameter.)
- Use the minimum screw major diameter for the assembly shift calculations

The Tolerance Stackup report is shown in Fig. 16.25. The Tolerance Stackup sketch and a detail of the worst-case results are shown in Fig. 16.26.

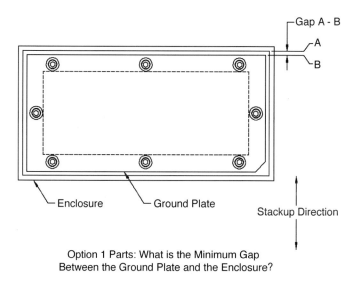

Option 1 Parts: What is the Minimum Gap
Between the Ground Plate and the Enclosure?

**FIGURE 16.24** Example 16.5: Ground plate to enclosure gap, Option 1.

Tolerance Stack                                                                                         Release 1.2a

| | | | | | | | | | | |
|---|---|---|---|---|---|---|---|---|---|---|

**Program:** Electronics Packaging Program AV-11

| | | | | **Stack Information:** | |
|---|---|---|---|---|---|
| **Product:** | **Part Number** | **Rev** | **Description** | **Stack No:** | Example 16.5 |
| | 12345678-001 | A | Ground Plate Enclosure Assembly. Option 1 w Surfaces as Datum Features B & C | **Date:** | 07/04/02 |
| | | | | **Revision** | A |
| **Problem:** | Edges of Ground Plate must not Touch Walls of Enclosure | | | | |
| **Objective:** | Option 1: Determine if Ground Plate Contacts Enclosure Walls | | **Direction:** | Y Axis |
| | | | | **Author:** | BR Fischer |

| Description of Component / Assy | Part Number | Rev | Item | Description | + Dims | - Dims | Tol | Percent Contrib | Dim / Tol Source & Calcs |
|---|---|---|---|---|---|---|---|---|---|
| Enclosure | 12345678-002 | A | 1 | Profile Edge Along Pt A | | | ± 0.5000 | 18% | Profile 1, A, B, C |
| | | | 2 | Datum Feature Shift | | | ± 0.0000 | 0% | N/A - Datum Features not Features of Size |
| | | | 3 | Dim: Edge of Enclosure - Datum B | 78.0000 | | ± 0.0000 | 0% | 78 Basic on Dwg |
| | | | 4 | Dim: Datum B - CL M4 Holes | | 8.5000 | ± 0.0000 | 0% | 8.5 Basic on Dwg |
| | | | 5 | Position: M4 Holes | | | ± 0.3000 | 11% | Position dia 0.6 @ MMC A, B, C |
| | | | 6 | Bonus Tolerance | | | ± 0.0900 | 3% | = (3.422 - 3.242)/2 |
| | | | 7 | Datum Feature Shift: | | | ± 0.0000 | 0% | N/A - Datum Features not Features of Size |
| Ground Plate | 12345678-004 | A | 8 | Assembly Shift (Mounting Holes$_{MC}$ - F$_{MC}$) / 2 | | | ± 0.8660 | 32% | = [(6.4 + 0.15) - 3.82]/2 |
| | | | 9 | Position: Dia 5.4 ±0.15 Holes | | | ± 0.3250 | 12% | Position dia 0.65 @ MMC A |
| | | | 10 | Bonus Tolerance | | | ± 0.1500 | 5% | = (0.15 + 0.15)/2 |
| | | | 11 | Datum Feature Shift: | | | ± 0.0000 | 0% | N/A - Datum Features not Features of Size |
| | | | 12 | Dim: CL Dia 5.4 Holes - Datum B | 6.0000 | | ± 0.0000 | 0% | 6 Basic on Dwg |
| | | | 13 | Dim: Datum B - Edge of Ground Plate | | 73.0000 | ± 0.0000 | 0% | 73 Basic on Dwg |
| | | | 14 | Profile Edge Along Pt B | | | ± 0.5000 | 18% | Profile 1, A, B, C |
| | | | 15 | Datum Feature Shift: | | | ± 0.0000 | 0% | N/A - Datum Feature s not Features of Size |

| | | | | |
|---|---|---|---|---|
| | Dimension Totals | 84.0000 | 81.5000 | |
| | Nominal Distance: Pos Dims - Neg Dims = | | 2.5000 | |

| | Nom | Tol | Min | Max |
|---|---|---|---|---|
| **RESULTS:** | | | | |
| Arithmetic Stack (Worst Case) | 2.5000 | ± 2.7300 | -0.2300 | 5.2300 |
| Statistical Stack (RSS) | 2.5000 | ± 1.2143 | 1.2857 | 3.7143 |
| Adjusted Statistical 1.5*RSS | 2.5000 | ± 1.8214 | 0.6786 | 4.3214 |

**Notes:** - M4 Screw Dimensions: Major Dia: 4/3.82   - M4 Tapped Hole Dimensions: Minor Dia: 3.422/3.242
- Used Min & Max M4 Tapped Hole Minor Diameters to Calculate Bonus Tolerance on Line 6 because the Positional Tolerance Applies to the Minor Diameter.
- Used smallest screw major dia in Assembly Shift Calculations on line 8.
- The positional tolerances on the clearance holes and the M4 holes are larger because they are toleranced relative to the edge surfaces, and manufacturing said that was the best they could do.
- The larger positional tolerances required the clearance holes to be larger, due to the Fixed Fastener Formula. This increased the Assembly Shift calculated in line 8.

**Assumptions:**
- Although threads are typically assumed to be self centering, the Positional Tolerance applies to the Minor Diameter of the M4 holes. Use the min/max Minor Dia to calculate the Bonus Tolerance on line 6.

**Suggested Action:**
- May want to holes as locators instead of edges.  See Examples 16.6 and 16.7.

**FIGURE 16.25**  Tolerance Stackup report for Example 16.5.

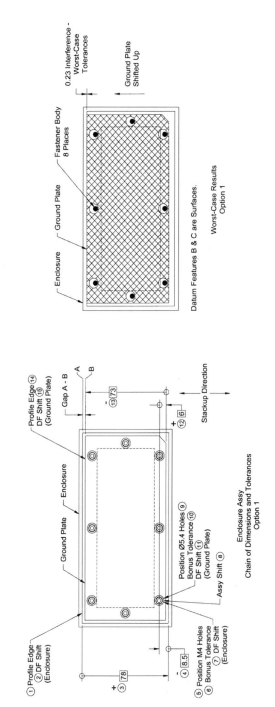

**FIGURE 16.26** Tolerance Stackup sketch and detail for Example 16.5.

The Tolerance Stackup sketch and the detail are included as page 2 of the Tolerance Stackup report.

*Results*. This is the first of three examples that study the same problem, each using a different dimensioning and tolerancing scheme for the ground plate and the enclosure. This example uses the Option 1 ground plate and enclosure, in which surfaces are specified as the secondary and tertiary datum features. To many this seems to be the simplest dimensioning and tolerancing scheme of the three, but it is the least functional, and leads to the greatest overall variation. It must be stated that the values in these three dimensioning and tolerancing schemes are not quite equivalent—they are close, however. For the purpose of these examples they help to show that the dimensioning and tolerancing scheme can have a big impact on the variation between important features.

Using this dimensioning and tolerancing scheme, the worst-case Tolerance Stackup result shows that there is a potential interference of 0.23 mm between the ground plate and the enclosure—the ground plate contacts the enclosure. The result indicates interference because it is a negative number—a positive result indicates clearance in this example.

Notice that the values for the positional tolerance and the associated bonus tolerance on lines 5, 6, 9 and 10 are included in the Tolerance Stackup. Using this dimensioning and tolerancing technique requires these tolerances to be included in the Tolerance Stackup. Together these make up 31% of the total tolerance in the Tolerance Stackup, so there is still some room for improvement.

### Example 16.6.   Ground Plate to Enclosure Gap Study: Option 2 Parts

Determine if the ground plate contacts the inside walls of the enclosure.
Extra data:

- M4 tapped hole Dims: minor diameter $= 3.242$–$3.422$.
- M4 $\times$ 8 socket head cap screw dims: major diameter $= 3.82$–$4$.

Solve Example 16.6 as follows:

- Stackup direction is along the $Y$ axis (see Fig. 16.27).
- Use Option 2 parts for this example.
- Use minimum and maximum minor diameter for bonus tolerance calculations on the M4 tapped holes in the enclosure. (This is because the positional tolerance is specified on the minor diameter.)
- Use the minimum screw major diameter for the assembly shift calculations

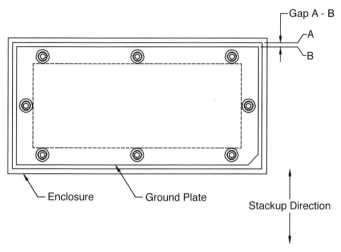

Option 2 Parts: What is the Minimum Gap
Between the Ground Plate and the Enclosure?

**FIGURE 16.27**   Example 16.6: Ground plate to enclosure gap, Option 2.

The Tolerance Stackup report is shown in Fig. 16.28. The Tolerance Stackup sketch and a detail of the worst-case results are shown in Fig. 16.29. The Tolerance Stackup sketch and the detail are included as page 2 of the Tolerance Stackup report.

*Results.*   This is the second of three examples that study the same problem, each using a different dimensioning and tolerancing scheme for the ground plate and the enclosure. This example uses the Option 2 ground plate and enclosure. The pattern of eight ⌀5 ± 0.15 clearance holes are specified as the secondary datum feature on the ground plate, and the minor diameters of the pattern of eight M4 tapped holes are specified as the secondary datum feature on the enclosure. This dimensioning and tolerancing scheme reflects the function of the mating parts better and leads to less overall variation than was seen when using the Option 1 parts. Using the minor diameters of the tapped holes as datum features was not a functional decision; it was done to allow the use of a functional gage for inspection. Even though a nonfunctional concession was made to facilitate inspection, this example still results in less overall variation, which again better reflects the functional requirements of the assembly.

Using this dimensioning and tolerancing scheme, the worst-case Tolerance Stackup result shows that there is a minimum clearance of 0.17 mm

Tolerance Stack                                                                        Release 1.2a

| Program: | Electronics Packaging Program AV-11 |
|---|---|

Product: Part Number 12345678-001  Rev A  Description Ground Plate Enclosure Assembly: Option 2 w 8 Holes as Datum Feature B

Problem: Edges of Ground Plate must not Touch Walls of Enclosure

Objective: Option 2: Determine if Ground Plate Contacts Enclosure Walls

**Stack Information:**
Stack No: Example 16.6
Date: 07/04/02
Revision: A
Direction: Y Axis
Author: BR Fischer

| Description of Component / Assy | Part Number | Rev | Item | Rev | Description | + Dims | - Dims | Tol | Percent Contrib | Dim / Tol Source & Calcs |
|---|---|---|---|---|---|---|---|---|---|---|
| Enclosure | 12345678-002 | A | 1 | | Profile Edge Along Pt A | | | ± 0.5000 | 21% | Profile 1, A, Bm |
| | | | 2 | | Datum Feature Shift: (DF$_B$ @ LMC - DFS$_B$)/2 | | | ± 0.2900 | 12% | = [3.422 - (3.242 - 0.4)]/2   (Shift within Minor Dia) |
| | | | 3 | | Dim: Edge of Enclosure - Datum B | 8.5000 | | ± 0.0000 | 0% | 8.5 Basic on Dwg |
| | | | 4 | | Position: DF$_B$ M4 Holes | | | ± 0.0000 | 0% | N/A (See Assumption #1) |
| | | | 5 | | Bonus Tolerance | | | ± 0.0000 | 0% | N/A (See Assumption #1) |
| | | | 6 | | Datum Feature Shift: (DF$_B$ @ LMC - DFS$_B$)/2 | | | ± 0.0000 | 0% | N/A - DF$_A$ not a Feature of Size |
| Ground Plate | 12345678-004 | A | 7 | | Assembly Shift: (Mounting Holes$_{LMC}$ - F$_{LMC}$)/2 | | | ± 0.6650 | 29% | = [(5 + 0.15) - 3.82]/2 |
| | | | 8 | | Position: DF$_B$ Dia 5 ± 0.15 Holes | | | ± 0.0000 | 0% | N/A (See Assumption #1) |
| | | | 9 | | Bonus Tolerance | | | ± 0.0000 | 0% | N/A (See Assumption #1) |
| | | | 10 | | Datum Feature Shift: (DF$_B$ @ LMC - DFS$_B$)/2 | | | ± 0.0000 | 0% | N/A - DF$_A$ not a Feature of Size |
| | | | 11 | | Dim: Datum B - Edge of Ground Plate | 6.0000 | | ± 0.0000 | 0% | 6 Basic on Dwg |
| | | | 12 | | Profile Edge Along Pt B | | | ± 0.5000 | 21% | Profile 1, A, Bm |
| | | | 13 | | Datum Feature Shift: (DF$_B$ @ LMC - DFS$_B$)/2 | | | ± 0.3750 | 16% | = [(5 + 0.15) - (5 - 0.15 - 0.45)]/2 |

Dimension Totals    8.5000    6.0000
Nominal Distance: Pos Dims - Neg Dims =    2.5000

**RESULTS:**

| | Nom | Tol | Min | Max |
|---|---|---|---|---|
| Arithmetic Stack (Worst Case) | 2.5000 | ± 2.3300 | 0.1700 | 4.8300 |
| Statistical Stack (RSS) | 2.5000 | ± 1.0803 | 1.4197 | 3.5603 |
| Adjusted Statistical: 1.5*RSS | 2.5000 | ± 1.6204 | 0.8796 | 4.1204 |

Notes:
- M4 Screw Dimensions: Major Dia 4/3.82    - M4 Tapped Hole Dimensions: Minor Dia: 3.422/3.242
- Used min and max screw thread minor dia in Datum Feature Shift Calculations on line 2.
- Used smallest screw major dia in Assembly Shift Calculations on line 7.

Assumptions:
1 The Positional Tolerances applied to the Secondary Datum Feature B holes on the Enclosure and the Ground Plate have no effect on the Tolerance Stackup - nor do the associated Bonus Tolerances. This is because as the Secondary Datum Features, the Datums derived from these holes are the basis from which all measurements are made in the Direction of the Tolerance Stackup. If the Datum Features were produced toward one extreme within their Tolerance Zones, the Datum Reference Frame derived from the Datum Features and all related features would be biased in the same direction.

Suggested Action:
- May want to use two holes as locators instead of all eight. See Example 16.7.

**FIGURE 16.28**  Tolerance Stackup report for Example 16.6.

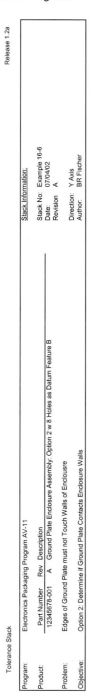

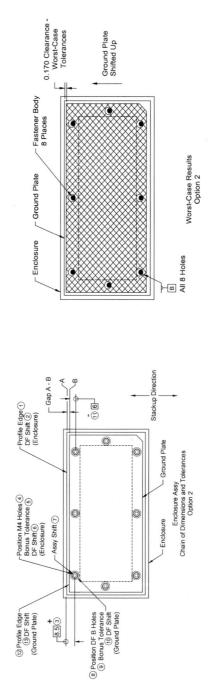

**FIGURE 16.29** Tolerance Stackup sketch and detail for Example 16.6.

between the ground plate and the enclosure—the ground plate does not contact the enclosure.

Notice that the values for the positional tolerance and the associated bonus tolerance on lines 4, 5, 8, and 9 are not included in the Tolerance Stackup—the values have been set to zero and "N/A (See Assumption #1)" has been placed in the "Dim/Tol Source & Calcs" column. Surprisingly, using this dimensioning and tolerancing technique makes these tolerances inconsequential to the Tolerance Stackup result. The reason is as follows: the pattern of holes is specified as the secondary datum feature. As the secondary datum feature, these holes are the basis from which all related tolerances are measured in the direction of the Tolerance Stackup. To put it another way, wherever the holes go, the rest of the features follow. So, using this technique has eliminated four tolerances from the Tolerance Stackup. This is a function of the rules of GD&T and can best be visualized by picturing the parts staged on the appropriate functional gage.

However, remember an MMC or LMC material condition modifier should accompany the datum feature reference in the feature control frame when the datum feature is a pattern of features of size. In this example that is precisely what we have: the secondary datum feature is a pattern of holes, which are features of size. That is why the datum feature B holes are referenced at MMC in the profile tolerance feature control frames. The MMC material condition modifier associated with the datum feature B reference creates a condition where datum feature shift is possible. Notice that values for datum feature shift have been included in the Tolerance Stackup on lines 2 and 13 following the profile tolerances applied to the ground plate and the enclosure. Together these make up 28% of the total tolerance in the Tolerance Stackup; so there is still some room for improvement.

### Example 16.7. Ground Plate to Enclosure Gap Study: Option 3 Parts

Determine if the ground plate contacts the inside walls of the enclosure.
     Extra data:

*   M4 tapped hole Dims: minor diameter = 3.242–3.422.
*   M4 × 8 socket head cap screw Dims: major diameter = 3.82–4.

Solve Example 16.7 as follows:

*   Stackup direction is along the $Y$ axis (see Fig. 16.30).
*   Use Option 3 parts for this example.
*   Use minimum and maximum minor diameter for bonus tolerance calculations on the M4 tapped holes in the enclosure. (This is because the positional tolerance is specified on the minor diameter.)

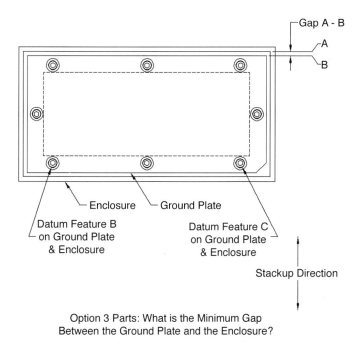

Option 3 Parts: What is the Minimum Gap
Between the Ground Plate and the Enclosure?

**FIGURE 16.30** Example 16.7: Ground plate to enclosure gap, Option 3.

- Use the minimum screw major diameter for the assembly shift calculations.

The Tolerance Stackup report is shown in Fig. 16.31. The Tolerance Stackup sketch and a detail of the worst-case results are shown in Fig. 16.32. The Tolerance Stackup sketch and the detail are included as page 2 of the Tolerance Stackup report.

*Results.* This is the third of three examples that study the same problem, each using a different dimensioning and tolerancing scheme for the ground plate and the enclosure. This example uses the Option 3 ground plate and enclosure. The lower left $\varnothing 4.5 \pm 0.1$ clearance hole is specified as the secondary datum feature, and the lower right $\varnothing 4.5 \pm 0.1$ clearance hole is specified as the tertiary datum feature on the ground plate. The minor diameter of the lower left M4 tapped hole is specified as the secondary datum feature, and the minor diameter of the lower right M4 tapped hole is specified as the tertiary datum feature on the enclosure. Of the three options, this dimensioning and tolerancing scheme reflects the function of the mating parts best and leads to the least overall variation. Using the minor diameters

Tolerance Stack 1                                                                         Release 1.2a

| | | | | | |
|---|---|---|---|---|---|
| Program: | Electronics Packaging Program AV-11 | | | **Stack Information:** | |
| Product: | Part Number 12345678-001  Rev A  Description: Ground Plate Enclosure Assembly, Option 3 w Single Holes as Datum Features B and C | | | Stack No: Example 16.7 | |
| | | | | Date: 07/04/02 | |
| | | | | Revision: A | |
| Problem: | Edges of Ground Plate must not Touch Walls of Enclosure | | | | |
| Objective: | Option 3: Determine If Ground Plate Contacts Enclosure Walls | | | Direction: Y Axis | |
| | | | | Author: BR Fischer | |

| Description of Component / Assy | Part Number | Rev | Item | Description | + Dims | − Dims | Tol | Percent Contrib | Dim / Tol Source & Calcs |
|---|---|---|---|---|---|---|---|---|---|
| Enclosure | 12345678-002 | A | 1 | Profile: Edge Along Pt A | | | ± 0.5000 | 36% | Profile 1, A, B, C |
| | | | 2 | Datum Feature Shift | | | ± 0.0000 | 0% | N/A - Datum Features Referenced RFS |
| | | | 3 | Dim: Edge of Enclosure Basic Loc Top Hole | 8.5000 | | ± 0.0000 | 0% | 8.5 Basic on Dwg |
| | | | 4 | Dim: Basic Loc Top Hole - Basic Loc Mid Hole | 30.5000 | | ± 0.0000 | 0% | 30.5 Basic on Dwg |
| | | | 5 | Dim: Basic Loc Mid Hole - Datums B & C | 30.5000 | | ± 0.0000 | 0% | 30.5 Basic on Dwg |
| | | | 6 | Perpendicularity: DF$_B$ M4 Hole; Position: DF$_C$ M4 Hole | | | ± 0.0000 | 0% | N/A (See Assumption #1) |
| | | | 7 | Bonus Tolerance | | | ± 0.0000 | 0% | N/A (See Assumption #1) |
| | | | 8 | Datum Feature Shift | | | ± 0.0000 | 0% | N/A - DF$_A$ not FOS; DF$_B$ Ref'd RFS |
| Ground Plate | 12345678-004 | A | 9 | Assembly Shift: (Mounting Holes$_{LMC}$ - F$_{LMC}$)/2 | | | ± 0.3900 | 28% | = [(4.5 + 0.1) - 3.82]/2 |
| | | | 10 | Perpendicularity: DF$_B$ Hole; Position: DF$_C$ Hole | | | ± 0.0000 | 0% | N/A (See Assumption #1) |
| | | | 11 | Bonus Tolerance | | | ± 0.0000 | 0% | N/A (See Assumption #1) |
| | | | 12 | Datum Feature Shift (DF$_B$ @ LMC - DF$_{Sa}$)/2 | | | ± 0.0000 | 0% | N/A - DF$_A$ not FOS; DF$_B$ Ref'd RFS |
| | | | 13 | Dim: Datums B & C - Basic Loc Mid Hole | | 30.5000 | ± 0.0000 | 0% | 30.5 Basic on Dwg |
| | | | 14 | Dim: Basic Loc Mid Hole - Basic Loc Top Hole | | 30.5000 | ± 0.0000 | 0% | 30.5 Basic on Dwg |
| | | | 15 | Dim: Basic Loc Top Hole - Edge of Ground Plate | | 6.0000 | ± 0.0000 | 0% | 6 Basic on Dwg |
| | | | 16 | Profile: Edge Along Pt B | | | ± 0.5000 | 36% | Profile 1, A, B, C |
| | | | 17 | Datum Feature Shift | | | ± 0.0000 | 0% | N/A - Datum Features Referenced RFS |

Dimension Totals: + Dims 69.5000   − Dims 67.0000
Nominal Distance: Pos Dims - Neg Dims = 2.5000

**RESULTS:**

| | Nom | Tol | Min | Max |
|---|---|---|---|---|
| Arithmetic Stack (Worst Case) | 2.5000 | ± 1.3900 | 1.1100 | 3.8900 |
| Statistical Stack (RSS) | 2.5000 | ± 0.8075 | 1.6925 | 3.3075 |
| Adjusted Statistical 1.5*RSS | 2.5000 | ± 1.2113 | 1.2887 | 3.7113 |

Notes:
- M4 Screw Dimensions: Major Dia: 4/3.82   - M4 Tapped Hole Dimensions: Minor Dia: 3.422/3.242
- Used smallest screw major dia in Assembly Shift Calculations on line 9.

Assumptions:
1  The Perpendicularity and Positional Tolerances applied to the Secondary and Tertiary Datum Feature B & C holes on the Enclosure and Ground Plate have no effect on the Tolerance Stackup - nor do their Bonus Tolerances. This is because as the Secondary and Tertiary Datum Features, the Datums derived from these holes are the basis from which all measurements are made in the Direction of the Tolerance Stackup. If the Datum Features were produced toward one extreme within their Tolerance Zones, the Datum Reference Frame derived from the Datum Features and all related features would be biased in the same direction.

Suggested Action:

**FIGURE 16.31**  Tolerance Stackup report for Example 16.7.

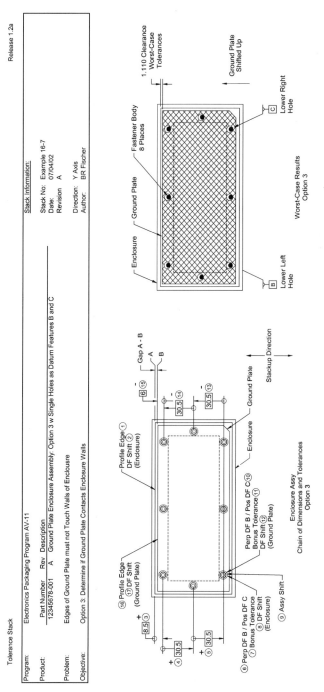

**FIGURE 16.32** Tolerance Stackup sketch and detail for Example 16.7.

of the tapped holes as datum features was not a functional decision; it was done to allow the use of a functional gage for inspection. Even though a nonfunctional concession was made to facilitate inspection, this example still results in less overall variation, which again better reflects the functional requirements of the assembly.

Specifying one of the holes to be the secondary datum feature and another of the holes to be the tertiary datum feature offers several advantages. First, an agreement may be reached with manufacturing to allow tighter tolerances on these two holes and looser tolerances to be specified on the other six holes. Notice that the size and the size tolerance of the datum feature B and C holes have been reduced in this example. This will minimize assembly shift. Second, specifying a single hole as the secondary datum feature and a single hole as the tertiary datum feature instead of a pattern of holes allows the datum features to be referenced RFS in a feature control frame. That means there is no datum feature shift for the profile tolerances that reference these datum features. Lastly, this technique allows the other six holes to be made larger (in the case of the clearance holes), as they no longer play a role in locating the ground plate to the enclosure—their role is now to merely allow a fastener to pass through to hold the part in place. Given that the size of the other six clearance holes has been increased, the fixed-fastener formula can be used to verify that the positional tolerance of the six clearance holes and the mating tapped holes can be increased.

There is a potential drawback to this method however, as the assembly method must be carefully coordinated with the tolerancing scheme. The fasteners must be started through the two datum feature holes first, as they have the tightest fit. If the fasteners were started through the other larger holes first and tightened, it is likely that the fasteners would interfere with the smaller holes. It is critical that the assembly personnel understand the requirement to follow the necessary assembly sequence.

Using this dimensioning and tolerancing scheme, the worst-case Tolerance Stackup result shows that there is a minimum clearance of 1.11 mm between the ground plate and the enclosure: the ground plate does not contact the enclosure.

Notice that the values for the perpendicularity tolerance and the associated bonus tolerance for the datum feature B holes and the positional tolerance and associated bonus tolerance for the datum feature C holes on lines 6, 7, 10, and 11 are not included in the Tolerance Stackup—the values have been set to zero and "N/A (See Assumption #1)" has been placed in the "Dim/Tol Source & Calcs" column. As with the previous Example, using this dimensioning and tolerancing technique makes these tolerances inconsequential to the Tolerance Stackup result. The reason is as follows: the holes specified as the secondary and tertiary datum features are the basis from which all related tolerances are measured in the direction of the Tolerance Stackup. To put it another way, wherever these holes go, the rest of the features follow. Using this technique has eliminated six tolerances from the Tolerance Stackup. This is a function of the rules of GD&T and can best be visualized by picturing the parts staged on the appropriate functional gage.

The secondary and tertiary datum features may be referenced in a feature control frame RFS because a single hole was specified for each. Notice that values for datum feature shift have been set to zero and labeled "N/A" in the Tolerance Stackup on lines 2 and 17 following the profile tolerances applied to the ground plate and the enclosure.

The Option 3 parts lead the least variation of the three methods.

# 17

## Calculating Component Tolerances Given a Final Assembly Tolerance Requirement

Sometimes a final assembly tolerance requirement is known, and tolerances must be determined that will allow the final requirement to be met. This is commonly encountered where assembly level or finished product level objectives have been set. For example, automotive and truck body panels must meet predetermined design and manufacturing objectives for quality and fit. The final assembly tolerancing requirements must be met when all the subcomponents are assembled.

Complex assemblies such as vehicle bodies are usually toleranced using a combination of what-if tolerancing and computer statistical variational modeling software. Iterations are performed until an achievable combination of component tolerances are shown to yield an acceptable statistical result. Component tolerances must be selected that are within known manufacturing process capabilities for the analysis to be meaningful. Where it is shown that the overall assembly tolerance cannot be met by assigning realistic component tolerances, the design geometry must be altered to work with a larger tolerance.

Design geometry may be altered by using oversized holes or slots for adjustment at assembly or in combination with tighter geometry coordinated with assembly fixtures. Other methods include changing mating relationships, such as changing butt joints to lap joints, changing surface geometry to make misalignment less obvious, using shims at assembly, reducing the number of parts, or redimensioning the parts to reduce the number of tolerances contributing to the accumulated total.

Different industries and assembly preferences drive different solutions to this dilemma. Industries where manual assembly methods are prevalent and the skill and care of the assemblers can be relied upon often use oversized holes and slots as an easy solution. Here the assembler manually adjusts each part to a near optimal position before tightening fasteners or welding. Industries where automated assembly or assembly line methods are prevalent typically cannot rely on the assembler to make fine adjustments at final assembly. Parts must work even if assembled in the worst possible manner. Typically these designs must be altered to allow for worst-case assembly. Factors include part weight and gravity, awkwardness of handling large parts, assembly line speed, and turnover of workforce.

This "what-if" method also works well with simple Tolerance Stackups. Guesses at the tolerances can be entered into a spreadsheet and the results studied. Once a satisfactory result is obtained, the study is complete.

In Fig. 17.1 a final assembly tolerance of ≤2.5 mm is given and the part tolerances are to be determined. A spreadsheet with iterative calculations are shown in Fig. 17.2, in which it is assumed that all parts have the same tolerance value and that an adjusted RSS Tolerance Stackup result will be used.

Another more precise technique is to use the Goal Seek function in Microsoft Excel, which allows the analyst to determine the required part

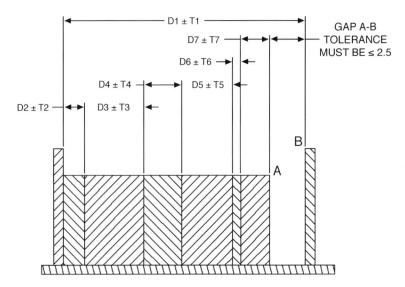

**FIGURE 17.1**   Simple assembly with assembly tolerance goal.

Tolerance Stackup

| Product: | Sample Assembly with Gap Tolerance Only | | Date: | 07/04/02 |
|---|---|---|---|---|

Problem:   Component Dimensions must be Toleranced to Achieve Assy Tolerance Requirement

Objective:   Determine the Tolerance for the  of Each Dimension

Revision:   A
Direction: X Axis

Guess #1   Try +/-0.5 tolerance for each Dimension

| Component / Assy | Item | Description | Tol | Tol Source & Calcs |
|---|---|---|---|---|
| Dimension 1 | 1 | Tolerance T1 | +/- 0.5 | Guess #1 |
| Dimension 2 | 2 | Tolerance T2 | +/- 0.5 | " |
| Dimension 3 | 3 | Tolerance T3 | +/- 0.5 | " |
| Dimension 4 | 4 | Tolerance T4 | +/- 0.5 | " |
| Dimension 5 | 5 | Tolerance T5 | +/- 0.5 | " |
| Dimension 6 | 6 | Tolerance T6 | +/- 0.5 | " |
| Dimension 7 | 7 | Tolerance T7 | +/- 0.5 | " |

|  | Arithmetic Stack (Worst Case) | +/- 3.500 |  |
|---|---|---|---|
|  | Statistical Stack (RSS) | +/- 1.323 |  |
|  | Adjusted Stack: 1.5*RSS | +/- 1.984 | ◄──────── Too small - Tolerances may be larger |

Guess #2   Try +/-0.75 tolerance for each Dimension

| Component / Assy | Item | Description | Tol | Tol Source & Calcs |
|---|---|---|---|---|
| Dimension 1 | 1 | Tolerance T1 | +/- 0.75 | Guess #2 |
| Dimension 2 | 2 | Tolerance T2 | +/- 0.75 | " |
| Dimension 3 | 3 | Tolerance T3 | +/- 0.75 | " |
| Dimension 4 | 4 | Tolerance T4 | +/- 0.75 | " |
| Dimension 5 | 5 | Tolerance T5 | +/- 0.75 | " |
| Dimension 6 | 6 | Tolerance T6 | +/- 0.75 | " |
| Dimension 7 | 7 | Tolerance T7 | +/- 0.75 | " |

|  | Arithmetic Stack (Worst Case) | +/- 5.250 |  |
|---|---|---|---|
|  | Statistical Stack (RSS) | +/- 1.984 |  |
|  | Adjusted Stack: 1.5*RSS | +/- 2.976 | ◄──────── Too large - Tolerances must be smaller |

Guess #3   Try +/-0.63 tolerance for each Dimension

| Component / Assy | Item | Description | Tol | Tol Source & Calcs |
|---|---|---|---|---|
| Dimension 1 | 1 | Tolerance T1 | +/- 0.63 | Guess #3 |
| Dimension 2 | 2 | Tolerance T2 | +/- 0.63 | " |
| Dimension 3 | 3 | Tolerance T3 | +/- 0.63 | " |
| Dimension 4 | 4 | Tolerance T4 | +/- 0.63 | " |
| Dimension 5 | 5 | Tolerance T5 | +/- 0.63 | " |
| Dimension 6 | 6 | Tolerance T6 | +/- 0.63 | " |
| Dimension 7 | 7 | Tolerance T7 | +/- 0.63 | " |

|  | Arithmetic Stack (Worst Case) | +/- 4.410 |  |
|---|---|---|---|
|  | Statistical Stack (RSS) | +/- 1.667 |  |
|  | Adjusted Stack: 1.5*RSS | +/- 2.500 | ◄──────── Satisfies Overall Requriement |

Use +/-0.63mm tolerance for each Dimension

**FIGURE 17.2**   Spreadsheet with iterative solution for simple assembly.

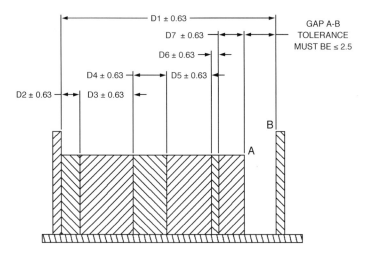

**FIGURE 17.3**  Simple assembly with iterative tolerances.

tolerance value without iteration. Using this function the Tolerance Analyst can set the desired assembly tolerance value and ask the program to iterate a contributing tolerance value to find the exact solution. This is a very powerful tool.

The tolerances derived in the above spreadsheet are used for the components in the assembly. The simple assembly is shown in Fig. 17.3 with

Formula:

$$PT = \frac{TOL}{ADJ \sqrt{n}}$$

Where:
n    = Number of Parts in Assembly
TOL  = Overall or Gap Tolerance (Given)
PT   = Part Tolerance (Calculated)
ADJ  = RSS Adjustment Factor

Note:
Use ADJ = 1 for straight RSS Stackup
          (no adjustment factor)

Sample Problem:
Example in Figure 17.1 Solved:

$$PT = \frac{TOL}{ADJ \sqrt{n}}$$

$$PT = \frac{2.5}{1.5 \sqrt{7}}$$

$$\boxed{PT = 0.63}$$

**FIGURE 17.4**  Adjusted RSS part allocation formula.

RSS Part Allocation Formula:

$$PT = \frac{TOL}{\sqrt{n}}$$

Where:
n   = Number of Parts in Assembly
TOL = Overall or Gap Tolerance (Given)
PT   = Part Tolerance (Calculated)

Adjusted RSS Part Allocation Formula:

$$PT = \frac{TOL}{ADJ\sqrt{n}}$$

Where:
n   = Number of Parts in Assembly
TOL = Overall or Gap Tolerance (Given)
PT   = Part Tolerance (Calculated)
ADJ = RSS Adjustment Factor

Note:
Use ADJ = 1 for straight RSS Stackup
    (no adjustment factor)
(Reduces to RSS Part Allocation Formula)

Derivation from RSS Formula:

$$\sqrt{PT_1^2 + PT_2^2 + \ldots PT_n^2} = TOL$$

$$\sqrt{n\,PT^2} = TOL$$

$$\sqrt{n}\sqrt{PT^2} = TOL$$

$$\sqrt{n}\;PT = TOL$$

$$PT = \frac{TOL}{\sqrt{n}}$$

Where:
All Part Tolerances are Equal:
$$PT_1 = PT_2 = \ldots PT_n$$
$$PT_1^2 + PT_2^2 + \ldots PT_n^2 = n\,PT^2$$

Derivation from Adjusted RSS Formula:

$$ADJ\sqrt{PT_1^2 + PT_2^2 + \ldots PT_n^2} = TOL$$

$$ADJ\sqrt{n\,PT^2} = TOL$$

$$ADJ\sqrt{n}\sqrt{PT^2} = TOL$$

$$ADJ\sqrt{n}\;PT = TOL$$

$$\sqrt{n}\;PT = \frac{TOL}{ADJ}$$

$$PT = \frac{TOL}{ADJ\sqrt{n}}$$

Where:
All Part Tolerances are Equal:
$$PT_1 = PT_2 = \ldots PT_n$$
$$PT_1^2 + PT_2^2 + \ldots PT_n^2 = n\,PT^2$$
ADJ = RSS Adjustment Factor

**FIGURE 17.5**   Derivation of adjusted RSS part allocation formula.

the iteratively calculated tolerance values. In this example, the same tolerance was applied to each part. Different tolerances for each part may be used with this method of tolerance assignment as well, inserting different tolerance value guesses into the spreadsheet for each part. It is more likely that the parts in most Tolerance Stackups will require different tolerances.

Where multiple parallel part features are be assigned the same tolerance as in the previous examples, a simpler approach may be to use the formula in Fig. 17.4. The formula works for RSS Tolerance Stackups and for adjusted RSS Tolerance Stackups. The sample problem in Fig. 4 shows that using the same values as in the previous examples, the formula yields the same results.

Derivation of the adjusted RSS allocation formula and the RSS allocation formula are shown in Fig. 17.5. The only difference between the formulas is that the ADJ coefficient variable is not included in the RSS part allocation formula. The RSS part allocation formula is actually a special form of the adjusted RSS part allocation formula. If the adjustment factor (ADJ) was set equal to 1, the adjusted RSS part allocation formula reduces to the RSS part allocation formula. These formulas offer a simpler way to calculate the values for a set of equal-value tolerances. The result of these formulas and the statistical and adjusted statistical results from the Tolerance Stackup report form would be the same given the same inputs and the same RSS adjustment factor.

# 18

## Floating-Fastener and Fixed-Fastener Formulas and Considerations

*Floating fastener* and *fixed fastener* are terms describing two possible relationships between the corresponding features in mating parts. These features include clearance holes, tight-fitting holes, tapped holes, slots, pins, studs, keys, keyways, etc.

An example of a floating-fastener situation is where a bolt passes though clearance holes in mating parts, perhaps terminating in a hex nut.

An example of a fixed-fastener situation is where a bolt passes through a clearance hole in one part and threads into a tapped hole in the mating part. Another example is where a part has clearance holes that fit over threaded studs protruding from the mating part. Another example is where a part has tight-fitting locating holes that fit over locating pins pressed into the mating part.

The tolerancing for each situation is determined by the relationship of the fastener, pin, or shaft to the holes in each part.

### FLOATING-FASTENER SITUATION

Definition: Where internal features, such as holes, in one or more parts must clear a common external feature, such as a fastener or a shaft, is referred to as a *floating-fastener situation*. A common application is where a fastener passes through clearance holes in mating parts. This is common for applications using nuts and bolts or when determining hole sizes for shims and washers.

Corollary: The holes do not locate the fastener in a floating-fastener situation. The fastener is free to "float" within the holes. All the holes must do is stay out of the way.

An example of a floating-fastener relationship in mating parts can be seen in Fig. 18.1, which shows a section through two mating parts with matching patterns of clearance holes. Note that the diameters of the holes may be different in each part. In this example, the function of the holes is to allow fasteners to pass so the parts can be fastened together. It is also important that the holes are not so large that there is no longer adequate bearing surface for the head of the bolts and nuts. The holes should be as small as they can be to maximize bearing surface. The floating-fastener formula allows the designer to determine the minimum size the holes can be and still allow the fasteners to pass at the worst case.

*Floating-fastener formula*:

$$H = F + T$$

where

$H$ = minimum clearance hole diameter (MMC)

$F$ = maximum fastener diameter (MMC)

$T$ = clearance hole positional tolerance at MMC

in considered part

Figure 18.2 shows a drawing of mating parts with positional tolerances and floating-fastener calculations. The holes in each part are different sizes.

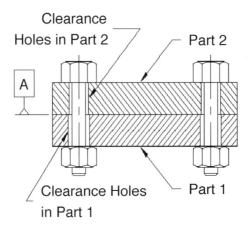

**FIGURE 18.1**  Floating fastener: Section thru mating parts.

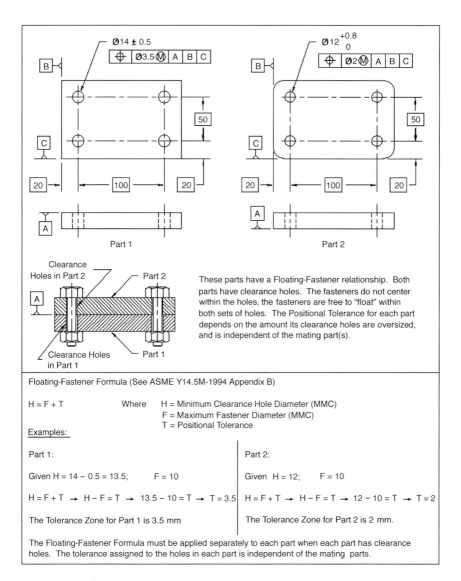

**FIGURE 18.2** Floating-fastener situation: Drawings of parts and calculations.

The drawings are shown at the top of the figure, and the calculations are shown at the bottom. In this example, the floating-fastener formula was used to calculate the positional tolerance allowed for the clearance holes. The floating-fastener formula may also be used to calculate the minimum allowable hole diameter or the maximum allowable fastener diameter—the variable to be calculated is based on which variables are known. If the fastener is already selected and the positional tolerance has already been determined, then the formula is used to solve for the minimum hole diameter. If the hole diameter has already been selected and the fastener diameter is known, then the formula is used to solve for the positional tolerance.

In floating-fastener situations, the positional tolerance for the clearance holes in each part is calculated separately. The functional requirement for this application is that the fastener must pass through the clearance holes in each part. One way to think of it is that the edges of the holes must not block the passage of the fastener. The absolute minimum clearance hole diameter is the fastener's maximum diameter, which would require a positional tolerance of zero at MMC on the holes. To put it another way, the virtual condition of the clearance holes must be equal to or greater than the maximum fastener diameter. This last statement is not an absolute rule, but it is good design practice. Violation of this last statement requires very careful consideration, and may change the problem from a floating-fastener situation to a fixed-fastener situation.

As shown in Fig. 18.3, the fastener passes through worst-case clearance holes in both parts. The clearance hole diameters ($H$) must allow for their variation in orientation and location, due to their respective positional tolerances ($T$). From these considerations we derive the floating-fastener formula above.

It is important to note that each respective interfacial surface between the mating parts must be specified as the primary datum feature on each part. As such, the datum plane along the interfacial surface establishes the orientation of the positional tolerance zones for the holes through each part. It is assumed that the datum plane on both parts is the same plane, that is, the datum planes are coplanar. Consequently, the tolerance zones in each part are colinear, and the formulas are valid.

The tolerance zones on a part would have additional orientation error if a different feature or surface was chosen as the primary datum feature for the positional tolerance applied to the holes. This additional orientation error would have to be accounted for when determining the required size of the holes or the allowable positional tolerance value. This applies to the floating-fastener and the fixed-fastener formulas. A similar case can be seen in Examples 16.2 and 16.3 in Chapter 16. Remember in those examples the top surface of the enclosure was not the primary datum feature for the inside walls of the enclosure, which led to an apparent foreshortening of the opening.

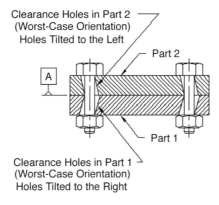

Clearance Holes in Part 2 —
(Worst-Case Orientation)
Holes Tilted to the Left

Part 2

A

Part 1

Clearance Holes in Part 1
(Worst-Case Orientation)
Holes Tilted to the Right

Holes Tilted Worst-Case
in Both Parts

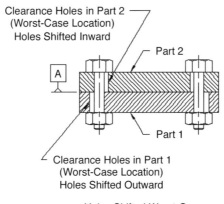

Clearance Holes in Part 2 —
(Worst-Case Location)
Holes Shifted Inward

Part 2

A

Part 1

Clearance Holes in Part 1
(Worst-Case Location)
Holes Shifted Outward

Holes Shifted Worst-Case
in Both Parts

**FIGURE 18.3** Floating fastener: Holes tilted and shifted worst-case.

## Example 18.1

Determine the minimum clearance hole diameter for the following situation, where

$H =$ minimum clearance hole diameter (MMC)

$F =$ maximum fastener diameter (MMC)

$T =$ positional tolerance of clearance hole at MMC
in considered part

Given

$$F = \text{M8 bolt} = \text{8-mm maximum OD}$$

$$T = \text{positional tolerance on clearance holes} = \varnothing 1 \text{ mm at MMC}$$

solve $H = F + T$ for $H$:

$$H = 8 + 1 = 9$$

The minimum clearance hole diameter $= \varnothing 9$ mm.

Calculate the nominal clearance hole diameter, where

$H$    = minimum clearance hole diameter
        (from floating-fastener calculation)

$ST$   = applicable $\pm$ size tolerance on clearance holes

$H_{\text{nom}}$ = nominal hole diameter

From the result of the previous calculation, we see that the minimum clearance hole diameter $= 9$ mm. Also assume an equal bilateral size tolerance ($ST$) is given of $\pm 0.5$ mm for the holes.
solve $H_{\text{nom}} = H + ST$ for $H_{\text{nom}}$:

$$H_{\text{nom}} = 9 + 0.5 = 9.5$$

The nominal clearance hole diameter $= \varnothing 9.5$. The hole specification on the drawing is $\varnothing 9.5 \pm 0.5$.

In the above illustrations it is assumed the parts are held in place—that they do not shift relative to each other. Only the holes are free to move, and their movement is relative to the datum reference frame on each part, within the specified positional tolerance.

Using the floating-fastener formula ensures that the virtual condition of the holes allows the fastener to pass. In most applications, parts may shift relative to one another about the fasteners at assembly, which is assembly shift. To review, assembly shift is due to the clearance between the holes and the fastener. When parts shift about their fasteners, there is greater variation from their nominal location than the fixed- and floating-fastener formulas accommodate. This shifting is irrelevant where there is only one hole in each part. When fastening parts through a pattern of holes (more than hole), care must be taken not to shift the parts to align one or more holes at the expense of the other holes. Fasteners should be started through all the holes in a pattern before any one is tightened, thus ensuring that the part was not overly shifted to align any of the holes. In cases where the assembly methods do not guarantee or allow all fasteners to be started before any one is tightened, the floating-fastener formula should not be used, and a traditional Tolerance Stackup should be performed.

Gravity always affects the location of vertically oriented parts, especially if the parts are large or heavy. Gravity pulls parts downward against the fasteners unless there is some other means of holding the parts in place. An example where gravity affects parts at assembly can be seen in Fig. 18.4.

In this example, the force of gravity has caused worst-case assembly shift between both parts and the fasteners. Perhaps the maximum distance shown in the figure is critical. If it is not acceptable to allow gravity to shift to the parts as shown, the parts should be assembled using a fixture, the parts should be redesigned such that surfaces align the parts vertically, or the parts should be manually adjusted at assembly.

Figure 18.5 shows a common problem that occurs when all the fasteners are not started through all the holes simultaneously. The hanger and the bracket have mating patterns of four holes in this example, and positional tolerances applied on their respective detail drawings (not shown). The floating-fastener formula was used to calculate the required diameters and positional tolerances for the holes in each part. The size and location of the holes in both parts is within specifications.

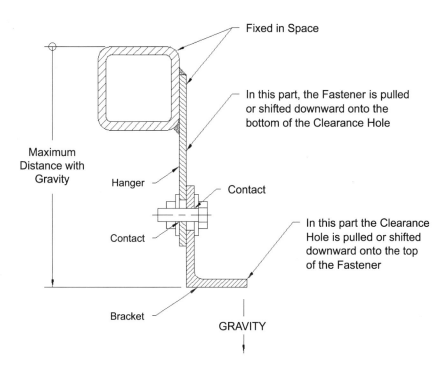

**FIGURE 18.4** Floating fastener: Hanger Assembly with Effect of Gravity.

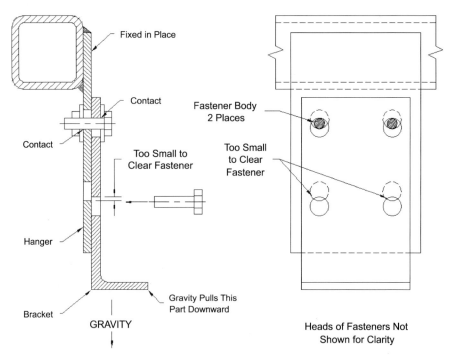

**FIGURE 18.5** Floating fastener: Assembly sequence violates floating-fastener formula (with gravity).

In this example, the fasteners were inserted and tightened into the upper holes first, and the bracket was allowed to slide down against the fasteners. This pulled the fasteners down against the upper holes in the hanger. Consequently, this added two positional tolerances and two occurrences of assembly shift to the location of the lower holes.

The positional error of the upper holes in both parts and the assembly shift in both parts contributes to the total error in the location of the lower holes. The fasteners cannot fit into the lower holes even though the floating-fastener formula was used. The assembler has to loosen the upper fasteners and readjust the parts to allow the lower fasteners to fit, change the assembly procedure to start all the fasteners at the same time, or a different Tolerance Stackup must be done to determine how large the lower holes must be to accommodate the total variation allowed by the assembly process.

Figure 18.6 shows two assembly methods for a pair of mating parts with clearance holes, (Figs. 18.6A and B). The holes in each part have been produced within their positional tolerance but are at the extremes of their tol-

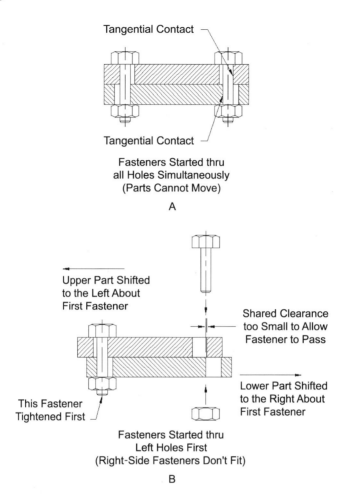

**FIGURE 18.6** Floating fastener: Assembly sequence violates floating-fastener formula (horizontal).

erance zones. The holes in the upper part are located inward within their tolerance zones, and the holes in the lower part are located outward within their tolerance zones. The fasteners are started through all the holes before any are tightened in Fig. 18.6A. Even with the worst-case positional error all of the fasteners can pass through the mating holes—this agrees with the results of properly applying the floating-fastener formula. Notice in this example at worst-case the parts cannot move relative to one another—the holes contact the fasteners in such a way as to disallow part movement. The parts are locked up.

Figure 18.6B shows what happens if the fasteners on the left are inserted and tightened first. Remember in this example the holes in each as-produced part are biased in opposite directions. Figure 18.6B shows that the parts are shifted before the fasteners on the left are tightened. In this example, the positional tolerance and assembly shift of the left-hand holes in both parts are added to the right-hand holes, in effect adding four tolerances (two positional tolerances and two assembly shifts) to the right-hand holes. If the positional tolerances were specified at MMC, two occurrences of bonus tolerance would also be added.

Although in these examples it seems easy to change the assembly procedure, in cases where all the fasteners cannot be accessed simultaneously, where parts are very heavy, very large, or just awkward and difficult to handle, it may not be possible to start all the fasteners simultaneously. A common tolerancing mistake occurs when engineering personnel use the floating- and fixed-fastener formulas assuming the assembly procedure will start all fasteners simultaneously, but the assembly procedure does not agree.

For example, the author was consulting with a firm designing large, heavy parts to be assembled into a large frame structure. There were several large cast and machined cross members with flanges and clearance holes on each end that were to be assembled in between two frames with matching flanges and tapped holes. The design team assumed that all the cross members would be put into place between the frame members, the fasteners would be started though the 150 or so clearance holes, and then the fasteners would be tightened.

A trip to the assembly facility proved us wrong. The left-hand flange on the first cross member was bolted down first. The frame on the right-hand side of the cross member was shifted to line up with the right-side mating holes and tightened. The left-hand flange on the second cross member was then bolted down. When they attempted to fasten the flange on the other side, the holes were completely misaligned. The positional tolerances and assembly shift from the holes on both sides of the first cross member and the left-hand side of the second cross member were added to the total tolerance on the second cross member's right-hand holes—six tolerances were added to the total tolerance on the holes in the right-hand flange of the second cross member. From design's point of view, this was the absolute worst possible assembly process. From the assembly personnel's point of view, this was the only way they could assemble the parts, given the size and weight of the parts and the tools available. The ultimate solution was to assemble the parts in a fixture and to start all fasteners before tightening any.

Remember, it is absolutely necessary to validate the assumptions made about the assembly process during the design process. If a design and its Tolerance Stackup are based on the assumption that all fasteners will be started simultaneously, and they are not, then the Tolerance Stackup results do not represent reality. The assembly process must be changed, the design must be changed, or the Tolerance Stackup must be changed to match the actual assembly process. In cases where the assembly process is unknown, it may be a good idea to solve the Tolerance Stackup using several assembly models.

## FIXED-FASTENER SITUATION

Definition: Where external features, such as pins or studs, are fixed in place in one part and pass though internal features, such as clearance holes, in a mating part is referred to as a *fixed-fastener situation*. A common application is where two or more parts are fastened together, and the fasteners are fixed in one part, and the other parts have clearance holes. The fastener may be "fixed" by a number of methods, such as by pressing a pin or a stud into a hole, welding studs onto a part, or threading a fastener into a tapped hole or weldnut.

Corollary: The fastener cannot move relative to one of the parts in a fixed-fastener situation. It is commonly assumed that a bolt or screw threaded into a tapped hole is fixed in place. Although there may be some movement allowed between mating threads, most Tolerance Stackups assume the fastener and the threaded hole are coaxial. Note: In very critical applications it may be necessary to calculate the amount of clearance and coaxiality error between the fastener and the threaded hole.

An example of a fixed-fastener relationship in mating parts can be seen in Fig. 18.7, which shows a section through two mating parts. The upper part (part 2) has a pattern of clearance holes and the lower part (part 1) has a matching pattern of threaded holes. In this example, the function of the clearance holes is to allow fasteners to pass into the threaded holes. It is also important that the clearance holes are not so large that there is no longer adequate bearing surface for the head of the bolts. The holes should be as

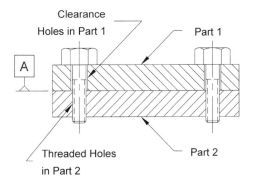

**FIGURE 18.7** Fixed fastener: Section through mating parts.

small as they can be to maximize bearing surface. The fixed-fastener formula allows the designer to determine the minimum size the holes can be and still allow the fasteners to pass into the threaded holes at the worst case.

*Fixed-fastener formula*:

$$H = F + T_1 + T_2$$

where

$H$ = minimum clearance hole diameter (MMC)

$F$ = Maximum fastener diameter (MMC)

$T_1$ = clearance hole positional tolerance at MMC

$T_2$ = threaded hole positional tolerance at MMC

Figure 18.8 shows a drawing of mating parts with positional tolerances and fixed-fastener calculations. Part 1 has a pattern of clearance holes, and part 2 has a matching pattern of threaded holes. The drawings are shown at the top of the figure and the calculations are shown at the bottom. In this example, the fixed-fastener formula was used to calculate the positional tolerance allowed for both sets of holes. The value for the positional tolerance applied to the clearance holes does not have to be the same as the value applied to the threaded holes. In this example there was 2 mm available for the positional tolerance applied to both parts. The 2 mm available was split between the parts as follows: a 1.2-mm positional tolerance zone was specified for the clearance holes and a 0.8-mm positional tolerance zone was specified for the threaded holes. The fixed-fastener formula may also be used to calculate the minimum allowable clearance hole diameter or the maximum allowable fastener diameter—the variable to be calculated is based on which variables are known. If the fastener is already selected and the positional tolerances have already been determined, then the formula is used to solve for the minimum clearance hole diameter. If the clearance hole diameter has already been selected and the fastener diameter is known, then the formula is used to solve for the positional tolerances.

The fixed-fastener formula presented in this section requires that a projected tolerance zone is specified for the positional tolerance applied to the threaded or press-fit holes. The height of the projected tolerance zone should be equal to or greater than the maximum thickness of the mating part(s). See Chapter 10 in this text, Section 5.5 and Appendix B3 of ASME Y14.5M-1994 for more information on projected tolerance zones.

Projected tolerance zones are not the most popular specifications that manufacturing and inspection personnel encounter on drawings. The means of validating compliance with a projected tolerance zone specification using conventional (physical) inspection techniques can be cumbersome, time-

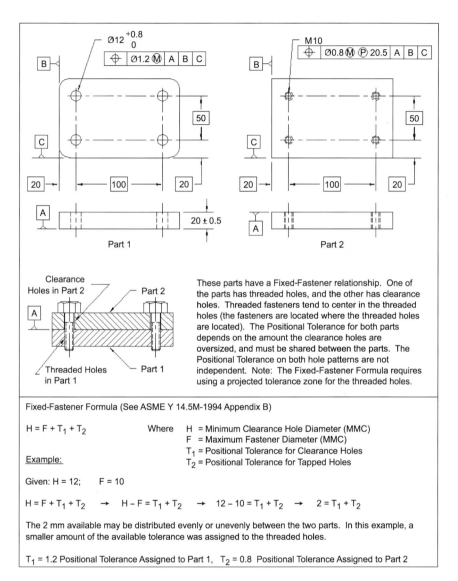

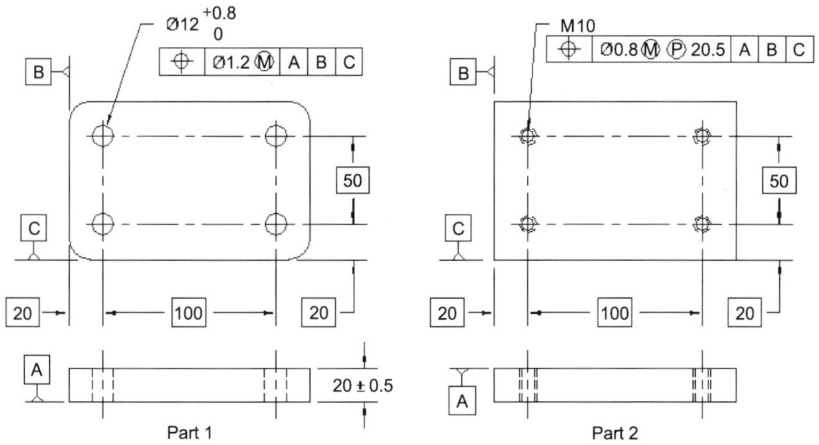
Ø12 +0.8 0
Ø1.2 Ⓜ A B C
B
C
50
20
100
20
A
20 ± 0.5
Part 1
M10
Ø0.8 Ⓜ Ⓟ 20.5 A B C
B
C
50
20
100
20
A
Part 2

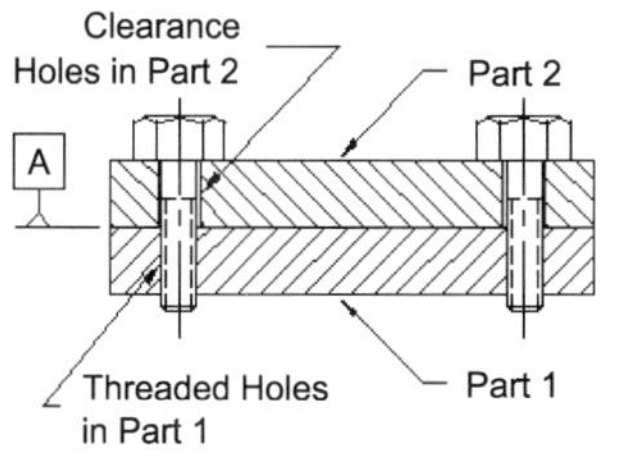
Clearance Holes in Part 2
Part 2
A
Threaded Holes in Part 1
Part 1

These parts have a Fixed-Fastener relationship. One of the parts has threaded holes, and the other has clearance holes. Threaded fasteners tend to center in the threaded holes (the fasteners are located where the threaded holes are located). The Positional Tolerance for both parts depends on the amount the clearance holes are oversized, and must be shared between the parts. The Positional Tolerance on both hole patterns are not independent. Note: The Fixed-Fastener Formula requires using a projected tolerance zone for the threaded holes.

Fixed-Fastener Formula (See ASME Y 14.5M-1994 Appendix B)

$H = F + T_1 + T_2$

Where
- H = Minimum Clearance Hole Diameter (MMC)
- F = Maximum Fastener Diameter (MMC)
- $T_1$ = Positional Tolerance for Clearance Holes
- $T_2$ = Positional Tolerance for Tapped Holes

<u>Example:</u>

Given: H = 12; F = 10

$H = F + T_1 + T_2 \rightarrow H - F = T_1 + T_2 \rightarrow 12 - 10 = T_1 + T_2 \rightarrow 2 = T_1 + T_2$

The 2 mm available may be distributed evenly or unevenly between the two parts. In this example, a smaller amount of the available tolerance was assigned to the threaded holes.

$T_1 = 1.2$ Positional Tolerance Assigned to Part 1, $T_2 = 0.8$ Positional Tolerance Assigned to Part 2

**FIGURE 18.8** Fixed-fastener situation: Drawings of parts and calculations.

consuming, and therefore more expensive than validating compliance of a nonprojected tolerance zone. Often this involves threading plug gages into each threaded hole and verifying the positional tolerance on a mandrel that projects the required distance outside the part. This extra effort tends to cause grief with some manufacturing and inspection personnel, especially where their organization is insensitive to why it is important, but concerned about the extra time required to perform such tasks. Where virtual inspection methods are used, such as a CMM, validating compliance with a projected tolerance zone specification should take no more time than if a projected tolerance zones was not specified. Whether projected tolerance zones are difficult or easy to inspect should not be the primary concern, however. Projected tolerance zones are necessary to ensure functional requirements are met.

In fixed-fastener situations, both parts must be toleranced together, as the location of the tapped hole affects the location of the fastener. The functional requirement for this application is that the fastener must pass unobstructed through the clearance hole into the threaded hole. As shown in Fig. 18.9, the fastener is located by the threaded hole and in a sense follows the threaded hole—wherever the threaded hole ends up, the fastener is centered within it. The clearance hole diameter ($H$) must be sized to allow for the variation in the orientation and location of the fastener allowed by the threaded hole's positional tolerance ($T_2$). The clearance hole diameter ($H$) must also allow for its own variation in orientation and location, allowed by its positional tolerance ($T_1$). From these considerations we derive the fixed-fastener formula above. Unlike the floating-fastener formula where the positional tolerance value $T$ for the clearance holes in each part is calculated independently, the positional tolerance values $T_1$ and $T_2$ in the fixed-fastener formula are dependent variables. In the fixed-fastener situation the amount of clearance between the maximum fastener diameter and the minimum clearance hole diameter equals the total available to be shared by both tolerances $T_1$ and $T_2$.

It is important to note that each respective interfacial surface between the mating parts must be specified as the primary datum feature on each part. As such, the datum plane along the interfacial surface establishes the orientation of the positional tolerance zones for the holes through each part. It is assumed that the datum plane on both parts is the same plane, that is, the datum planes are coplanar. Consequently, the tolerance zones in each part are colinear, and the formulas are valid.

The tolerance zones on a part would have additional orientation error if a different feature or surface was chosen as the primary datum feature for the positional tolerance applied to the holes. This additional orientation error would have to be accounted for when determining the required size of the holes or the allowable positional tolerance value. This applies to the floating-fastener and the fixed-fastener formulas.

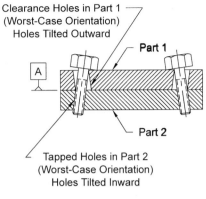

Clearance Holes in Part 1
(Worst-Case Orientation)
Holes Tilted Outward

Part 1

A

Part 2

Tapped Holes in Part 2
(Worst-Case Orientation)
Holes Tilted Inward

Holes Tilted Worst-Case
in Both Parts

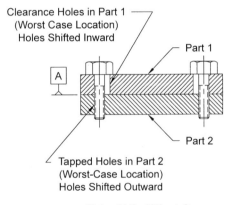

Clearance Holes in Part 1
(Worst Case Location)
Holes Shifted Inward

Part 1

A

Part 2

Tapped Holes in Part 2
(Worst-Case Location)
Holes Shifted Outward

Holes Shifted Worst-Case
in Both Parts

**FIGURE 18.9**  Fixed fastener: Holes tilted and shifted worst-case.

## Example 18.2

Determine the minimum clearance hole diameter for the following situation, where

$H$ = minimum clearance hole diameter (MMC)

$F$ = maximum fastener diameter (MMC)

$T_1$ = positional tolerance of clearance hole at MMC

$T_2$ = positional tolerance of threaded hole at MMC

Given

$\quad F = $ M10 bolt $= 10$-mm maximum OD

$\quad T_1 = $ positional tolerance of clearance hole $= \varnothing 1$ mm at MMC

$\quad T_2 = $ positional tolerance of threaded hole $= \varnothing 1.5$ mm at MMC

solve $H = F + T_1 + T_2$ for $H$:

$\quad H = 10 + 1 + 1.5 = 12.5$

The minimum clearance hole diameter $= 12.5$ mm.

$\quad$ Calculate the nominal clearance hole diameter, where

$\quad H \quad = $ minimum clearance hole diameter
$\quad\quad\quad$ (from fixed-fastener calculation)

$\quad ST \quad = $ applicable $\pm$ size tolerance on clearance holes

$\quad H_{nom} = $ nominal hole diameter

From the result of the previous calculation, we see that the minimum clearance hole diameter $= 12.5$ mm. Also assume an equal bilateral size tolerance $(ST)$ is given of $\pm 0.3$ mm for the holes.

$\quad$ Solve $H_{nom} = H + ST$ for $H_{nom}$:

$\quad H_{nom} = 12.5 + 0.3 = 12.8$

The nominal clearance hole diameter $= \varnothing 12.8$. The hole specification on the drawing is $\varnothing 12.8 \pm 0.3$.

$\quad$ In the above illustrations, it is assumed the parts are held in place, that they do not shift relative to each other. Only the holes are free to move, and their movement is relative to the datum reference frame on each part, within the specified positional tolerance.

$\quad$ Using the fixed-fastener formula ensures that the virtual condition of the clearance holes allows the fastener to pass into the worst-case tapped holes. As mentioned earlier, for the fixed-fastener formula to be valid, a projected tolerance zone must be specified for the tapped holes' positional tolerance.

$\quad$ In most applications, parts may shift relative to one another about the fasteners, which is due to the clearance between the holes and the fastener at assembly. This is called *assembly shift*. As with the floating-fastener formula, the fixed-fastener formula requires all fasteners in a pattern to be started before any one is tightened. Tightening any one fastener before the other fasteners are inserted into the holes could allow the parts to shift relative to one another, invalidating the results of the fixed-fastener formula for the remaining holes. Refer to the material at the end of the floating-fastener section that discusses this assembly issue.

# 19

# Fit Classifications

Generally speaking, there are three types of fit between cylindrical parts. These are *clearance* fits, *transition* fits, and *interference* fits. These are standard fit classifications each is based on how the mating parts interact. There are U.S. and international standards and systems governing these fit classifications. Information on these standard systems can be found in *Machinery's Handbook* * or in documents from the applicable standards governing bodies.

These fit classifications do not take into account positional error between parts; the part features are assumed to be coaxial. Typically these fits are used for shafts into bearings, pressing pins into holes, or similar applications.

The fit classification standards include tables of standardized fits, each offering slightly more or less relative clearance or interference. Given a nominal size, the designer determines the parts' functional requirement, and selects the appropriate fit. The fits in each table are grouped to address a certain set of conditions (such as high speed rotation or light press fit). Nominal sizes are listed with corresponding upper and lower limits for the shaft and hole. The upper and lower limits are applied to nominal shaft and hole, leading to the desired fit.

Note: In several fit tables, the hole and shaft are derived from the same nominal size. For example, given a 10-mm nominal and a clearance fit, the hole tolerances may be listed as $+0.10/+0.05$, and the shaft tolerances listed

---

* Machinery's Handbook. 23rd Ed. Erik Oberg, Franklin D. Jones and Holbrook L. Horton, New York, Industrial Press, 1988.

as −0.02/-0.10. Notice that the tolerances for the hole are both + tolerances, and the tolerances for the hole are both − tolerances. This convention should not be used on drawings: never tolerance a feature with two positive tolerances, such as 0.25 +0.2/+0.1, or two negative tolerances, such as 1.5 − 0.5/−0.75. Features such as holes on drawings should be toleranced using limit dimensioning, unilateral, equal bilateral, or unequal bilateral tolerances. Examples of acceptable tolerancing include

| | | |
|---|---|---|
| 0.27 | | |
| 0.26 | 0.27 0/−0.1 | 0.26 + 0.1/0 |
| 0.265 ± 0.005 | 0.262 + 0.008/−0.002 | 0.267 + 0.003/−0.007 |

The charts use this tolerancing scheme (allowing both limits to be positive or negative) as a convenience. Where fits are specified using letter designations it is appropriate to size both the hole and the shaft at the same nominal, with limits as specified by the specified fit. In these cases, the actual fit designations would be placed adjacent to the nominal size, and reference to the fit standard would be made by note.

Any fit between a female and male feature of size may be classified as clearance, transition, or interference, regardless if the fit was selected from a standard chart. This is true for all features of size, which includes width features as well as cylindrical features. Examples of width features of size are keys and keyways.

For the sake of simplicity, the following discussion will be in terms of a shaft passing through a hole.

## CLEARANCE FITS

A clearance fit must always have clearance between the shaft and the hole. The maximum size shaft will fit into the minimum size hole with clearance. This means that the hole is always larger than the shaft. Typically the functional requirement is that the fit allows rotation or guarantees clearance for other purposes.

## TRANSITION FITS

A transition fit may have clearance or interference between the shaft and the hole. This means that the hole may be larger than the shaft or the hole may be smaller than the shaft. Typically the functional requirement is that the fit is tight; whether there is a small amount of clearance or interference is immaterial.

## INTERFERENCE FITS (FORCE FITS)

An interference fit must always have interference between the shaft and the hole. The minimum size shaft will fit into the maximum size hole with interference. This means that the hole is always smaller than the shaft. Typically the functional requirement is for a press fit, guaranteeing that the shaft will not break loose from the hole.

It is important to remember that these fit classifications are discussed in terms of an external feature that is fit into an internal feature. They assume that the external feature is oriented and located coaxial with the internal feature. In many parts the orientation and location tolerances on mating features will affect the fit between the features. The designer must take the orientation and location tolerances into account along with the size tolerances when determining the fit between features in an assembly.

# 20

# Form Tolerances in Tolerance Stackups

Form tolerances are not included in most linear Tolerance Stackups. In most cases, the form of features in the chain of Dimensions and Tolerances has little or no effect on the result of the Tolerance Stackup, as the form tolerance is almost always smaller than the location tolerance. The Tolerance Stackup problem is idealized, and these tolerances are not included in the chain of Dimensions and Tolerances. Usually there is little risk in omitting form tolerances from the Tolerance Stackup.

The location of features is typically the most important characteristic of features in linear Tolerance Stackups, which is why position and profile tolerances are more commonly included in Tolerance Stackups than form tolerances.

The orientation of features may also be important, but orientation tolerances in Tolerance Stackups are also not as common as location tolerances—whether orientation tolerances are included in the chain of Dimensions and Tolerances is determined on a case-by-case basis. Even though they are usually far less important than location tolerances in Tolerance Stackups, orientation tolerances are typically more important than form tolerances in Tolerance Stackups.

The reason that form tolerances are of less concern in linear Tolerance Stackups is because most Tolerance Stackups are done to find a minimum or maximum distance, and in the majority of cases the form or shape of a feature has little to no effect on the distance being studied. As stated above, the location of features in the Tolerance Stackup has the greatest effect on the distance being studied. This is because when features in the Tolerance Stackup are located at extreme positions within their tolerance zones they have the greatest effect on the distance being studied—the worst-case distance is seen when the features in the Tolerance Stackup are at their worst-case locations. Again, usually their form has little or no effect on this worst-case condition.

However, form tolerances may play a role in Tolerance Stackups. As stated earlier, their effect in most cases is probably miniscule, but in some cases variation in the form of a feature can have a dramatic effect on the Tolerance Stackup. For a form tolerance to have a significant effect on a Tolerance Stackup, the form of one or more features in the chain of Dimensions and Tolerances must vary in a particular way. To put it in different terms, the form of a surface will only affect the Tolerance Stackup if its form varies a particular way. In many cases, similar variation must occur on the mating surfaces of mating parts to see the worst-case condition.

## DATUM FEATURE FORM TOLERANCES

Datum features, like all features, must be toleranced if a part is to be completely defined. Their allowable variation must be quantified.

As primary datum features are independent and the basis of all subsequent feature relationships, they are typically not toleranced relative to other features (all other features are toleranced relative to them). A form tolerance is most often applied to primary datum features that are not features of size, such as planar features.

Planar primary datum features are usually toleranced using flatness. Multiple coplanar or offset parallel planar surfaces are usually toleranced using profile of a surface. Curved or contoured surfaces (single or multiple surfaces) are usually toleranced using profile of a surface as well. Primary datum features of size may also have form tolerances applied, such as cylindricity or straightness, but such form tolerances will probably not affect the result of a Tolerance Stackup.

The form tolerance applied to primary datum features may be included in Tolerance Stackups if it contributes to the total possible variation between the features being studied. Where parts are functionally dimensioned and toleranced, planar mating surfaces are commonly specified as primary datum features. The form tolerances applied to each mating surface should be considered if the interface is part of the Tolerance Stackup. This is due to the difference between the method used to simulate the primary datum feature at inspection and the geometry of the actual parts at assembly.

When a primary planar datum feature is simulated in inspection, an "ideally flat" surface is used as the datum feature simulator. Certain "high points" of the as-produced datum feature contact corresponding points on the datum feature simulator. The form of the surface is measured relative to the datum feature simulator or simulated datum, which is the tangent plane along the surface of the datum feature simulator.

In assembly, the as-produced part with its planar primary datum feature is mounted against another as-produced part. It is unlikely that the same high points on the surface will contact the corresponding high points on the mating part's primary datum feature surface. Therefore, the part may not sit in the same location or orientation against the mating part as it did against the datum feature simulator. When there is no load applied to the interface, the maximum amount of this difference in location is equivalent to the form tolerance specified for the feature.

Form and orientation tolerances applied to secondary and tertiary datum features may also be considered in Tolerance Stackups where applicable. They are treated similarly to the primary datum features described above, except their condition is slightly more complex because they are located and/or oriented to higher precedence datums.

Form tolerances can affect the result of a Tolerance Stackup in two ways:

- As translational variation only, such as where parts are very rigid or where they are not subjected to forces that may deform the interfacial surfaces at assembly
- As rotational variation projected out to a linear displacement, such as where thin-walled or sheet metal parts are subjected to loads that may deform the interfacial surfaces at assembly and the rotational displacement causes other features on the parts to deform

Both of these cases are discussed in detail in the following material. For the worst case to occur, both scenarios require somewhat improbable combinations of geometric form error on both mating surfaces, but it is important to undestand how form tolerances may affect a Tolerance Stackup. It is the Tolerance Analyst's responsibility to decide whether to include these tolerances in the Tolerance Stackup, so it is critical that the Tolerance Analyst understands how form tolerances may play a role in Tolerance Stackups.

## FORM TOLERANCES TREATED AS ADDING TRANSLATIONAL VARIATION ONLY

The parts shown in Fig. 20.1 mate along planar surfaces. Detail drawings with dimensions and tolerances of these parts can be seen in Figs. 20.2 and 20.3. If we use the techniques learned earlier in this text to perform a Tolerance

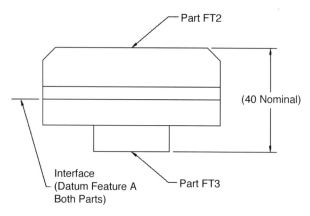

**FIGURE 20.1**   Form tolerances: Translation assembly.

Stackup between the upper and lower surfaces, the chain of Dimensions and Tolerances starts at the upper surface (marked point A) and passes through the interface and down to the lower surface (marked point B). The Tolerance Stackup sketch for this problem can be seen in Fig. 20.4, and the associated Tolerance Stackup report can be seen in Fig. 20.5.

Using the techniques learned earlier in this text we see that the flatness tolerances specified for the datum feature A surfaces on each part are not included in the chain of Dimensions and Tolerances in the Tolerance Stackup sketch or the Tolerance Stackup report. The belief is that for all intents and purposes, the datum plane for each of the two surfaces will be the same plane. This is an oversimplification, as the problem has been idealized.

Remember, the variation allowed for all aspects of every feature on a part must be directly or indirectly defined. A flatness tolerance has been specified for each datum feature A surface to quantify how much its form can vary, to clearly state how flat it must be to pass inspection and still work at

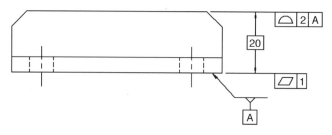

**FIGURE 20.2**   Form tolerances: Upper part drawing (FT2).

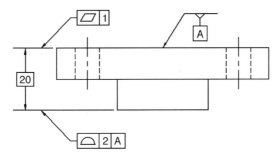

**FIGURE 20.3** Form tolerances: Lower part drawing (FT3).

assembly. The flatness tolerance allows the surface to bow, to warp, to have any type of form error within its tolerance zone. In this example, each datum feature A surface must be flat within 1 mm. The datum feature surfaces will not be perfectly flat; so the allowable form error must be stated.

A bit of review of from basic GD&T is in order. To simulate primary datum plane A from datum feature A, a minimum of three points of contact are required between the datum feature and its datum feature simulator.

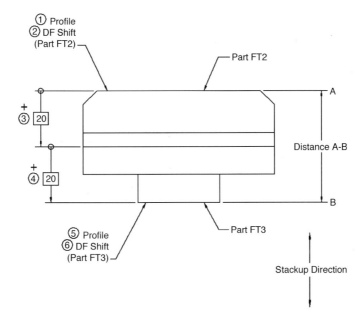

**FIGURE 20.4** Form tolerances: Tolerance Stackup sketch for FT1.

Tolerance Stack                                                                                    Release 1.2a

| Program: | Tolerance Analysis Textbook | | | | | | | | |
|---|---|---|---|---|---|---|---|---|---|
| Product: | Part Number | Rev | Description | | | | Stack No | Figure 20.5 | |
| | FT1 | A | Sample Assembly FT1 for Form Tolerance Chapter | | | | Date | 07/04/02 | |
| Problem: | Overall Height of Assembly Is Critical | | | | | | Revision | A | |
| Objective: | Determine the Minimum and Maximum Height of the Assembly | | | | | | Direction: | Vertical | |
| | | | | | | | Author | BR Fischer | |

Stack Information:

Description of
Component / Assy

| Upper Part | Part Number | Rev | Item | Description | + Dims | – Dims | Tol | Percent Contrib | Dim / Tol Source & Calcs |
|---|---|---|---|---|---|---|---|---|---|
| | FT2 | A | 1 | Profile Upper Surface (Point A) | | | ± 1.0000 | 50% | Profile 2, A |
| | | | 2 | Datum Feature Shift | | | ± 0.0000 | 0% | N/A – Datum Feature A not a Feature of Size |
| | | | 3 | Dim Upper Surface – Datum A | 20.0000 | | ± 0.0000 | 0% | 20 Basic on Dwg |
| Lower Part | FT3 | A | 4 | Dim Datum A – Lower Surface | 20.0000 | | ± 0.0000 | 0% | 20 Basic on Dwg |
| | | | 5 | Profile Lower Surface (Point B) | | | ± 1.0000 | 50% | Profile 2, A |
| | | | 6 | Datum Feature Shift | | | ± 0.0000 | 0% | N/A – Datum Feature A not a Feature of Size |

| | | | |
|---|---|---|---|
| Dimension Totals | 40.0000 | 0.0000 | |
| Nominal Distance Pos Dims – Neg Dims = | 40.0000 | | |

RESULTS:

| | Nom | Tol | Min | Max |
|---|---|---|---|---|
| Arithmetic Stack (Worst Case) | 40.0000 | ± 2.0000 | 38.0000 | 42.0000 |
| Statistical Stack (RSS) | 40.0000 | ± 1.4142 | 38.5858 | 41.4142 |
| Adjusted Statistical 1.5*RSS | 40.0000 | ± 2.1213 | 37.8787 | 42.1213 |

Notes:    - The Flatness Tolerances Specified for Datum Feature A on the parts are not included in the Tolerance Stackup.

Assumptions:    - The Flatness Tolerances Specified for the Datum Feature A Surfaces does not add Variation to the Tolerance Stackup.

Suggested Action:

**FIGURE 20.5** Form tolerances: Tolerance Stackup report for FT1.

Typically these three points are assumed to be the highest points on the surface, the three points that protrude farthest from the part. This is true for both parts. Figure 20.6 shows an imperfect as-produced part FT2 staged against its datum feature simulator, and Fig. 20.7 shows an imperfect as-produced part FT3 staged against its datum feature simulator. The relationship between datum feature A and datum feature simulator A is shown along with the profile tolerance zone in both figures.

Notice that the high points of datum feature A contact the datum feature simulator in Figs. 20.6 and 20.7. Also notice that there are gaps where datum feature A does not touch the datum feature simulator.

Again from basic GD&T, the profile tolerance applied to the upper surface in the figures is basically related to datum plane A. At inspection, out of necessity the profile tolerance zone is related to simulated datum plane A, which is the tangent plane along the high points of the datum feature simulator. For the sake of simplicity, we may assume this plane is the same plane as the plane along the high points of the datum feature itself. This is a simplification, but it is often difficult to predict and quantify the potential mismatch between these two planes. In all but the most extreme cases this mismatch is insignificant. When the upper surface in Fig. 20.6 is measured to determine if it lies within its profile tolerance zone, the profile tolerance zone is basically located from datum plane A. The upper surface is measured to determine if it is within the profile tolerance limits, which are

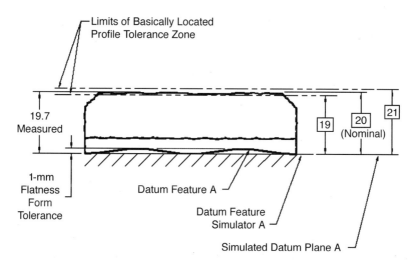

**FIGURE 20.6**  Form tolerances: Upper part with variation. As-produced part FT2 at inspection.

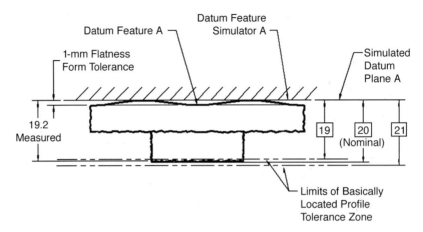

**FIGURE 20.7** Form tolerances: lower part with variation. As-produced part FT3 at inspection.

from 9 mm to 11 mm from datum plane A. The same is true for the mating part.

To ignore the form tolerance on these mating surfaces in a Tolerance Stackup requires that the same high points that contact the datum feature simulator contact the mating part at assembly. Remember the datum feature simulator is assumed to be perfect, perfectly flat in these examples. The datum feature A surfaces on the mating parts are not perfectly flat. In fact, the flatness tolerance has been specified to clearly state how much they can vary from perfectly flat. It is very unlikely that the form of the mating part surface will be as near perfect as the form of the datum feature simulator, and it is equally unlikely that the high points that contacted the datum feature simulator will contact the mating part. There will be some mismatch between the datum feature A surfaces of the as-produced parts at assembly—to be certain, the high points of the parts will probably not contact each other. This mismatch can be seen in Fig. 20.8, where the two imperfect mating parts have been brought together.

For the form tolerance to have its full effect as a translational displacement in a linear Tolerance Stackup, the form error of both mating surfaces must be mirror images of one another, geometric inverses if you will. Consider the form of the as-produced datum feature A surfaces of the parts in Figs. 20.6 and 20.7. The as-produced surfaces are shown with a sine-wave (sinusoidal) shape, with peaks and valleys at some interval. The peaks on one part align with the valleys on the mating part. As seen in Fig. 20.8, the high points of the surfaces will not contact each other when these parts are brought together at

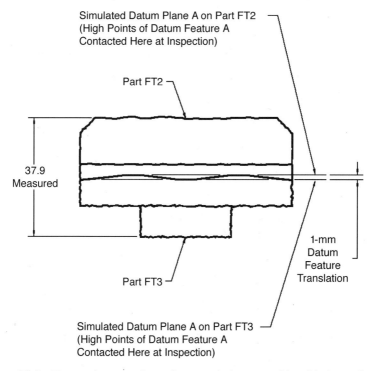

Simulated Datum Plane A on Part FT2
(High Points of Datum Feature A
Contacted Here at Inspection)

Part FT2

37.9
Measured

1-mm
Datum
Feature
Translation

Part FT3

Simulated Datum Plane A on Part FT3
(High Points of Datum Feature A
Contacted Here at Inspection)

**FIGURE 20.8**   Form tolerances: Imperfect translation assembly with 1-mm flatness.

assembly. In this extreme example, the surface imperfections are the same and 180° out of phase with one another. In this example, the as-assembled relationship of the features is 1 mm different than the as-inspected relationship of the features—the datum feature surfaces have nested or translated 1 mm within each other.

This can only happen when the form tolerance values are the same on both parts and the form error is such that the geometry of both surfaces align as shown. The same situation would happen if the form error of one mating surface was convex and the other was concave. In both cases, the high points that touched the datum feature simulator do not touch the high points of the mating part.

## Probability

It is important to realize that the form error shown in these examples is for three-dimensional parts. To achieve this translational error between the as-

assembled parts, the form error would have to have exactly the right shape in three-dimensions, like the concave and convex example. The sinusoidal shape shown above would have to be three-dimensional as well, extending in both the $X$ and $Y$ directions along the entire interface.

In my opinion, it is very unlikely that this sort of variation is encountered in most mating part applications. The likelihood of this perfectly inverse form error occurring on both mating parts is very remote, close to zero. However, it is possible. As in Chapter 8 which deals with statistics, again we are faced with choosing between the possible and the probable. I also believe that the likelihood of the high points of both mating surfaces contacting each other at assembly is also close to zero.

## FORM TOLERANCES TREATED AS ADDING ROTATIONAL VARIATION

Form tolerances can also add to a Tolerance Stackup rotationally. The deformation of mating features during assembly may cause them to rotate, causing other features to rotate and translate in the direction of the Tolerance Stackup. This deformation is possible due to the form error of the mating surfaces and forces applied at assembly. The following example discusses the potential effect of the form tolerances applied to mating datum features.

Consider the assembly shown in Fig. 20.9. Two parts are bolted together as an assembly, a brace assembly. The brace assembly consists of two parts

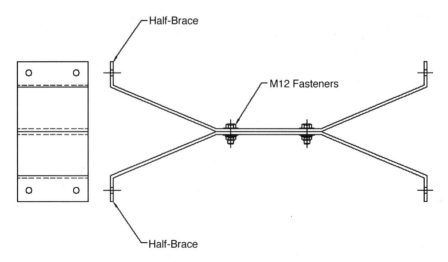

**FIGURE 20.9**   Form tolerances: Brace assembly.

and fasteners. Both parts are the same part (the half-brace) and bolted back to back. At the next higher level this assembly is assembled into a frame, as shown in Fig. 20.10.

The clearance holes in the brace assembly must allow fasteners to pass through the clearance holes in the frame at final assembly. At first glance, this appears to be a floating-fastener problem, as fasteners pass through clearance holes in both mating parts. However, the forces applied as the half-braces are bolted together may deform the parts, adding to the positional tolerance applied to the holes on the detail drawing. The detail drawing for the half-brace is shown in Fig. 20.11. For this example, assume the half-brace is a stamped part.

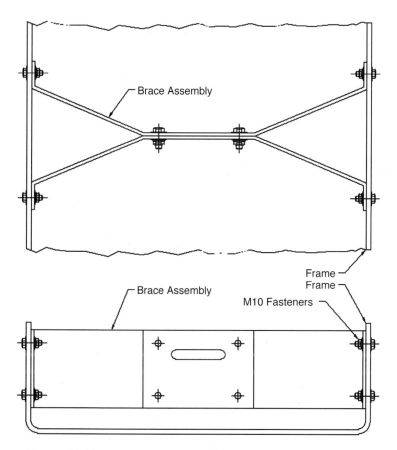

**FIGURE 20.10** Form tolerances: Brace assembly installed in frame.

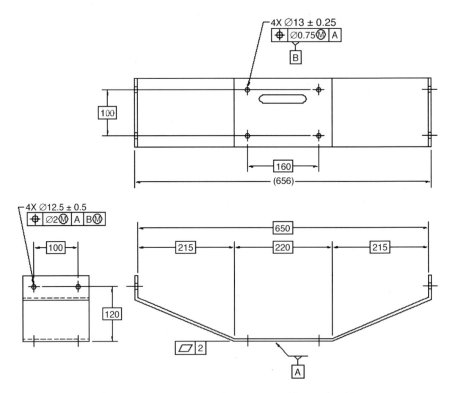

**FIGURE 20.11**   Form tolerances: Half-brace detail.

The half-brace has been toleranced to match its function. The primary datum feature (datum feature A) is the mounting surface between the half-braces at assembly. A 2-mm flatness tolerance has been specified for this surface—according to the manufacturing engineer, this is as tight as the manufacturing process can reliably hold. The designer wanted to hold this surface within a tighter flatness tolerance, but the process can not be improved. The ⌀12.5 ± 0.5 holes in the side flanges mate with holes in the frame and must allow M10 bolts to pass, as shown in Fig. 20.10. The 2-mm positional tolerance applied to the holes in the side flanges of the half-brace also reflects the process capability limits and cannot be any tighter.

A Tolerance Stackup shall be done to determine the total location tolerance of the ⌀12.5 ± 0.5 holes in the half-brace assembly. The holes were sized using the floating-fastener formula. If the formula was valid in this application, the worst-case assembly would allow M10 bolts to pass. However, in this application, the flatness tolerance applied to datum feature A of the half-braces affects the location of the holes by adding rotational variation to their location. The Tolerance Stackup will show that the holes must be much larger to allow fasteners to pass. Figures 20.12 through 20.17 show how the flatness tolerance applied to the half-braces affects the location of the holes during the assembly process.

An imperfect as-produced half-brace is shown in Fig. 20.12. All features on the part are within tolerance. Datum feature A is shown convex with maximum form error, completely bowed within its 2-mm flatness tolerance zone. The form error is such that the edges of the datum feature A surface are 2 mm above the middle of the surface. The part is stabilized on the datum feature simulator, and each side is adjusted or shimmed until the variation of datum feature A is approximately equal on both sides. Notice that there is a gap between the datum feature and the datum feature simulator on both sides. When the parts are fastener together at assembly, these gaps will close, and the parts will deform.

The flatness tolerance applied to each datum feature A surface creates a tolerance zone that consists of two parallel planes 2 mm apart. All points of the surface must lie within this tolerance zone. The form error of the surface may take on any form, be it sinusoidal, concave, convex, vee-shaped, or jagged; all that is required is that all points of the surface are within the form

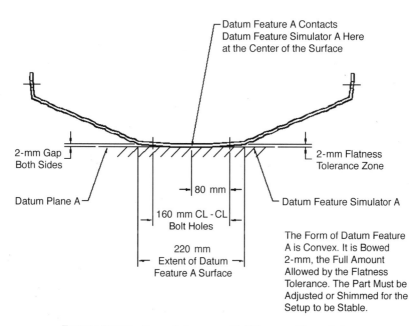

**FIGURE 20.12** Form tolerances: Half-brace with variation.

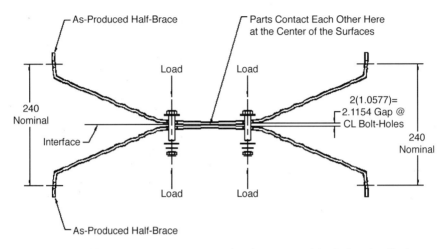

**FIGURE 20.13**  Form tolerances: Imperfect brace assembly (before loading).

tolerance zone. For this example the form error of the surface is convex. It is assumed that the form error is uniformly distributed across the surface. Because the bolt holes are not at the edges of the surface and the surface is assumed to have convex form error, and the mating surfaces will be pulled together at the bolt holes, the full 2-mm flatness tolerance value will not be used in the calculations. As will be shown, the approximate form error at the

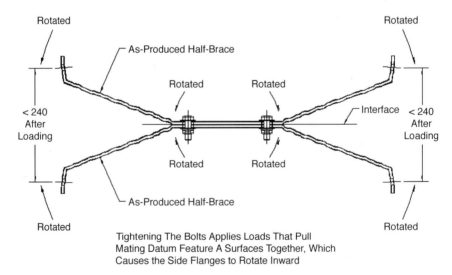

**FIGURE 20.14**  Form tolerances: Imperfect brace assembly (after loading).

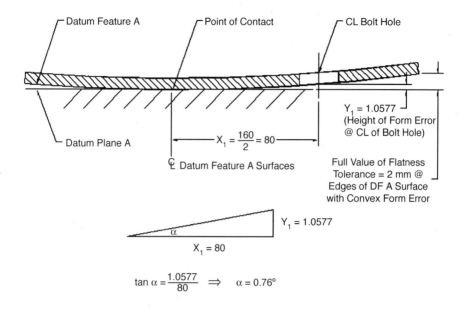

**FIGURE 20.15** Form tolerances: Enlarged view with triangle 1.

bolt holes is only 1.0577 mm. If the surface was produced with a different form error, e.g., vee-shaped, the approximate value of the form error at the bolt holes would be larger, approximately 1.45 mm. For our calculations, however, we will use the 1.0577 form error from the convex form error.

Two imperfect as-produced half-braces are shown at assembly in Fig. 20.13. The fasteners have been inserted but have not yet been tightened. Axial loads will pull the half-braces together as the fasteners are tightened. For the sake of this example, it is assumed the bolt forces bring the mating surfaces together at the bolts—the distance between the bolt holes and their relationship to the middle of datum feature A will be used in the calculations. As seen in Fig. 20.13, the $\varnothing 12.5 \pm 0.5$ holes in each pair of side flanges are approximately 240 mm apart before the bolt forces are applied. The bolt forces will pull the mating surfaces together, closing the gap on both sides of the interface. These gaps are the sum of each pair of 2-mm flatness tolerances on the interfacial surfaces.

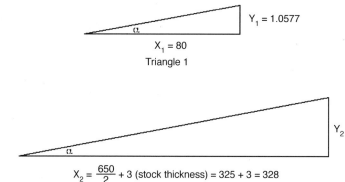

$$Y_1 = 1.0577$$

$$X_1 = 80$$

Triangle 1

$$Y_2$$

$$X_2 = \frac{650}{2} + 3 \text{ (stock thickness)} = 325 + 3 = 328$$

Triangle 12

Solve For $Y_2$:

$$\frac{Y_1}{X_1} = \frac{Y_2}{X_2} \implies \frac{1.0577}{80} = \frac{Y_2}{328}$$

$$Y_2 = 1.0577(328/80) = 4.34$$

Note:

$Y_1$ is less than the full 2-mm value of the Flatness tolerance. With convex form error, the vertical variation of the Datum Feature A surfaces is only 1.0577 at the centerline of the bolt holes. It is assumed that the bolt forces will pull the surfaces together along the CL of the bolts.

**FIGURE 20.16**   Form tolerances: Like triangles projecting the rotation out to the side flanges.

It will also be assumed that the parts deform in a uniform manner. The deformation between the point of contact on datum feature A and where the bolt forces are applied will cause the part features farther out to rotate uniformly, as if that portion of the part were rigid.

Figure 20.14 shows the half-brace assembly after the bolts have been tightened. The axial loads have pulled the mating surfaces together, and the side flanges have rotated inward a corresponding amount.

As the mating surfaces of the half-braces are deformed, they cause the edges of the interfacing surfaces and side flanges to rotate through an angle. It is relatively easy to calculate this angle, and just as easy to determine the amount this rotation affects the holes in the side flanges. This problem will be solved in the same manner as in Chapter 15, which used *like triangles* to convert the rotation of one set of features into the linear translation of another set of features. Figure 20.15 shows an enlarged detail view of the half-brace against datum feature simulator A. In this example, datum feature A of each

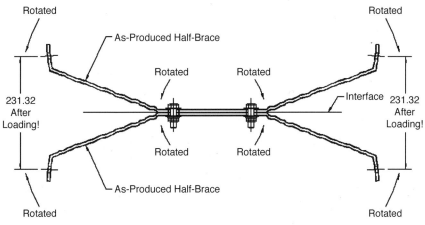

The Distance Between the Side Flange Hole was
Reduced by 4.34 mm on Both Parts. The Effect of the
Positional Tolerance on the Holes is not included.

**FIGURE 20.17** Form tolerances: Brace assembly after loading with dimensions.

half-brace may deviate 1.0577 mm, which is the height of the form error at the bolt holes. Again it is assumed that when the bolts are tightened, each surface will be pulled through the gap until the gap is closed and the surfaces touch. It is also assumed the gap will close along the centerline of the bolts, which is where the forces are applied. The triangle shown in Fig. 20.15 represents the amount each half-brace will rotate as the bolts are tightened. One last assumption is that the bolt forces are large enough to fully close the gaps.

Using like triangles, the angle of rotation about datum feature A will be projected out to the $\varnothing 12.5 \pm 0.5$ holes in the side flanges. Triangle 1 represents the variation in datum feature A: $X_1$ is the horizontal distance from the point of contact at the center of datum feature A to the axis of the bolt holes; $Y_1$ is the vertical height of the gap at the centerline of the bolts allowed by the flatness tolerance on datum feature A. Triangle 2 in Fig. 20.16 has the same angle as triangle 1. $X_2$ on triangle 2 is the horizontal distance from the center of datum feature A to the outside of the side flanges, which is 328 mm (325 basic to inside edge + 3 mm stock thickness to outside edge). The problem is solved for the $Y_2$ distance, which represents the linear translation created by projecting the rotation of datum feature A.

Using like triangles, the value of $Y_2$ is 4.34 in Fig. 20.16. Remember, this is the variation found in one half-brace. This variation would be added to the Tolerance Stackup twice, once for each part, as both parts would deform the same in this scenario. Figure 20.17 shows that the as-assembled worst-case distance between the $\oslash 12.5 \pm 0.5$ holes in the side flanges has been reduced by $2(4.34) = 8.68$ mm worst case! This is only a function of the deformation of the datum feature A surfaces and does not include the positional tolerance applied to the holes in the side flanges. If the as-produced shape of the mating datum feature A surfaces was different, such as vee-shaped or wedge-shaped, the variation would be larger. In fact, with vee-shaped geometry the distance between the holes in side flanges would be 11.82 mm worst case.

A comment is in order here about rigid vs. nonrigid parts: in many applications, 3-mm thick sheet metal parts as shown in this example are treated as rigid parts, at least from a dimensioning and tolerancing point of view. If the half-braces were treated as nonrigid parts and inspected with forces that approximate the bolt loads encountered at assembly, much of the potential problem described above would disappear. There would still be potential for error as there would be a difference between the near perfect form of the datum feature simulator and the imperfect mating part, but the overall effect form tolerances would be much less severe, as the inspection method would more closely match the as-installed condition of the parts.

Typically when the effect of form tolerances in Tolerance Stackups is treated as adding rotational variation there are assembly forces involved, that is, the form tolerances are applied to mating surfaces that are subjected to loading and deformation at assembly. This is not always the case, but it is the most likely scenario in most mechanical assemblies.

An example of where interfacial surfaces are not loaded and rotational variation is possible can be seen in your kitchen cabinet. If you consider a stack of cereal bowls and the interface between each pair of bowls, you will see that the upper bowl can rotate within the bowl beneath it. If a Tolerance Stackup was done to determine the height of the stack of bowls, this rotation would affect the result. This is more a function of the geometry of the mating surfaces than the form tolerances applied to them, but it provides an example of geometry where rotation is possible without loading. Another similar geometric example is a pair of spherical washers, which are used to level equipment during installation.

Tolerance Stackups where form tolerances are treated as adding translational variation only are usually only subjected to the force of gravity, which does not appreciably deform the parts in most assemblies. It is also the same force the part was subjected to when it was inspected; so there should not be a difference at assembly due to gravity.

The translational variation caused by form tolerances acts in one direction in the example shown in Figs. 20.1 to 20.8. When treated as purely translational tolerances, the form tolerances shown in the first example only serve to reduce the distance between the surfaces under consideration. Given the geometry of the rotational example shown in Figs. 20.9 through 20.17, tightening the bolts is far more likely to move the side flange holes closer together than farther apart. It is also possible, however, that if the form error of the datum feature A surfaces was concave instead of convex, the surfaces would rotate outward while the bolts were fastened (see Fig. 20.18). It is possible that the placement of the bolt holes out near the edges of the datum feature A surface could reduce any potential rotational displacement outward caused by tightening the bolts. If the bolts that tightened the half-braces together were only located at the middle of the datum feature A surface, it would be far more likely that the rotational displacement would only move the holes outward. This can be seen in Fig. 20.19.

## Probability

As with the translation-only example, it is important to realize that the form error shown in these examples is for three-dimensional parts. To encounter this rotational error between the as-assembled parts, the form error would have to have exactly the right shape in three dimensions. The entire mating

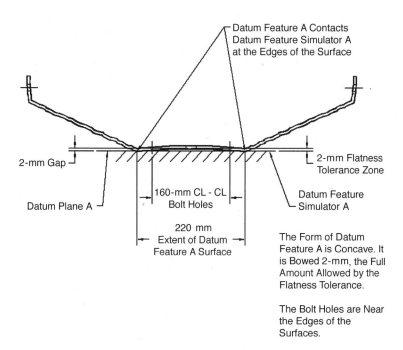

**FIGURE 20.18**  Form tolerances: Half-brace with concave form error.

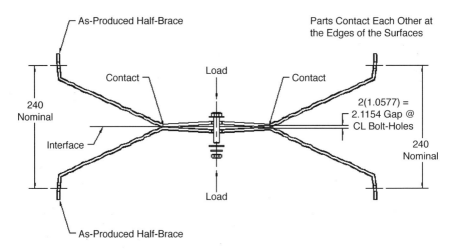

**FIGURE 20.19** Form tolerances: Brace assembly with concave form error (before loading).

datum features of both parts would have to be bowed in exactly the right manner.

In my opinion, it is unlikely that the full effect of this sort of variation is encountered in most mating part applications. The likelihood of the full effect of this form error occurring on both mating parts is very remote. However, it is possible. It is very likely that the same or similar form error would be seen in mating parts where both parts are the same part as in this example. If the stamping die used to make the parts was warping datum feature A and both parts were from the same production run, the form of datum feature A on both parts would probably be very similar. Whether the form error of datum feature A was at this worst-case extreme is another issue, and much less likely. As in Chapter 8, which deals with statistics, again we are faced with choosing between the possible and the probable.

Determining whether to treat the form tolerance as adding translational or rotational variation and how to include it in the Tolerance Stackup presents us with a problem. There are four things to consider:

- Whether form tolerances should be included in the Tolerance Stackup.
- Whether the variation allowed by form tolerances should be treated as translation or as rotation.
- How to include the form tolerances in the Tolerance Stackup.
- How to quantify the potential effect of the form tolerances.

## WHETHER FORM TOLERANCES SHOULD BE INCLUDED IN THE TOLERANCE STACKUP

Whether a form tolerance is included in a Tolerance Stackup is left to the discretion of the Tolerance Analyst. As stated earlier, the likelihood of both surfaces being perfectly misshapen is very remote. This improbability may lead the Tolerance Analyst to omit the form tolerances from the Tolerance Stackup. The likelihood of the high points of both surfaces aligning and touching is also very remote. This may lead the Tolerance Analyst to include the form tolerances in the Tolerance Stackup. The decision ultimately must be based on the sensitivity of the design, understanding of the manufacturing process, and the risk associated with including or not including the form tolerances in the Tolerance Stackup.

There is risk on both sides. If the form tolerances are not included in the chain of Dimensions and Tolerances, and some mismatch occurs at assembly, there may be greater variation in a critical distance than predicted by the Tolerance Stackup. Likewise, if the form tolerances are included in the chain of Dimensions and Tolerances and no mismatch occurs at assembly, there may be less variation in a critical distance than predicted by the Tolerance Stackup.

Generally speaking, form tolerances are less likely to add translational error than rotational error to a Tolerance Stackup. In Figs. 20.1 to 20.8 the variation allowed by form tolerances is treated as purely translational. It is assumed in these examples that the mismatch between the datum features at assembly acts purely in the direction of the Tolerance Stackup and that the simulated datum planes remain parallel but separated. To state it in different terms, the datum reference frames of the parts are assumed to remain parallel but shifted by the amount of the form tolerance. This can be seen as the distance between simulated datum planes A for the mating parts in Fig. 20.8.

It is also possible that the variation allowed by form tolerances may result in rotational error, that is, that the mismatch between the datum reference frames on the mating parts are at an angle to one another rather than being parallel. As shown in the material covering rotation of parts within the Tolerance Stackup, rotational error can be much greater than translational error when projected over a distance.

## WHETHER THE VARIATION ALLOWED BY FORM TOLERANCES SHOULD BE TREATED AS TRANSLATION OR AS ROTATION

As stated in the previous section, the variation allowed by form tolerances may be treated as translation or as rotation. It requires a lot of factors to be in place for either case to affect a Tolerance Stackup, such as matching geometric imperfection in both mating parts.

Most likely the mismatch between parts will be a combination of translation and rotation. Given the fact that we don't have the time to play around with a Tolerance Stackup and try 1000 iterations with different combinations of translation and rotation, it is likely that our best attempt would be to solve a Tolerance Stackup once using translation and once using rotation. The sort of iteration required to get a feel for the most likely worst-case combination of translation and rotation is handled well by 3D statistical tolerance modeling programs such as vis-VSA or 3DCS. Once the data is entered, these programs run a series of iterations that address many, many translational and rotational possibilities. This is very difficult for a person to do, but it is very easy for a computer program. However, these programs only offer statistical results, so the problem should be solved linearly if a worst-case result is needed.

It is possible to solve the Tolerance Stackup twice if desired, by treating the allowable form error as translational in one study and rotational in another. This approach is similar to the approach taken in Chapter 15.

Form tolerances treated as adding purely translational variation in Tolerance Stackups only serve to reduce the overall gap or distance being studied—they act in one direction only. Form tolerances treated as adding rotational variation in Tolerance Stackups may act only in one direction, or they may act in both the positive and negative directions, depending on the geometry of the parts.

## HOW TO INCLUDE FORM TOLERANCES IN THE TOLERANCE STACKUP

Look again at Fig. 20.8 which shows the assembled imperfect as-produced parts from the first example. Notice that there is 1-mm mismatch between datum A for both parts—if the surfaces were perfectly flat, they would not be able to nest within each other, and there would be no difference between how the features were inspected and how they assemble. This 1-mm mismatch is due to the flatness tolerances specified for the surfaces. With careful consideration, we see that the most these two flatness tolerances can add to the Tolerance Stackup is 1 mm total. This is because the most either datum feature can be displaced from its location against the datum feature simulator is the amount of its form tolerance. Furthermore, since both datum feature flatness tolerances are the same value (1 mm), the total amount both flatness tolerances add to the Tolerance Stackup is 1 mm.

In fact, in cases where the flatness tolerances on both mating surfaces are not equal, the amount added to the Tolerance Stackup is the smaller of the

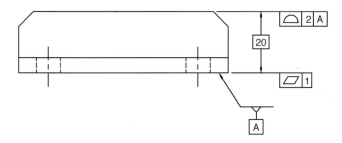

**FIGURE 20.20** Form tolerances: Upper part FT20 drawing (same as FT2).

two flatness tolerances. This can be seen in Figs. 20.20 through 20.24. Even though the flatness tolerance on datum feature A on part FT20 is 1 mm, it is not possible for the datum feature to translate the full 1 mm at assembly because the flatness tolerance on the mating part in Fig. 20.22 is smaller. The most the datum feature can be displaced is limited to the smaller 0.5-mm flatness tolerance specified for the mating surface.

The effects of the form tolerance on each part are considered together where the form tolerance is treated as adding translational variation only. The effect of the mating form tolerances in only added once to the Tolerance Stackup.

The effects of the form tolerances on each part are considered independently in cases where the tolerances are treated as adding rotational var-

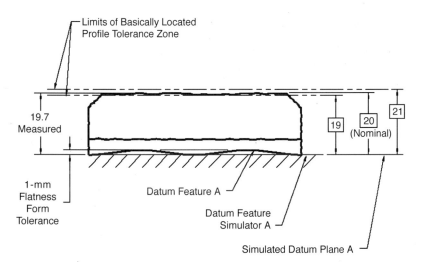

**FIGURE 20.21** Form tolerances: Upper part FT20 with variation (same as FT2).

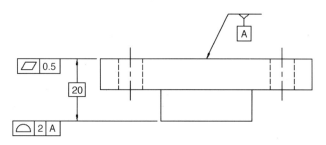

**FIGURE 20.22** Form tolerances: Lower part FT22 drawing with 0.5 flatness.

iation. The effect of each mating form tolerance is added to the Tolerance Stackup.

Depending on the geometry of the mating surfaces and how the form tolerance is treated, the form tolerance may only affect the Tolerance Stackup in one direction, or it may affect the Tolerance Stackup in both the positive and negative directions. If the form tolerance is treated as adding translational variation only, it can only reduce the Tolerance Stackup result. In that case the effect of the form tolerance is only seen in the negative direction. In such cases a negative dimension is added on the same line as the equal bilateral equivalent form tolerance in the Tolerance Stackup report. The form tolerance is numbered as it is encountered in the chain of Dimensions and Tolerances in the Tolerance Stackup sketch, and a negative sign is placed adjacent to its item number to highlight that it acts in the negative direction. Form tolerances that only affect the Tolerance Stackup in one direction are

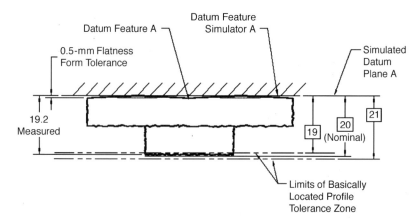

**FIGURE 20.23** Form tolerances: Lower part FT22 with 0.5 flatness with variation.

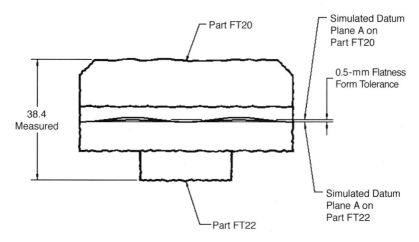

**FIGURE 20.24** Form tolerances: Imperfect translation assembly with 0.5 flatness.

treated differently than other geometric tolerances, as is evidenced by the negative dimension and the negative sign associated with the tolerance's item number. An example can be seen in Figs. 20.25 and 20.26. See the following material for more detailed instructions.

## FORM TOLERANCES TREATED AS ADDING TRANSLATIONAL VARIATION ONLY

Form tolerances treated as adding translational variation only are included in the Tolerance Stackup as follows:

- If the form tolerances of the mating surfaces are the same value,
  - Only one of the form tolerances on the mating surfaces should be included in the Tolerance Stackup.
  - Convert the form tolerance to its equal bilateral equivalent. (Divide the form tolerance value by 2.)
  - Enter the equal bilateral form tolerance into the Tolerance Stackup report as it is encountered in the chain of Dimensions and Tolerances.
  - Enter a value in the negative direction dimension column equal to half the value of the form tolerance. The dimension value is entered on the same line as the form tolerance in the Tolerance Stackup report. (This is a mean shift.) (Because the form tolerances only act in the negative direction and the form tolerance

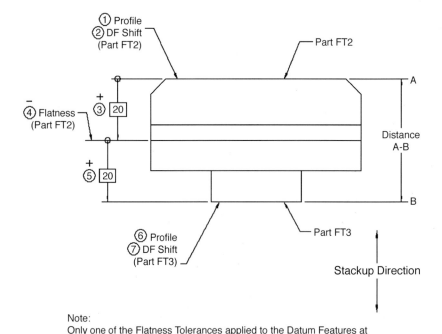

Note:
Only one of the Flatness Tolerances applied to the Datum Features at
the interface is included in the Tolerance Stackup. Since the Tolerances
are the same value, it doesn't matter which one is used.

**FIGURE 20.25** Form tolerances: Tolerance Stackup sketch for FT1 with form
tolerance included.

has been converted to an equal bilateral format, a mean shift of
1/2 the form tolerance value must be subtracted from the Tolerance Stackup. This is done by adding the dimension in negative
column.)

Figure 20.25 shows a Tolerance Stackup sketch, and Fig.
20.26 shows a Tolerance Stackup report for parts FT2 and FT3.

• If the form tolerances of the mating surfaces are not the same value,

  ○ Only one of the form tolerances on the mating surfaces should be
    included in the Tolerance Stackup.

  ○ Use the smaller form tolerance value in the Tolerance Stackup.

  ○ Convert the smaller form tolerance to its equal bilateral
    equivalent. (Divide the form tolerance value by 2.)

  ○ Enter the equal bilateral form tolerance into the Tolerance
    Stackup report as it is encountered in the chain of Dimensions and
    Tolerances.

Tolerance Stack

Release 1.2a

| Program: | Tolerance Analysis Textbook | | | | | | |

| Product: | Part Number | Rev | Description |
| | FT1 | A | Sample Assembly FT1 for Form Tolerance Chapter |

Problem: Overall Height of Assembly Is Critical

Objective: Determine the Minimum and Maximum Height of the Assembly

**Stack Information:**

| | |
|---|---|
| Stack No | Figure 20.26 |
| Date | 07/04/02 |
| Revision | A |
| Direction: | Vertical |
| Author | BR Fischer |

| Description of Component / Assy | Part Number | Rev | Item | Description | + Dims | − Dims | Tol | Percent Contrib | Dim / Tol Source & Calcs |
|---|---|---|---|---|---|---|---|---|---|
| Upper Part | FT2 | A | 1 | Profile Upper Surface (Point A) | | | ± 1.0000 | 40% | Profile 2, A |
| | | | 2 | Datum Feature Shift | | | ± 0.0000 | 0% | N/A – Datum Feature A not a Feature of Size |
| | | | 3 | Dim Upper Surface - Datum A | 20.0000 | | ± 0.0000 | 0% | 20 Basic on Dwg |
| | | | 4 | Flatness Tolerance (Datum Feature A) | | 0.5000 | ± 0.5000 | 20% | Flatness T on Dwg - See Notes and Assumptions |
| Lower Part | FT3 | A | 5 | Dim Datum A - Lower Surface | 20.0000 | | ± 0.0000 | 0% | 20 Basic on Dwg |
| | | | 6 | Profile Lower Surface (Point B) | | | ± 1.0000 | 40% | Profile 2, A |
| | | | 7 | Datum Feature Shift | | | ± 0.0000 | 0% | N/A – Datum Feature A not a Feature of Size |

| | | | |
|---|---|---|---|
| Dimension Totals | 40.0000 | 0.5000 | |
| Nominal Distance: Pos Dims − Neg Dims = | 39.5000 | | |

**RESULTS:**

| | Nom | Tol | Min | Max |
|---|---|---|---|---|
| Arithmetic Stack (Worst Case) | 39.5000 | ± 2.5000 | 37.0000 | 42.0000 |
| Statistical Stack (RSS) | 39.5000 | ± 1.5000 | 38.0000 | 41.0000 |
| Adjusted Statistical 1.5*RSS | 39.5000 | ± 2.2500 | 37.2500 | 41.7500 |

Notes:
- The Flatness Tolerance Specified for Datum Feature A on Part FT2 is included in the Tolerance Stackup.
Since both Flatness tolerance values are the same, it doesn't matter which one is included in the Tolerance Stackup.

Assumptions:
- It is assumed that the Flatness tolerances Specified for the Datum Feature A Surfaces adds Translational-only Variation to the Tolerance Stackup. Only one of the Flatness tolerances are included in the Tolerance Stackup. A Mean Shift dimension of 1/2 the Tolerance Value has been added to the Tolerance Stackup with the Form tolerance on Line 4, because a Translation-only Form tolerance only affects the Tolerance Stackup in the negative direction.

Suggested Action:

**FIGURE 20.26** Form tolerances: Tolerance Stackup report for FT1 with form tolerances.

○  Enter a value in the negative direction dimension column equal to half the value of the form tolerance. The dimension value is entered on the same line as the form tolerance in the Tolerance Stackup report. (This is a mean shift.) (Because the form tolerances only act in the negative direction and the smaller form tolerance has been converted to an equal bilateral format, a mean shift of 1/2 the smaller form tolerance value must be subtracted from the Tolerance Stackup. This is done by adding the dimension in negative column.)

Figure 20.27 shows a Tolerance Stackup sketch, and Fig. 20.28 shows a Tolerance Stackup report for parts FT20 and FT22.

## FORM TOLERANCES TREATED AS ADDING ROTATIONAL VARIATION

Form tolerances treated as adding rotational variation are included in the Tolerance Stackup as follows:

*   If the rotational variation allowed by the form tolerance acts in one direction only,
    ○  The effect of the form tolerances applied to both mating surfaces should be included in the Tolerance Stackup.

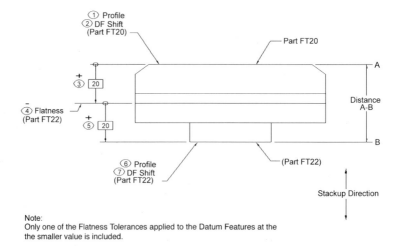

**FIGURE 20.27**  Form tolerances: Tolerance Stackup sketch for FT20 and 22 with form tolerances.

Tolerance Stack

Release 1.2a

| Program: | Tolerance Analysis Textbook | | |
|---|---|---|---|
| Product: | Part Number | Rev | Description |
| | - | A | Sample Assembly for Form Tolerance Chapter with Unequal Form Tolerance Values |
| Problem: | Overall Height of Assembly Is Critical | | |
| Objective: | Determine the Minimum and Maximum Height of the Assembly | | |

Stack Information:

Stack No.: Figure 20.28
Date: 07/04/02
Revision: A

Direction: Vertical
Author: BR Fischer

| Description of Component / Assy | Part Number | Rev | Item | Description | + Dims | – Dims | Tol | Percent Contrib | Dim / Tol Source & Calcs |
|---|---|---|---|---|---|---|---|---|---|
| Upper Part | FT20 | A | 1 | Profile Upper Surface (Point A) | | | ± 1.0000 | 44% | Profile 2, A |
| | | | 2 | Datum Feature Shift | | | ± 0.0000 | 0% | N/A - Datum Feature A not a Feature of Size |
| | | | 3 | Dim. Upper Surface - Datum A | 20.0000 | | ± 0.0000 | 0% | 20  Basic on Dwg |
| Lower Part | FT22 | A | 4 | Flatness Tolerance (Datum Feature A) | | 0.2500 | ± 0.2500 | 11% | Flatness 0.5 on Dwg - See Notes and Assumptions |
| | | | 5 | Dim. Datum A - Lower Surface | 20.0000 | | ± 0.0000 | 0% | 20  Basic on Dwg |
| | | | 6 | Profile Lower Surface (Point B) | | | ± 1.0000 | 44% | Profile 2, A |
| | | | 7 | Datum Feature Shift | | | ± 0.0000 | 0% | N/A - Datum Feature A not a Feature of Size |

| | + Dims | – Dims | Tol |
|---|---|---|---|
| Dimension Totals | 40.0000 | 0.2500 | |
| Nominal Distance: Pos Dims - Neg Dims = | | 39.7500 | |

**RESULTS:**

| | Nom | Tol | Min | Max |
|---|---|---|---|---|
| Arithmetic Stack (Worst Case) | 39.7500 | ± 2.2500 | 37.5000 | 42.0000 |
| Statistical Stack (RSS) | 39.7500 | ± 1.4361 | 38.3139 | 41.1861 |
| Adjusted Statistical: 1.5*RSS | 39.7500 | ± 2.1542 | 37.5958 | 41.9042 |

Notes:
- The Flatness Tolerance Specified for Datum Feature A on Part FT22 is included in the Tolerance Stackup because it is smaller than the Flatness Tolerance value for Part 20

Assumptions:
- It is assumed that the Flatness tolerances Specified for the Datum Feature A Surfaces adds Translational-only Variation to the Tolerance Stackup.  Only one of the Flatness tolerances are included in the Tolerance Stackup.  A Mean Shift dimension of 1/2 the Tolerance Value has been added to the Tolerance Stackup with the Form tolerance on Line 4, because a Translation-only Form tolerance only affects the Tolerance Stackup in the negative direction.

Suggested Action:

**FIGURE 20.28**  Form tolerances: Tolerance Stackup report for FT20 and 22 with form tolerances.

○  The rotational variation allowed by the form tolerance applied to each surface must be calculated.

○  The projected linear displacement allowed by the rotational variation should be calculated using like triangles for the form tolerance applied to each surface. (Use the techniques described earlier in this chapter.)

○  Convert each projected linear displacement to its equal bilateral equivalent. (Divide the projected linear displacement value by 2.)

○  Enter the equal bilateral equivalent for both form tolerances into the Tolerance Column in the Tolerance Stackup report as they are encountered in the chain of Dimensions and Tolerances.

○  Add a dimension value on the same line in the Tolerance Stackup report as the form tolerance as follows:

  ▪  If the projected linear translation allowed by the form tolerance makes the considered gap or distance smaller, enter the dimension value in the negative direction dimension column. The dimension value is equal to half the value of the projected linear translation. (This is a mean shift. Because the form tolerances only act in the negative direction and the form tolerance has been converted to an equal bilateral format, a mean shift of $1/2$ the form tolerance value must be subtracted from the Tolerance Stackup. This is done by entering the dimension value in negative direction dimension column.)

  ▪  If the projected linear translation allowed by the form tolerance makes the considered gap or distance larger, enter the dimension value in the positive direction dimension column. The dimension value is equal to half the value of the projected linear translation. (This is a mean shift. Because the form tolerances only act in the positive direction and the effect of the projected linear displacement has been converted to an equal bilateral format, a mean shift of $1/2$ the projected linear displacement value must be added to the Tolerance Stackup. This is done by entering the dimension value in positive direction dimension column.)

Figure 20.29 shows a Tolerance Stackup sketch, and Fig. 20.30 shows a Tolerance Stackup report for the brace assembly FT9. In this example, the form tolerances are assumed to only make the distance between the holes in the side flanges smaller.

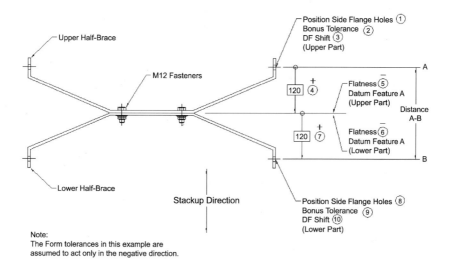

**FIGURE 20.29** Form tolerances: Tolerance Stackup sketch for FT9 with form tolerances in one direction.

If the form tolerances were assumed to only make the distance between the holes in the side flanges larger, the mean shift dimensions would be entered in the positive dimension column.

- If the rotational variation allowed by the form tolerance acts in both the positive and negative directions:
  - The effect of the form tolerances applied to both mating surfaces should be included in the Tolerance Stackup.
  - The rotational variation allowed by the form tolerance applied to each surface must be calculated.
  - The projected linear displacement allowed by the rotational variation should be calculated using like triangles for the form tolerance applied to each surface. (Use the techniques described earlier in this chapter.)
  - The projected linear displacement is used as its equal bilateral equivalent. (This is because the projected linear displacement can act in both the positive and negative directions.)
  - Enter the equal bilateral equivalent into the Tolerance column in the Tolerance Stackup report as it is encountered in the chain of Dimensions and Tolerances. (No mean shift is required if the

Tolerance Stack                                                                                    Release 1.2a

| Program: | Tolerance Analysis Textbook | | | | | Stack Information |
|---|---|---|---|---|---|---|
| Product: | Part Number | Rev | Description | | | Stack No. | Figure 20.30 |
| | FT9 | A | Sample Assembly for Form Tolerance Chapter with Linear Displacement from Rotational Variation | | | Date | 07/04/02 |
| | | | | | | Revision | A |
| Problem: | When Bolts are Tightened Datum Feature A Surfaces will Deform and The Side Flange Bolt Holes will Be Displaced | | | | | Direction | Vertical |
| Objective | Determine the Variation Between the Holes in The Side Flanges in the Vertical Direction | | | | | Author | BR Fischer |

| Description of Component / Assy | Part Number | Rev | Item | Description | + Dims | − Dims | Tol | Percent Contrib | Dim / Tol Source & Calcs |
|---|---|---|---|---|---|---|---|---|---|
| Half-Brace (Upper Part) | FT11 | A | 1 | Position Holes in Upper Side Flanges | | | ± 1.0000 | 14% | Position dia 2 @ MMC, A, B @ MMC |
| | | | 2 | Bonus Tolerance | | | ± 0.5000 | 7% | = (0.5 + 0.5)/2 |
| | | | 3 | Datum Feature Shift | | | ± 0.0000 | 0% | N/A - DF$_B$ Shift is Perpendicular to Tolerance Stackup |
| | | | 4 | Dim. CL Side Flange Holes - Datum A | 120.0000 | | ± 0.0000 | 0% | 120 Basic on Dwg |
| | | | 5 | Flatness Tolerance (Datum Feature A) | | 2.1700 | ± 2.1700 | 30% | Flatness 1 on Dwg   - 4.34 Projected (See Notes) |
| Half-Brace (Lower Part) | FT11 | A | 6 | Flatness Tolerance (Datum Feature A) | | 2.1700 | ± 2.1700 | 30% | Flatness 1 on Dwg   - 4.34 Projected (See Notes) |
| | | | 7 | Dim  Datum A - CL Side Flange Holes | 120.0000 | | ± 0.0000 | 0% | 120 Basic on Dwg |
| | | | 8 | Position Holes in Upper Side Flanges | | | ± 1.0000 | 14% | Position dia 2 @ MMC, A, B @ MMC |
| | | | 9 | Bonus Tolerance | | | ± 0.5000 | 7% | = (0.5 + 0.5)/2 |
| | | | 10 | Datum Feature Shift | | | ± 0.0000 | 0% | N/A - DF$_B$ Shift is Perpendicular to Tolerance Stackup |

|  | Dimension Totals | 240.0000 | 4.3400 |
|---|---|---|---|
|  | Nominal Distance Pos Dims - Neg Dims = | 235.6600 | |

**RESULTS:**

| | Nom | Tol | Min | Max |
|---|---|---|---|---|
| Arithmetic Stack (Worst Case) | 235.6600 | ± 7.3400 | 228.3200 | 243.0000 |
| Statistical Stack (RSS) | 235.6600 | ± 3.4522 | 232.2078 | 239.1122 |
| Adjusted Statistical 1.5*RSS | 235.6600 | ± 5.1783 | 230.4817 | 240.8383 |

Notes: - The Flatness Tolerance Specified for Datum Feature A on Both Parts is included in the Tolerance Stackup.

Assumptions: - It is assumed that the form error allowed by the Flatness tolerances specified for both Datum Feature A Surfaces allows the surfaces to rotate when the bolts are tightened. When the Bolts are tightened, the Datum Feature A surfaces will deform and rotate until the surfaces meet along the nominal interface  The Datum Feature A surfaces are assumed to rotate to close the gap between them. This rotation is projected out to the Side Flanges, which rotate through the same angle. Both parts are subject to the same deformation as the Bolts are tightened, so both the effect of both Flatness tolerances are included in the Tolerance Stackup. The Form tolerances are assumed to only act to make the distance between the holes smaller, which is why negative dimensions have been included on the same line as the Flatness tolerances. These dimensions are Mean Shift dimensions.

Suggested Action: - Add a note to the Half-Brace detail drawing to inspect the part with forces applied. The forces should approximate the forces encountered at assembly; this will negate most of the effect of the projected Form error.

**FIGURE 20.30**  Form tolerances: Tolerance Stackup report for FT9 with form tolerances in one direction.

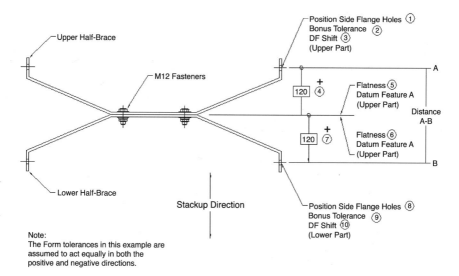

**FIGURE 20.31**  Form tolerances: Tolerance Stackup sketch for FT9 with form tolerances in both directions.

effect of the form tolerance allows the same variation in both the negative and positive directions.)

Figure 20.31 shows a Tolerance Stackup sketch, and Fig. 20.32 shows a Tolerance Stackup report for the brace assembly FT9. In this example, the form tolerances are assumed to act in both the negative and positive directions, making the distance between the holes in the side flanges smaller and larger.

It is a good idea to explain how the form tolerance variation was treated by adding a note to the Tolerance Stackup report in the Notes or Assumptions block. This is critical for anyone interpreting the Tolerance Stackup report. The Tolerance Analyst should state whether the form tolerance was treated as adding translational or rotational variation and whether the form tolerance affects the Tolerance Stackup in one or both the positive and negative directions.

## HOW TO QUANTIFY THE POTENTIAL EFFECT OF THE FORM TOLERANCES

Again we are faced with the dilemma of determining how much a tolerance is likely to contribute to a Tolerance Stackup. At one extreme lies the worst case. Regardless of probability (or improbability), all of the dimensions and

Tolerance Stack                                                                                Release 1.2a

| Program: | Tolerance Analysis Textbook | | | | | |
|---|---|---|---|---|---|---|

| | Part Number | Rev | Description | | | |
|---|---|---|---|---|---|---|
| Product: | FT9 | A | Sample Assembly for Form Tolerance Chapter with Linear Displacement from Rotational Variation |
| Problem: | | | When Bolts are Tightened Datum Feature A Surfaces will Deform and The Side Flange Bolt Holes will Be Displaced |
| Objective: | | | Determine the Variation Between the Holes in The Side Flanges in the Vertical Direction |

**Stack Information**

Stack No    Figure 20.32
Date        07/04/02
Revision    A

Direction:  Vertical
Author:     BR Fischer

| Description of Component / Assy | Part Number | Rev | Item | Description | + Dims | − Dims | Tol | Percent Contrib | Dim / Tol Source & Calcs |
|---|---|---|---|---|---|---|---|---|---|
| Half-Brace (Upper Part) | FT11 | A | 1 | Position - Holes in Upper Side Flanges | | | ± 1.0000 | 9% | Position dia 2 @ MMC, A, B @ MMC |
| | | | 2 | Bonus Tolerance | | | ± 0.5000 | 4% | = (0.5 + 0.5)/2 |
| | | | 3 | Datum Feature Shift | | | ± 0.0000 | 0% | N/A - DF₈ Shift is Perpendicular to Tolerance Stackup |
| | | | 4 | Dim: CL Side Flange Holes - Datum A | 120.0000 | | ± 0.0000 | 0% | 120 Basic on Dwg |
| | | | 5 | Flatness Tolerance (Datum Feature A) | | | ± 4.3400 | 37% | Flatness 1 on Dwg - 4.34 Projected (See Notes) |
| Half-Brace (Lower Part) | FT11 | A | 6 | Flatness Tolerance (Datum Feature A) | | | ± 4.3400 | 37% | Flatness 1 on Dwg - 4.34 Projected (See Notes) |
| | | | 7 | Dim: Datum A - CL Side Flange Holes | 120.0000 | | ± 0.0000 | 0% | 120 Basic on Dwg |
| | | | 8 | Position - Holes in Upper Side Flanges | | | ± 1.0000 | 9% | Position dia 2 @ MMC, A, B @ MMC |
| | | | 9 | Bonus Tolerance | | | ± 0.5000 | 4% | = (0.5 + 0.5)/2 |
| | | | 10 | Datum Feature Shift | | | ± 0.0000 | 0% | N/A - DF₈ Shift is Perpendicular to Tolerance Stackup |

| | | | |
|---|---|---|---|
| Dimension Totals | 240.0000 | 0.0000 | |
| Nominal Distance: Pos Dims - Neg Dims = | 240.0000 | | |

**RESULTS:**

| | Nom | Tol | Min | Max |
|---|---|---|---|---|
| Arithmetic Stack (Worst Case) | 240.0000 | ± 11.6800 | 228.3200 | 251.6800 |
| Statistical Stack (RSS) | 240.0000 | ± 6.3381 | 233.6619 | 246.3381 |
| Adjusted Statistical 1.5*RSS | 240.0000 | ± 9.5071 | 230.4929 | 249.5071 |

Notes:    - The Flatness Tolerance Specified for Datum Feature A on Both Parts is included in the Tolerance Stackup.

Assumptions: - It is assumed that the form error allowed by the Flatness tolerances specified for both Datum Feature A Surfaces allows the surfaces to rotate when the bolts are tightened.
When the Bolts are tightened, the Datum Feature A surfaces will deform and rotate until the surfaces meet along the nominal interface. The Datum Feature A surfaces are assumed to rotate to close the gap between them. This rotation is projected out to the Side Flanges, which rotate through the same angle. Both parts are subject to the same deformation as the Bolts are tightened, so both the effect of both Flatness tolerances are included in the Tolerance Stackup. The Form tolerances are assumed to act equally in both the positive and negative directions, making the distance between the holes either larger or smaller.

Suggested Action: - Add a note to the Half-Brace detail drawing to inspect the part with forces applied. The forces should approximate the forces encountered at assembly; this will negate most of the effect of the projected Form error.

**FIGURE 20.32**   Form tolerances: Tolerance Stackup report for FT9 with form tolerances in both directions.

tolerances in the chain of Dimensions and Tolerances are assumed to be at their worst-case value, which leads to a worst-case result. At the other extreme is assuming that all of the dimensions will be at their nominal values. Obviously that makes no sense. Root-sum-square and adjusted root-sum-square were presented as statistical options that approximate the probabilistic result provided a certain set of factors are in place.

Regardless of whether the worst-case, root-sum-square, or adjusted root-sum-square methods are used, the Tolerance Analyst may still not want to include the full amount of the form tolerance value in the Tolerance Stackup. This reluctance may be due to understanding how unlikely it is for the mating surfaces to have exactly the right form error. In fact, you will notice that form tolerances are not included in the Tolerance Stackups presented in the other chapters of this text.

Although form tolerances are possible contributors to Tolerance Stackups, in many applications they probably play a very small role. If it is decided that the potential variation allowed by a form tolerance must be included in the Tolerance Stackup, one of two choices is available: the full possible variation may be added to the Tolerance Stackup as described earlier in this chapter, or the possible variation may be reduced, multiplied by a factor less than 1 that represents the Tolerance Analyst's best guess as to the probability of actually encountering the variation.

## RECAP

Form tolerances may add variation to Tolerance Stackups. The form tolerance may be treated as adding translational variation only or as adding rotational variation. If it is desired to split the form tolerance variation into some combination of translational and rotation components, it is suggested that the problem be solved using 3D statistical modeling software instead of manually.

The variation from form tolerances that is treated as adding translational variation only is calculated differently than the variation that is treated as adding rotational variation. Depending on the geometry of the parts in the Tolerance Stackup, the form tolerance variation may affect the Tolerance Stackup in one direction only, or it may affect the Tolerance Stackup is both the positive and negative directions.

The effect of form tolerances are added to the Tolerance Stackup report as described earlier in this chapter. A mean shift dimension value is added to the Tolerance Stackup where form tolerances only affect the Tolerance Stackup in one direction.

The full amount of the possible variation may be added to the Tolerance Stackup, or the amount may be reduced if desired to reflect the improbability of its occurrence. Alternatively, the variation allowed by the form tolerance may be discounted altogether and excluded from the Tolerance Stackup. These decisions should be made by the Tolerance Analyst.

It is a good idea to explain how the form tolerance variation was treated by adding a note to the Tolerance Stackup report. This includes whether the form tolerance was treated as adding translational or rotational variation, and if it affects the Tolerance Stackup in one or both the positive and negative directions.

The form of a surface may also be controlled by an orientation or a profile tolerance. These tolerances and their effect on the form of a surface may be the same as shown in this chapter with a flatness tolerance. Consequently, if desired, the form error allowed by orientation and profile tolerances may be included in a Tolerance Stackup similar to the methods presented in this chapter.

## CONCLUSION

All parts and part features are imperfect. Often, the amount of allowable variation must be calculated, between features on a part and between parts in an assembly. Tolerance Analysis and Tolerance Stackups are the only way to determine the allowable variation and whether parts will satisfy their dimensional objectives. Sometimes it is necessary to work from the top down, from the assembly down to the individual parts, letting the assembly requirements determine the part tolerances. Sometimes it is necessary to work from the bottom up, from the part tolerances up to the assembly.

Depending on the number of parts and the willingness to accept risk, a determination must be made whether worst-case or statistical Tolerance Stackups are appropriate. For complex three-dimensional Tolerance Stackups, computer statistical variational modeling software is necessary, as linear Tolerance Stackups are insufficient tools for the job.

The material presented in this text gives the Tolerance Analyst all the tools needed to solve a variety of Tolerance Stackups.

Remember, to become adept at performing Tolerance Stackups requires practice.

Keep at it and good luck.

# Index